Design and Optimization of Pharmaceutical Gels

Design and Optimization of Pharmaceutical Gels

Editors

Ying Huang
Zhengwei Huang
Xuanjuan Zhang

MDPI

Basel • Beijing • Wuhan • Barcelona • Belgrade • Novi Sad • Cluj • Manchester

Editors

Ying Huang
College of Pharmacy
Jinan University
Guangzhou
China

Zhengwei Huang
College of Pharmacy
Jinan University
Guangzhou
China

Xuanjuan Zhang
College of Pharmacy
Jinan University
Guangzhou
China

Editorial Office
MDPI
St. Alban-Anlage 66
4052 Basel, Switzerland

This is a reprint of articles from the Special Issue published online in the open access journal *Gels* (ISSN 2310-2861) (available at: www.mdpi.com/journal/gels/special_issues/Y22NN1L4RC).

For citation purposes, cite each article independently as indicated on the article page online and as indicated below:

Lastname, A.A.; Lastname, B.B. Article Title. *Journal Name* **Year**, *Volume Number*, Page Range.

ISBN 978-3-7258-0006-3 (Hbk)
ISBN 978-3-7258-0005-6 (PDF)
doi.org/10.3390/books978-3-7258-0005-6

Contents

About the Editors

Ying Huang

Ying Huang (Y. Huang), Ph.D., is an associate professor of pharmaceutics in College of Pharmacy, Jinan University. She obtained her doctor's degree from the School of Pharmaceutical Sciences, Sun Yat-sen University, in 2016, and her supervisor is Prof. Chuanbin Wu (Distinguished Expert of "Thousand Talents Program" in China, Fellow of American Association of Pharmaceutical Scientists). She obtained her bachelor's degree from the School of Pharmaceutical Sciences, Sun Yat-sen University, in 2011.

Y. Huang devotes herself to investigation of innovative drug delivery systems, e.g., pulmonary drug delivery systems and transdermal drug delivery systems. In recent years, she received several scientific fundings, including National Natural Science Foundation of China. She has published over 80 papers; among those, she published 53 SCI-indexed papers as the first/corresponding author. She possesses an H-index of 21 (by Jan, 2024). Y. Huang has also applied for 18 Chinese patents, including 15 authorized patents. She earned a lot of awards, including the Innovation and Promotion Award of China Industry-University-Research Institute Collaboration Association (1st prize, 9/10), the Science and Technology Progress Award of Guangdong Province (2nd prize, 6/10), and the Best Author of the Journal of Guangdong Pharmaceutical University, etc. In addition, she serves as the youth editorial of Chinese Pharmacological Bulletin (1st session) and the committee member of Excipient Committee of Guangdong Pharmaceutical Association (1st session).

Zhengwei Huang

Zhengwei Huang (Z. Huang), Ph.D., is an associate professor of pharmaceutics in College of Pharmacy, Jinan University. He obtained his doctor's degree from the School of Pharmaceutical Sciences, Sun Yat-sen University, in 2020, and his supervisor is Prof. Chuanbin Wu (Distinguished Expert of "Thousand Talents Program" in China; Fellow of Royal Society of Chemistry and American Association of Pharmaceutical Scientists). He obtained his bachelor's degree from the School of Pharmaceutical Sciences, Sun Yat-sen University, in 2015. In recent years, he has received several scientific fundings, including two from the National Natural Science Foundation of China. He has published over 60 papers; among them, he published 48 SCI-indexed papers as the first/corresponding author. He possesses an H-index of 16 (Google Scholar). Z. Huang has also applied for six Chinese patents, three issued. In addition, he serves as the Guangdong Science and Technology Consulting Expert and the Guangzhou Science and Technology Expert, as well as the Section Associate Editor member of *Eur J Med Res*, member of the Academic Editor Board of SCI-indexed journals *BMC Pharmacol Toxico* and *J Trop Med* and Early Career Board of *Chin Med*. He won First-Class Prize of National Tertiary Industry Science and Technology Innovation Award, and was elected as a Favored Author by Dove Press.

Xuanjuan Zhang

Xuejuan Zhang (X. Zhang), Ph.D., is an associate professor of pharmaceutics in College of Pharmacy, Jinan University. She obtained her doctor's degree from the School of Biomedical and Pharmaceutical Sciences, Guangdong University of Technology, in 2020, and her supervisors are Prof. Chuanbin Wu and Prof. Wen Tan (Distinguished Expert of "Thousand Talents Program" in China). She obtained her bachelor's degree from the School of Pharmaceutical Sciences, Sun Yat-sen University, in 2017. X. Zhang focuses on the basic research and industrial transformation of innovative pulmonary inhalable medicines. In recent years, she received several scientific fundings, including the National Natural Science Foundation of China. She has published over 40 papers; among trhem, she published 23 SCI-indexed papers as the first/corresponding author. X. Zhang has also applied for 11 Chinese patents, 8 of which were granted. In addition, she serves as a member of the Academic Editor Board of SCI-indexed journal *J Trop Med*, Early Career Board of *Chin Med* and *Journal of Exploratory Research in Pharmacology*. She also won the First-Class Prize of National Tertiary Industry Science and Technology Innovation Award.

Preface

The efficacy of many bioactive agents, including drugs, food supplements, and vaccines, is limited because of their poor chemical stability, low water solubility, and low oral bioavailability. For this reason, delivery vehicles are being developed to overcome these problems. In particular, gels have attracted significant attention in the fields of drug delivery systems (DDSs), such as sustained release, controlled release, targeted, and localDDSs, due to their high drug loading efficiency, high biocompatibility, and low toxicity.

Gels are three-dimensional, semi-solid systems consisting of polymeric matrices. The physicochemical properties of gels, such as their physical strength, viscosity, and self-healing ability, can be altered to meet the specific requirements of applications in various fields, such as drug and cell delivery, bioscaffolds, and the modeling of extracellular matrices. In particular, novel gel-based delivery systems (such as intelligent hydrogels, in situ gels, emulsion gels, nanogels, vesicular gels, and microgels) that have emerged in recent years can release drugs via specific biological or external stimuli, such as temperature, pH, enzymes, ultrasound, antigens, etc., to achieve precise and local drug delivery. Therefore, gels have broad clinical application prospects, and they are anticipated to provide new, effective, and robust strategies for the theranostics of diseases.

In this context, this Special Issue, entitled "Design and Optimization of Pharmaceutical Gels", in *Gels*, was established to shed light on gels in terms of material development, system construction, structural characterization, and the effect for disease treatment, since research on gels with high translational potential is particularly sought after. We received submissions from around the world, reflecting high cultural diversity.

Ying Huang, Zhengwei Huang, and Xuanjuan Zhang
Editors

gels

Editorial

Editorial on Special Issue "Design and Optimization of Pharmaceutical Gels"

Xuejuan Zhang [1,2,†], Ying Huang [1,2,†] and Zhengwei Huang [1,2,*]

[1] State Key Laboratory of Bioactive Molecules and Druggability Assessment, Jinan University, Guangzhou 511443, China; zhangxj02230223@jnu.edu.cn (X.Z.); huangy2007@jnu.edu.cn (Y.H.)
[2] College of Pharmacy, Jinan University, Guangzhou 511443, China
* Correspondence: huangzhengw@jnu.edu.cn; Tel.: +86-02039943117
† These authors contributed equally to this work.

1. Introduction

The efficacy of many bioactive agents, including drugs, food supplements, and vaccines, is limited because of their poor chemical stability, low water solubility, and low oral bioavailability [1]. For this reason, delivery vehicles are being developed to overcome these problems [2]. In particular, gels have attracted significant attention in the fields of drug delivery systems (DDSs) [3], such as sustained-release [4], controlled-release [5], targeted [6], and local [7] DDSs, due to their high drug loading efficiency, high biocompatibility, and low toxicity.

Gels are three-dimensional, semi-solid systems consisting of polymeric matrices [8]. The physicochemical properties of gels, such as their physical strength, viscosity, and self-healing ability, can be altered to meet the specific requirements of applications in various fields [9], such as drug and cell delivery [10], bioscaffolds [11], and the modeling of extracellular matrices [12]. In particular, novel gel-based delivery systems (such as intelligent hydrogels, in situ gels, emulsion gels, nanogels, vesicular gels, and microgels) that have emerged in recent years can release drugs via specific biological or external stimuli [13], such as temperature, pH, enzymes, ultrasound, antigens, etc., to achieve precise and local drug delivery [14]. Therefore, gels have broad clinical application prospects, and they are anticipated to provide new, effective, and robust strategies for the theranostics of diseases [15].

In this context, this Special Issue, entitled "Design and Optimization of Pharmaceutical Gels", in *Gels*, has been established to shed light on gels in terms of material development, system construction, structural characterization, and the effect for disease treatment, and research on gels with high translational potential is particularly sought after. We received submissions from around the world, reflecting high cultural diversity. In addition to this Editorial, this Special Issue comprises 12 items in total, including 9 articles and 3 reviews. It is encouraging that these contributions can boost the development of pharmaceutical gels, and the following content will provide a brief overview of them. It is important to note that the topics of contributions are roughly divided into three categories: hydrogels, nanogels, and gel reviews.

2. Overview of Contributions to Hydrogels

Hydrogels are widely used pharmaceutical gel platforms. In this Special Issue, four articles on hydrogels are included, which are briefly introduced as follows:

(1) An article from Korea, entitled "Development of Efficient Sodium Alginate/Polysuccinimide-Based Hydrogels as Biodegradable Acetaminophen Delivery Systems". In this article, a pharmaceutical sodium alginate/polysuccinimide-based hydrogel platform was developed for acetaminophen delivery, demonstrating good biodegradability

Citation: Zhang, X.; Huang, Y.; Huang, Z. Editorial on Special Issue "Design and Optimization of Pharmaceutical Gels". *Gels* **2024**, *10*, 38. https://doi.org/10.3390/gels10010038

Received: 21 December 2023
Revised: 27 December 2023
Accepted: 30 December 2023
Published: 1 January 2024

and mechanical properties. Systemic characterizations were performed, such as solid-state properties, rheological properties, degradation profile, drug loading and release, cytotoxicity, etc. This platform may have promising application aspects in pain and fever management.

(2) A collaborative article from India, Egypt, and Saudi Arabia, entitled "Piperine-Loaded In Situ Gel: Formulation, In Vitro Characterization, and Clinical Evaluation against Periodontitis". In this article, a pharmaceutical in situ gel platform was developed for piperine delivery, exhibiting sol-to-gel transformation behavior. Systemic characterizations were performed, such as drug–excipient compatibility, gelling profile, pH, viscosity, syringeability, solid-state properties, drug loading and release, etc. It is noteworthy that a randomized clinical trial was conducted for the designed formulation. This platform may have promising application aspects in periodontitis management.
(3) An article from Thailand, entitled "Response Surface Methodology for Optimization of Hydrogel-Forming Microneedles as Rapid and Efficient Transdermal Microsampling Tools". In this article, a pharmaceutical hydrogel-forming microneedle platform was developed for dermal interstitial fluid sampling, demonstrating interstitial fluid extraction effects. After utilizing Box–Behnken and central composite designs for the formulations, systemic characterizations were performed, such as morphology, swelling profile, mechanical strength, skin insertion, interstitial fluid extraction and recovery, etc. This platform may have promising application aspects in point-of-care testing.
(4) A collaborative article from Korea and India, entitled "pH Sensitive Drug Delivery Behavior of Palmyra Palm Kernel Hydrogel of Chemotherapeutic Agent". In this article, a pharmaceutical palmyra palm kernel hydrogel platform was developed for 5-fluorouracil delivery, exhibiting pH-responsive swelling capacities. Systemic characterizations were performed, such as swelling profile, drug loading and release, solid-state properties, etc. This platform may have promising application aspects in colon cancer management.

3. Overview of Contributions to Nanogels

Nanogels are defined as pharmaceutical gel platforms with nanometer sizes or those encapsulating nanoparticles. In recent years, nanogels have gained significant research interest. In this Special Issue, five articles on nanogels are included, which are briefly introduced below:

(1) A collaborative article from India and Saudi Arabia, entitled "Design and Development of a Topical Nanogel Formulation Comprising of an Unani Medicinal Agent for the Management of Pain". In this article, a pharmaceutical nanogel platform was developed for *Matricaria chamomilla* oil delivery, demonstrating enhancements in pain threshold and reductions in skin irritation capacities. Systemic characterizations were performed, such as pH, viscosity, spreadability, extrudability, texture, drug release, etc. This platform may have promising application aspects in migraine management.
(2) An article from Malaysia, entitled "Enhanced Osteogenesis Potential of MG-63 Cells through Sustained Delivery of VEGF via Liposomal Hydrogel". In this article, a pharmaceutical liposomal hydrogel platform was developed for vascular endothelial growth factor delivery, demonstrating elevated osteogenesis potential. Systemic characterizations were performed, such as morphology, porosity, solid-state properties, drug release, cell–hydrogel co-culture profile, etc. This platform may have promising application aspects in bone regeneration.
(3) A collaborative article from Saudi Arabia and Egypt, entitled "Numerical Optimization of Prednisolone-Tacrolimus Loaded Ultraflexible Transethosomes for Transdermal Delivery Enhancement; Box–Behnken Design, Evaluation, Optimization, and Pharmacokinetic Study". In this article, a pharmaceutical transethosomal gel platform was developed for prednisolone–tacrolimus co-delivery, exhibiting potential anti-inflammatory effects. After utilizing a Box–Behnken design for the formulations,

systemic characterizations were performed, such as drug loading and release, particle size, pH, spreadability, skin permeation, in vivo anti-inflammatory ability and pharmacokinetics, etc. This platform may have promising application aspects in the management of skin inflammatory conditions.

(4) An article from Poland, entitled "Emulsion-Based Gel Loaded with Ibuprofen and Its Derivatives". In this article, a pharmaceutical emulsion-based gel platform was developed for ibuprofen and derivatives thereof, demonstrating enhanced skin penetration capabilities. It is worth mentioning that the authors synthesized several derivatives of ibuprofen. Systemic characterizations were performed for the gel platform, such as morphology, stability, density, refractive index, viscosity, particle size, skin permeation, etc. This platform may have promising application aspects in the management of inflammatory skin diseases.

(5) An article from China, entitled "Design and Evaluation of Paeonol-Loaded Liposomes in Thermoreversible Gels for Atopic Dermatitis". In this article, a pharmaceutical liposome-in-gel platform was developed for paeonol delivery, exhibiting temperature-responsive gelation abilities. Systemic characterizations were performed, including liposome morphology, gelling profile, viscosity, pH, drug loading and release, in vitro and in vivo antioxidant activities, etc. This platform may have promising application aspects in atopic dermatitis management.

4. Overview of Contributions to Gel Reviews

In addition to the above interesting articles, three gel reviews are included in this Special Issue. The following is a brief introduction to them:

(1) A review from China, entitled "Utilization of Lyotropic Liquid Crystalline Gels for Chronic Wound Management". This review summarized and discussed gel-based pharmaceutical platforms suitable for chronic wound management. In particular, the authors argued that lyotropic liquid crystalline gels might be one of the most promising systems. The characteristics of lyotropic liquid crystalline gels were analyzed in detail. Of note, we are thrilled to announce that this review was contributed by the Guest Editor team of this Special Issue.

(2) A review from the USA, entitled "Enhancing Therapeutic Efficacy of Curcumin: Advances in Delivery Systems and Clinical Applications". This review summarized and discussed various drug delivery platforms suitable for curcumin delivery. In particular, the authors argued that gels might be one of the most promising systems. The composition of the reported curcumin gels was analyzed in detail.

(3) A review from Korea, entitled "Pectin Based Hydrogels for Drug Delivery Applications: A Mini Review". This (mini) review summarized and discussed representative pectin-based hydrogel pharmaceutical platforms. Interestingly, the pectin extraction methods were analyzed in detail. The authors also suggested that investigators in this field pay attention to the toxicity of such systems.

In summary, this Special Issue involved a number of interesting research articles and reviews that have represented the very recent progress of various aspects of pharmaceutical gels, including nanoscale, mesoscale, microscale, and macroscale gels. The indications for these gels also varied. These gels may offer new insights into the design, preparation, development, and application of pharmaceutical gels, and act as references for future studies. We truly appreciate the efforts of our authors, reviewers, and editors in disseminating invaluable knowledge. It is believed that open science in the research area of pharmaceutical gels has advanced, aligning well with the aims of MDPI Press.

Author Contributions: X.Z., writing—original draft preparation; Y.H., writing—original draft preparation; Z.H., writing—review and editing and conceptualization. All authors have read and agreed to the published version of the manuscript.

Funding: We are grateful for the funding from the General Program of Administration of Traditional Chinese Medicine of Guangdong Province (grant No.: 20241071) and the Guangzhou Science and Technology Plan Project (grant No.: 202201010589).

Institutional Review Board Statement: Not applicable.

Informed Consent Statement: Not applicable.

Data Availability Statement: No new data were created in this Special Issue.

Acknowledgments: We sincerely thank the editorial office (especially the assistant editor) for the kind assistance in preparing and editing this Special Issue.

Conflicts of Interest: The authors declare no conflicts of interest.

List of Contributions

1. Trinh, L.T.; Lim, S.; Lee, H.J.; Kim, I.T. Development of Efficient Sodium Alginate/Polysuccinimide-Based Hydrogels as Biodegradable Acetaminophen Delivery Systems. *Gels* **2023**, *9*, 980.
2. Sah, A.; Aggarwal, G.; Jain, G.K.; Zaidi, S.M.A.; Naseef, P.P.; Kuruniyan, M.S.; Zakir, F. Design and Development of a Topical Nanogel Formulation Comprising of a Unani Medicinal Agent for the Management of Pain. *Gels* **2023**, *9*, 794.
3. Gopalakrishna, P.K.; Jayaramu, R.A.; Boregowda, S.S.; Eshwar, S.; Suresh, N.V.; Abu Lila, A.S.; Moin, A.; Alotaibi, H.F.; Obaidullah, A.J.; Khafagy, E.-S. Piperine-Loaded In Situ Gel: Formulation, In Vitro Characterization, and Clinical Evaluation against Periodontitis. *Gels* **2023**, *9*, 577.
4. Tsai, M.H.; Megat Abdul Wahab, R.; Zainal Ariffin, S.H.; Azmi, F.; Yazid, F. Enhanced Osteogenesis Potential of MG-63 Cells through Sustained Delivery of VEGF via Liposomal Hydrogel. *Gels* **2023**, *9*, 562.
5. Alfadhel, M.M.; Zaki, R.M.; Aldosari, B.N.; Sayed, O.M. Numerical Optimization of Prednisolone–Tacrolimus Loaded Ultraflexible Transethosomes for Transdermal Delivery Enhancement; Box–Behnken Design, Evaluation, Optimization, and Pharmacokinetic Study. *Gels* **2023**, *9*, 400.
6. Agboola, A.A.; Nowak, A.; Duchnik, W.; Kucharski, Ł.; Story, A.; Story, G.; Struk, Ł.; Antosik, A.K.; Ossowicz-Rupniewska, P. Emulsion-Based Gel Loaded with Ibuprofen and Its Derivatives. *Gels* **2023**, *9*, 391.
7. Leanpolchareanchai, J.; Nuchtavorn, N. Response Surface Methodology for Optimization of Hydrogel-Forming Microneedles as Rapid and Efficient Transdermal Microsampling Tools. *Gels* **2023**, *9*, 306.
8. Wang, Y.; Yue, Y.; Jia, R.; Liu, X.; Cheng, Z.; Cheng, Y.; Xu, Y.; Xie, Z.; Xia, H. Design and Evaluation of Paeonol-Loaded Liposomes in Thermoreversible Gels for Atopic Dermatitis. *Gels* **2023**, *9*, 198.
9. Rao, K.-M.; Rao, K.S.V.K.; Palem, R.-R.; Uthappa, U.-T.; Ha, C.-S.; Han, S.-S. pH Sensitive Drug Delivery Behavior of Palmyra Palm Kernel Hydrogel of Chemotherapeutic Agent. *Gels* **2023**, *9*, 38.
10. Luo, P.; Shu, L.; Huang, Z.; Huang, Y.; Wu, C.; Pan, X.; Hu, P. Utilization of Lyotropic Liquid Crystalline Gels for Chronic Wound Management. *Gels* **2023**, *9*, 738.
11. Omidian, H.; Wilson, R.L.; Chowdhury, S.D. Enhancing Therapeutic Efficacy of Curcumin: Advances in Delivery Systems and Clinical Applications. *Gels* **2023**, *9*, 596.
12. Han, S.S.; Ji, S.M.; Park, M.J.; Suneetha, M.; Uthappa, U.T. Pectin Based Hydrogels for Drug Delivery Applications: A Mini Review. *Gels* **2022**, *8*, 834.

References

1. Rocha, B.; de Morais, L.A.; Viana, M.C.; Carneiro, G. Promising strategies for improving oral bioavailability of poor water-soluble drugs. *Expert Opin. Drug Dis.* **2023**, *18*, 615–627. [CrossRef] [PubMed]
2. Virmani, T.; Kumar, G.; Sharma, A.; Pathak, K.; Akhtar, M.S.; Afzal, O.; Altamimi, A.S.A. Amelioration of Cancer Employing Chitosan, Its Derivatives, and Chitosan-Based Nanoparticles: Recent Updates. *Polymers* **2023**, *15*, 2928. [CrossRef] [PubMed]

3. Kolawole, O.M.; Cook, M.T. In situ gelling drug delivery systems for topical drug delivery. *Eur. J. Pharm. Biopharm.* **2023**, *184*, 36–49. [CrossRef] [PubMed]
4. Liu, H.; Meng, X.Y.; Li, L.; Xia, Y.M.; Hu, X.Y.; Fang, Y. The incorporated hydrogel of chitosan-oligoconjugated linoleic acid vesicles and the protective sustained release for curcumin in the gel. *Int. J. Biol. Macromol.* **2023**, *227*, 17–26. [CrossRef] [PubMed]
5. Yan, S.Z.; Wu, S.Y.; Zhang, J.X.; Zhang, S.; Huang, Y.Y.; Zhu, H.P.; Li, Y.; Qi, B.K. Controlled release of curcumin from gelatin hydrogels by the molecular-weight modulation of an oxidized dextran cross-linker. *Food Chem.* **2023**, *418*, 135966. [CrossRef] [PubMed]
6. Lin, H.L.; Xie, L.K.; Lv, L.R.; Chen, J.R.; Feng, F.; Liu, W.Y.; Han, L.F.; Liu, F.L. Intranasally administered thermosensitive gel for brain-targeted delivery of rhynchophylline to treat Parkinson?s disease. *Colloid Surf. B* **2023**, *222*, 113065. [CrossRef] [PubMed]
7. Siafaka, P.I.; Bülbül, E.Ö.; Okur, M.E.; Karantas, I.D.; Okur, N.Ü. The Application of Nanogels as Efficient Drug Delivery Platforms for Dermal/Transdermal Delivery. *Gels* **2023**, *9*, 753. [CrossRef] [PubMed]
8. Li, Y.F.; Li, Z.; Lin, Q.; Yang, Y.W. Functional supramolecular gels based on pillar[*n*]arene macrocycles. *Nanoscale* **2020**, *12*, 2180–2200. [CrossRef] [PubMed]
9. Huang, T.; Tu, Z.C.; Shangguan, X.C.; Sha, X.M.; Wang, H.; Zhang, L.; Bansal, N. Fish gelatin modifications: A comprehensive review. *Trends Food Sci. Technol.* **2019**, *86*, 260–269. [CrossRef]
10. Wang, C.M.; Gong, Y.H.; Lin, Y.M.; Shen, J.B.; Wang, D.A. A novel gellan gel-based microcarrier for anchorage-dependent cell delivery. *Acta Biomater.* **2008**, *4*, 1226–1234. [CrossRef] [PubMed]
11. Yan, J.X.; Miao, Y.T.; Tan, H.P.; Zhou, T.L.; Ling, Z.H.; Chen, Y.; Xing, X.D.; Hu, X.H. Injectable alginate/hydroxyapatite gel scaffold combined with gelatin microspheres for drug delivery and bone tissue engineering. *Mater. Sci. Eng. C-Mater.* **2016**, *63*, 274–284. [CrossRef] [PubMed]
12. Zhang, Q.; Wang, P.D.; Fang, X.; Lin, F.; Fang, J.; Xiong, C.Y. Collagen gel contraction assays: From modelling wound healing to quantifying cellular interactions with three-dimensional extracellular matrices. *Eur. J. Cell Biol.* **2022**, *101*, 151253. [CrossRef] [PubMed]
13. Panja, S.; Adams, D.J. Stimuli responsive dynamic transformations in supramolecular gels. *Chem. Soc. Rev.* **2021**, *50*, 5165–5200. [CrossRef] [PubMed]
14. Webber, M.J.; Pashuck, E.T. (Macro)molecular self-assembly for hydrogel drug delivery. *Adv. Drug Deliv. Rev.* **2021**, *172*, 275–295. [CrossRef] [PubMed]
15. Cho, M.H.; Li, Y.; Lo, P.C.; Lee, H.R.; Choi, Y. Fucoidan-Based Theranostic Nanogel for Enhancing Imaging and Photodynamic Therapy of Cancer. *Nano-Micro Lett.* **2020**, *12*, 47. [CrossRef] [PubMed]

gels

Review

Utilization of Lyotropic Liquid Crystalline Gels for Chronic Wound Management

Peili Luo [1,†], Lei Shu [1,†], Zhengwei Huang [1,*], Ying Huang [1,*], Chuanbin Wu [1], Xin Pan [2] and Ping Hu [1]

1 College of Pharmacy, Jinan University, Guangzhou 511443, China; stu2022358310561pl@stu2022.jnu.edu.cn (P.L.); shulei@stu2021.jnu.edu.cn (L.S.); chuanbin_wu@126.com (C.W.); inzahu@hotmail.com (P.H.)
2 School of Pharmaceutical Sciences, Sun Yat-Sen University, Guangzhou 510006, China; mercurypan@foxmail.com
* Correspondence: huangzhengw@jnu.edu.cn (Z.H.); huangy2007@jnu.edu.cn (Y.H.)
† These authors contributed equally to this work.

Abstract: Management of chronic wounds is becoming a serious health problem worldwide. To treat chronic wounds, a suitable healing environment and sustained delivery of growth factors must be guaranteed. Different therapies have been applied for the treatment of chronic wounds such as debridement and photodynamic therapy. Among them, growth factors are widely used therapeutic drugs. However, at present, growth factor delivery systems cannot meet the demand of clinical practice; therefore new methods should be developed to meet the emerging need. For this reason, researchers have tried to modify hydrogels through some methods such as chemical synthesis and molecule modifications to enhance their properties. However, there are still a large number of limitations in practical use like byproduct problems, difficulty to industrialize, and instability of growth factor. Moreover, applications of new materials like lyotropic liquid crystalline (LLC) on chronic wounds have emerged as a new trend. The structure of LLC is endowed with many excellent properties including low cost, ordered structure, and excellent loading efficiency. LLC can provide a moist local environment for the wound, and its lattice structure can embed the growth factors in the water channel. Growth factor is released from the high-concentration carrier to the low-concentration release medium, which can be precisely regulated. Therefore, it can provide sustained and stable delivery of growth factors as well as a suitable healing environment for wounds, which is a promising candidate for chronic wound healing and has a broad prospective application. In conclusion, more reliable and applicable drug delivery systems should be designed and tested to improve the therapy and management of chronic wounds.

Keywords: chronic wound; growth factors; healing environment; delivery systems; lyotropic liquid crystalline

Citation: Luo, P.; Shu, L.; Huang, Z.; Huang, Y.; Wu, C.; Pan, X.; Hu, P. Utilization of Lyotropic Liquid Crystalline Gels for Chronic Wound Management. *Gels* **2023**, *9*, 738. https://doi.org/10.3390/gels9090738

Academic Editors: Pietro Matricardi and Esmaiel Jabbari

Received: 29 July 2023
Revised: 7 September 2023
Accepted: 9 September 2023
Published: 12 September 2023

1. Introduction

1.1. Chronic Wound

A chronic wound is defined as a wound that cannot be repaired in a normal, timely, and orderly manner to achieve anatomical and functional integrity within 8 weeks [1]. It is generally divided into five common types: pressure ulcer, venous ulcer, arterial ulcer, diabetic ulcer, and traumatic ulcer. Its etiology mainly includes chronic diseases, vascular problems, diabetes, neuropathy, malnutrition, old age, stress, infection, and edema [2]. It has become a major problem in the global medical field due to its high prevalence, long healing cycle, expensive treatment, and serious consequences such as amputation, and even death. In 2017, there were 6.7 million chronic wound patients in the United States, with an annual growth rate of more than 2%. The annual treatment cost is as high as USD 50 billion, and it is expected to increase yearly at a rate of 10% [3].

Wound healing is an orderly biological process with high complexity and precise regulation. The whole process is divided into four stages (Figure 1): hemostatic stage, inflammatory stage, proliferative stage, and remodeling stage [4,5]. (1) The hemostatic stage begins once blood leaks into the exposed wound and ends within the first minutes to hours of injury. Platelets aggregate and activate subendothelial collagen after the extrinsic clotting cascade is triggered and localized vasoconstriction mediators are released, leading to the formation of a hemostatic plug through the release of cytokines and growth factors. Hemostatic plug can also serve as a provisional matrix for many cell types involved in later stages of wound healing [6]. For non-bleeding chronic wounds, the hemostasis stage is generally omitted, but it is still an inherit stage of healing. (2) In inflammatory stage, platelets aggregate and leukocytes infiltrate into the wound site [7], aiming at building an immune barrier against microorganisms. Neutrophils perform phagocytosis by releasing proteolytic enzymes and oxygen-derived free-radical species to remove bacteria and xenobiotics from wounds [8]. They are removed from the wound before proceeding to the next stage of healing without causing tissue damage or enhancing inflammatory response. Then, macrophages appear in the wound and continue the process of phagocytosis [9]. (3) When it comes to the proliferative phase, fibroblasts play an important role. They proliferate and migrate to the wound, followed by changing into a myofibroblast phenotype and extending themselves to attach to fibronectin and collagen in the extracellular matrix. By synthesizing collagen, they build the integrity and strength of tissues [10]. In addition, endothelial cells and keratinocytes secret growth factors like keratinocyte growth factor 2 (KGF-2) and vascular endothelial growth factor (VEGF), leading to granulation tissue growth, neovascularization, and re-epithelization and promoting wound healing [11]. (4) As the final step, in remodeling stage, synthesis and breakdown of collagen as well as extracellular matrix remodeling take place continuously, which is regulated by inhibitory factors. The process is regulated by a number of factors like platelet-derived growth factor (PDGF), transforming growth factor-β (TGF-β), and fibroblast growth factor (FGF) [12]. Overall, the healing process involves the differentiation, migration, and proliferation of a variety of inflammatory and tissue cells, as well as the interaction between cells and structural proteins, cytokines, and protein kinases [4]. Any stage that hinders wound healing will affect the normal repair of the wound, and eventually form a chronic wound.

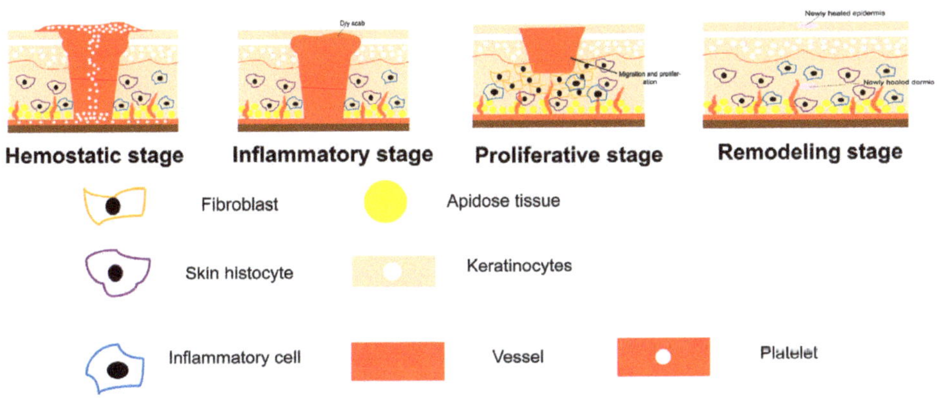

Figure 1. Inflammatory, proliferative, and remodeling stage during wound healing.

1.2. Healing Conditions

To treat chronic wounds, besides the treatment of primary diseases, an overall management of the wound is needed, which usually includes drug treatment, routine cleaning, local debridement, and infection control. During the treatment of chronic wounds, the local environment of the wound affects the whole healing process. Sufficient growth factors can promote the formation of granulation tissue and accelerate the re-epithelization process of wound, which is the main factor to prevent chronic wound formation and an important premise to promote wound repair [11]. Therefore, providing a suitable healing environment and stable delivery of growth factors are the keys to treat chronic wounds. The detailed underlying mechanisms are discussed as follows.

1.2.1. Suitable Healing Environment

The theories about the local environment needed for wound healing mainly include dry healing theory and wet healing theory. For many years, dry healing theory has been the mainstream guiding theory in clinical treatment. This theory presumes that wound healing requires the participation of atmospheric oxygen and a dry environment. Nevertheless, one study shows that atmospheric oxygen cannot be directly used by the wound tissue, and thus it has no practical significance for wound healing, and excessive oxygen even hinders wound healing [13]. At the same time, the dry healing environment makes the wound easy to dehydrate and scab, which is not conducive to the crawling of epithelial cells. The growth factors secreted by the body will lose their biological activity easily, resulting in slow healing [14]. In addition, the dry environment cannot prevent bacterial invasion or maintain the humidity and temperature of the wound, which is not favorable for wound healing. Wet healing theory considers that a specific wet microenvironment is necessary for the process of wound healing. Its principles include (Figure 2): (1) Reducing the oxygen pressure of the wound. The hypoxic healing environment is conducive to the formation of epithelial cells and collagen fibers, promoting the growth of fibroblasts, stimulating vascular proliferation, and improving the stability of growth factors [15]. (2) Maintaining a certain temperature and humidity of the wound can retain the release of wound exudates, and activate a variety of enzymes and enzyme activation factors (especially protease and urokinase). These enzymes can promote the degradation and absorption of necrotic tissue and fibrin, and avoid the formation of scabs [16]. (3) Avoiding air exposure to nerve endings can reduce pain. (4) Accelerating the migration of epithelized epidermal cells rapidly can shrink the wound [17], increase the formation of granulation tissue, and re-epithelize the wound [18]. (5) The hypoxic and slightly acidic environment, with pH ranging from 5.0–6.0 under closed conditions, can directly inhibit the growth of bacteria [19], which is favorable for the reproduction and function of leukocytes to prevent the penetration of bacteria and improve local immunity [20].

At present, the wet healing theory is preferably accepted by the academic community, compared to the dry healing theory. Some traditional wound dressing components like polysaccharides and vaseline are used to ensure that the wet conditions can be applied to the treatment [21,22]. Researchers have been devoted to developing new wound dressings, as well. For example, Rodrigues et al. developed a wound dressing using a hydrocellular functional material, which was found to be effective in diabetic wound healing. The material has shown good results due to the excellent balance of exudates removal, and a moist film on the wound can prevent the formation of dry scab and facilitate epithelialization of wound [23]. Agarwal et al. found that a curcumin-loaded polycaprolactone/polyvinyl alcohol-silk fibroin based electrospun nanofibrous mat can provide a moist microenvironment, to protect the diabetic wound from secondary infections, remove the wound exudate, control the biofilm, and promote tissue regeneration [24]. Based on these studies, an ideal delivery system for chronic wound application should be able to form a uniform and well-sealed moist film on the wound surface and create a moist and hypoxic local environment.

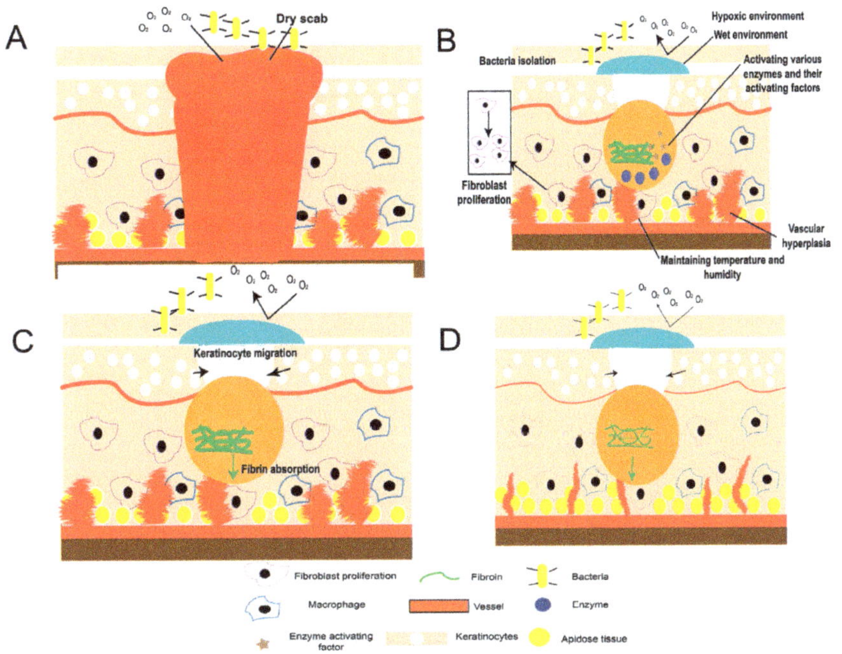

Figure 2. Wet healing theory: specific wet microenvironment can promote wound healing. (**A**) Initial inflammatory stage; (**B**) wet microenvironment protection; (**C**) keratinocyte migration; (**D**) regeneration outcome.

1.2.2. Sustained Delivery of Growth Factor

Growth factors are a class of bioactive proteins that significantly regulate cell growth and differentiation, which play important roles in the process of wound healing [25]. They have high efficiency and pleiotropy that can affect the processes of cell proliferation, migration, and extracellular matrix synthesis by regulating cell responses in the process of wound repair [26]. Although there have been many clinical trials on growth factor reparations, they were relatively small and single-centered. The results showed that the applications of growth factors to the intervention treatment of chronic wounds caused by burns, diabetic foot, and pressure ulcer can significantly accelerate the process of wound healing and improve the quality of healing [27–29]. The most widely used growth factor is human epidermal growth factor (hEGF). In addition, platelet-derived growth factor (PDGF) [30,31], basic fibroblast growth factor (bFGF) [32], and VEGF, have a considerable amount of applications [33].

The inactivation of these growth factors can be caused by different reasons, including physical changes such as salting-out and surface adsorption or chemical changes like breakage of bonds. Generally, chemical changes are the most important cause. Peptide bonds hydrolyze when the pH changes, while peptides are prone to aggregation at the isoelectric point. Temperature can cause changes in the conformation of peptides. As previously mentioned, different degradation reactions dominate under diverse conditions (such as temperature, pH, fluorescence, etc.) [34].

Specifically, an ideal delivery system to treat chronic wounds needs to have the following features: (1) Providing a moist and hypoxic environment suitable for wound healing; (2) releasing growth factors slowly and improving their stability [35]. In addition to these key attributes, it is expected to reduce the frequency of administration, be convenient to administrate, and finally improve the patient's compliance.

2. Growth Factor Delivery Systems to Heal Chronic Wounds

As previously discussed, both suitable healing environment and stable growth factor delivery are considered as prerequisites and crucial for wound healing. In this section, existing and innovative delivery systems are evaluated according to the two preconditions.

2.1. Existing Delivery Systems

In recent years, a variety of growth factor delivery systems for chronic wounds have been developed, most of which are in the form of solution or gel, as listed in Table 1 [36]. The majority of these delivery systems need to be applied and replaced frequently, a situation which is prone to cause pain and increase the probability of wound infection [37,38]. Solution and gel systems also have two critical bottlenecks—inability to maintain a suitable healing environment and stable growth factor delivery. After the solution is administered, because the wound is an open system and the drug could not be retained on the wound for a long time, it is impossible to control the formation of a hypoxic and moist microenvironment, or the isolation and protection effect on the wound. The gel is a normally crosslinked structure formed by the interlacing of polymer chains [39,40], which is not strong to control the release of growth factor, and is prone to cracking or falling due to dehydration caused by changes in external environmental humidity [41]. As a result, the clinical efficacy of current growth factor delivery systems for chronic wounds available in the market at present, is far from being ideal.

Table 1. Commercial formulations of growth factor drugs that are currently widely used in the clinic [36].

Growth Factor	Number of Amino Acid Residue	Commercial/Clinical Products for Wound Treatment	Dosage Form	Matrix Material
Epidermal growth factor	53	HEBERPROT-P® (Heber, Biotech, Havana, Cuba)	Solution	Normal saline
		EASYEF® (Daewoong Pharmaceutical, Seoul, Republic of Korea)	Solution	Normal saline
		REGEN-D™150 (Bharat Biotech International Limited, Hyderabad, India)	Gel	—
Platelet-derived growth factor	196–370	Regranex® (Smith and Nephew, Inc., London, UK)	Gel	Carboxymethyl cellulose
Basic fibroblast growth factor	155	Fiblast®Spray (Kaken Pharmaceutical Co., Ltd., Tokyo, Japan)	Solution	Normal saline
Vascular endothelial growth factor	121–206	Telbermin (Genentech, South San Francisco, CA, USA), phase I	Gel	—

2.2. Innovative Delivery Systems

New delivery systems need to be developed to provide a suitable healing environment and stable growth factor delivery. Modified natural polymer hydrogel and synthesized artificial polymer hydrogel are possible systems (Figure 3), as discussed below.

2.2.1. Hydrogels of Modified Natural Polymers

In order to provide a wet and hypoxic local environment suitable for wound healing, researchers have tried to use hydrogels that comprise natural polymers as drug carriers, such as sodium alginate, carrageenan, etc., which possess good biocompatibility and biodegradability and are less likely to cause immune rejection [42–44]. However, normal hydrogel possesses completely disordered polymeric network structures, with low mechanical strength, poor physical stability, and lack of self-healing property. It is difficult to maintain a hypoxic and moist environment for a wound consistently to meet the expectations of clinical use [45]. At the same time, this disordered network also has a limited stabilizing effect on polypeptide drugs such as growth factors. A sustained growth factor delivery cannot be obtained. Thus, investigators have improved the mechanical properties of natural polymer hydrogels by adding small organic molecules such as acrylamide and acrylic acids [46,47]. These molecules can react with the key functional groups of hydrogel components to form polymeric structures more orderly and create an anoxic and humid environment [48]. For example, Zhang et al. developed a new antibacterial hydrogel wound dressing that comprises poly(aminoethyl) modified chitosan, which not only can enhance its antibacterial activity, but also promote the formation and stabilization of the prepared hydrogel [49]. However, it is possible that the modification of natural polymers can lead to unexpected changes in biocompatibility when molecules react with functional groups or other functional groups are introduced in gels.

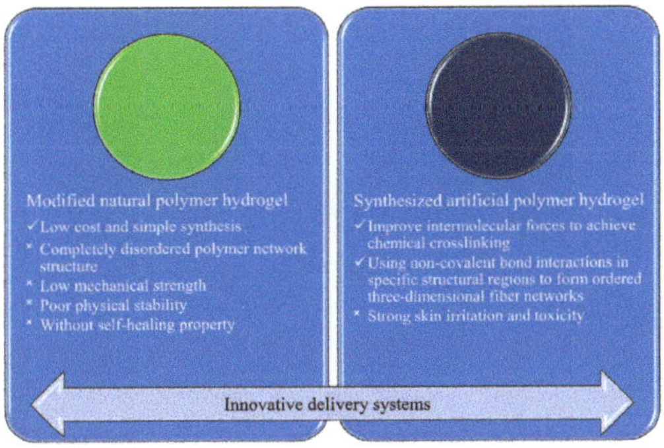

Figure 3. Comparison of two innovative delivery systems.

2.2.2. Hydrogels of Synthesized Polymers

Scientists have also tried to prepare hydrogels with chemically synthesized polymers. Usually, crosslinking agents, initiators, chelators, and chain transfer agents such as glutaraldehyde, N,N-methylene bis-acrylamide, metal ions, and ammonium persulfate, respectively, need to be added at the same time to control the polymerization kinetics, but the residue of the above substances normally causes unwanted strong skin irritation and toxicity [50,51]. Some studies have also achieved crosslinking by increasing intermolecular forces such as hydrogen bonds and electrostatic coupling, and used noncovalent bonding in specific structural regions to make the polymers self-assemble into an orderly three-dimensional fiber network [52,53]. Specifically, green chemical crosslinking methods like free-radical polymerization, reaction of complementary groups, and enzymatic reaction can be applied in various hydrogel systems, considering the biochemical features of the polymers [54,55]. For example, Raia et al. developed a composite hydrogel in which hyaluronic acid was covalently crosslinked with silk, and silk themselves crosslinked via the effect of horseradish peroxidase. This kind of hydrogel exhibited both mechanical integrity and hydrophilicity [56]. Synthetic hydrogels are very different from natural polymer hydrogels in the chemical composition and preparation process, which affect the mechanical properties and stability of gels [57].

3. Lyotropic Liquid Crystalline Might Be a Promising Candidate for Chronic Wound Healing

Although hydrogels of modified natural polymers and synthesized polymers are applied to provide a suitable environment and stable growth factor delivery, these two strategies may introduce byproducts that provoke safety issues and limit their clinical use [58]. Therefore, a better new delivery system should be developed. Lyotropic liquid crystalline (LLC) is an ideal candidate for chronic wound healing with its properties of good biosafety, ease of industrialization, and low cost [59,60].

LLC is a long-range ordered liquid crystal structure formed by self-assembly of amphiphilic molecules and solvents (Figure 4). There are two common types of LLC. One kind is prepared from alkyl-based materials like fatty acid salts, alkyl sulfonates, etc. The hydrophilic parts of these materials such as carboxyl and sulfonic groups are linked to a long hydrophobic group, forming a polar "head" and two hydrophobic "tails". The other type is prepared from phosphor-lipids such as egg/bean phospholipids and glycosphingolipids, with one polar "head" and two hydrophobic "tails" in the molecule. The hydrophobic groups in the molecule are usually arranged side-by-side [61]. Due to its spontaneous phase transition characteristics, ordered lattice structure, excellent drug loading, good biosafety and biocompatibility [62,63], release profile and stabilization of various drugs [64], it is expected to act as a prospect candidate delivery system for chronic wound treatment. LLC not only has a healing effect on the wound, but also can be useful as a nano drug carrier in implants [65,66].

Figure 4. Forms of lyotropic liquid crystalline.

LLC was featured with a unique spontaneous transition process from low-viscosity precursor to high-viscosity gel triggered by medium change. Its precursors are generally a layered liquid crystalline, which is formed by the superposition of bimolecular layers that comprise amphiphilic molecules, contributing to good fluidity. The polypeptide drugs (like growth factors) can be incorporated in the interlayers between the bimolecular layers. The precursors can be evenly smeared or sprayed on the wound surface when contacted with water, then a gel with ordered lattice structure is rapidly formed, transitioning from a layered phase to a cubic phase. Herein, a relatively closed aqueous channel is formed for loading growth factors and the formed LLC fits the non-smooth wound surface. Moreover, a hypoxic environment is created by the physical occultation, which can avoid the degradation of encapsulated drugs in contact with external water and oxygen molecules [67].

3.1. LLC Can Provide Sustained and Stable Growth Factor Delivery

If the diameter of the water channel is smaller than the molecular size of the polypeptide, the polypeptide cannot be completely encapsulated in the water channel, which may result in irregular drug release behavior and degradation [68]. In contrast, if the water channel is larger than the polypeptide, the effect of LLC on improving the stability of polypeptide drugs weakens. Only if the size of the water channel is similar to that of the polypeptide, the LLC lattice structure can embed the polypeptide nicely in the water channel, and the controlled release and stability of polypeptide can be expected. It is speculated that the size of the water channel can be adjusted to match the molecular size of the growth factors by optimizing the formulation, in order to improve the stability of the growth factors in the LLC system (Figure 5) [69].

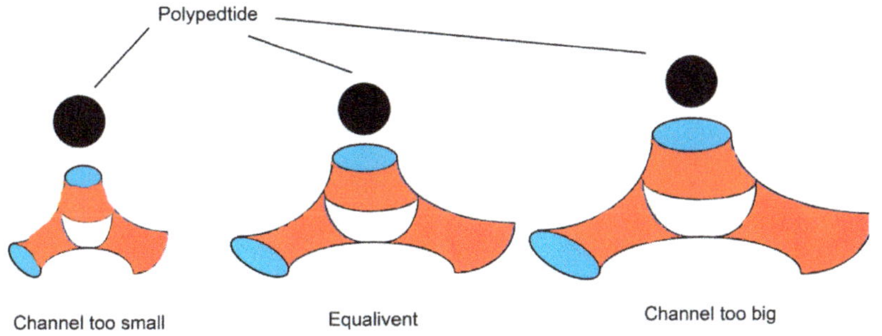

Figure 5. Relationship between the size of water channel and polypeptide to be embedded.

The release of drug from LLC is generally a concentration-mediated diffusion process. Specifically, the drug with higher concentrations in liquid crystalline carrier diffuses into the release medium with low concentrations. In general, the release of drugs from LLC conforms to Higuchi equation (Equation (1)) or Fick diffusion equation (Equation (2)) [70,71]:

$$\frac{Q}{A} = 2C_0\sqrt{\frac{Dt}{\pi}} \tag{1}$$

where Q/A represent the amount of drug diffusing into the receiving tank per unit diffusion area, C_0 is the initial drug concentration, D represents the drug apparent diffusion coefficient, and t means diffusion time of drug:

$$dS = -DF\frac{dC}{dx} \times dt \tag{2}$$

where dS represents the diffusion of substance (solute) in dt time, D is the diffusion coefficient, F is the diffusion area, and dC/dx represents the concentration gradient.

It is conjectured that through the sequential collapse of the lattice structure of LLC, the release rate of growth factor can be precisely regulated and the release mechanisms can be explained by Higuchi equation or Fick diffusion equation.

3.2. LLC Can Provide a Suitable Healing Environment for Wounds

Three major advantages of LLC can provide a suitable healing environment.

First, LLC can promote cell adhesion and proliferation. Its bimolecular layers comprise amphiphilic molecules that can provide a microenvironment similar to the cell membrane. Under physiological conditions, this biomimetic structure has good affinity and compatibility with the wound tissue [72], and acts as a skeleton or platform for tissue regeneration.

Second, LLC can provide a moist local environment for the wound. The water containing capacity of the system is due to its specific lattice structure. Common hydrogels (such as sodium alginate gel) lack a lattice structure [73], while the water channels within the LLC lattice structure have better water retention. The ordered lattice structure maintains a local environment with a certain temperature and humidity for the wound, which is favorable to activate a variety of enzymes and enzyme-activating factors, and promote the degradation and absorption of necrotic tissue and fibrin. In addition, the gel with certain mechanical strength formed can closely adhere to the uneven wound surface. LLC has low-oxygen permeability to isolate the external oxygen and reduce the oxygen pressure in the wound [74]. Nevertheless, the degradation products of LLC like glycerol have a moisturizing effect and oleic acid is non-irritating to the skin [75].

Third, LLC is a viscoelastic material. Upon the application of an external force, the lipid bilayer in the lattice structure has a spring-like effect, viz., it can respond to the external force by generating molecular compression and lattice rearrangement, and restore the original ordered structure. This effectively buffers and absorbs the impact force on the wound and protects the wound (Figure 6) [76]. Of note, the mechanical strength of LLC can be altered by the lattice structure. By controlling the lattice structure, the mechanical strength of the system can be adjusted. In summary, LLC not only can deliver the growth factor sustainably, but also maintain an environment that is beneficial for chronic wound healing. We call on more researchers to devote to this research and accelerate its translation and application.

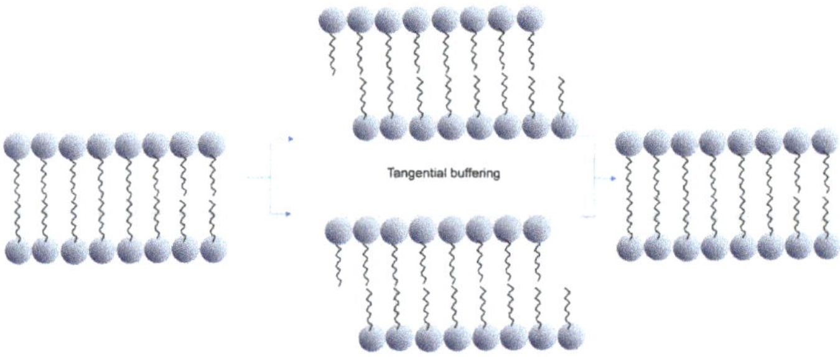

Figure 6. Bimolecular layers in the lyotropic liquid crystal exhibit the spring-like effect.

3.3. LLC Application in Wound Healing

LLC has made some progress as a wound healing dressing system, listed in Table 2. Zhou et al. constructed recombinant human epidermal growth factor (rhEGF)-containing lyotropic liquid crystalline precursor systems for chronic wound therapy, which exhibit good cargo stability and mechanical properties. It showed distinct promotion effects on wound closure, inflammatory recovery, and re-epithelization process in Sprague–Dawley rat models. Interestingly, different diameters of the internal water channels led to different in vitro release rates of rhEGF, and finally different therapeutic responses in cellulo and in vivo. Yue et al. investigated an LLC-based bacteria-resistant and self-healing spray dressing loaded with ε-polylysine, which can kill antibiotic-resistant bacteria efficiently, even the methicillin-resistant Staphylococcus aureus. The cubic cells of LLC could encapsulate PLL to improve its stability and induce a sustained release. The LLC precursor could spontaneously transit to a cubic phase gel once exposed to physiological fluid, which is endowed with mechanically responsive viscoelasticity that builds a robust and flexible defense for wounds. Chen et al. designed a hyaluronic acid combined LLC-based spray dressing loaded with the anti-fibrotic drug pirfenidone. The dressing possessed high levels of water absorption for exudate absorption. The self-assembled lattice nanostructures provide excellent mechanical protection to promote the healing process and steady pirfenidone release to exert a scar prophylaxis effect. In the deep partial thickness burn wound model they established, the dressing performs excellent healing effects, and pirfenidone-loaded dressing displayed an ideal prognosis with skin as smooth as healthy skin. The healing promotion of dressing was considered to be related to a clearly shortened inflammation phase, with contributions from water management and mechanical protection by the dressing. The scar prophylaxis of pirfenidone-loaded dressing was proven to be related to the regulation of collagen synthesis and degradation. It offered significant promise as a spray dressing for deep partial thickness burn injuries.

Table 2. Representing articles about wound healing using LLC.

No.	Main Materials	Type of Wound	Loading Drug	Reference
1	GMO	Chronic wound	rhEGF	[77]
2	GMO	Post-operative wound	ε-polylysine	[78]
3	Hyaluronic acid, GMO	Burn wound	Pirfenidone	[79]

4. Conclusions and Future Directions

Treatment of chronic wound is a serious health issue that is expanding around the world. Considering the different wound healing stages, providing a suitable healing environment and stable growth factor delivery are the keys to the treatment of chronic wounds. Existing delivery systems are difficult to meet all the requirements. Delivery systems like modified natural polymer hydrogel and synthesized artificial polymer hydrogel have been applied to heal chronic wounds. For natural polymers, small organic molecules can modify the gels to increase mechanical strength. Hydrogels that comprise chemically synthesized polymers can minimize their skin toxicity by green chemical crosslinking. However, they still have problems in biosafety and industrialization. Thus, there is a critical and urgent need to develop new systems which are able to fulfill the above demands. As an innovative delivery system, LLC meets the requirements of chronic wound healing. It builds a moist environment and its structure can keep the structure and activity of growth factors stable. The release of growth factor can be regulated in the LLC gel. Additionally, it shows advantages such as biocompatibility over other systems, which has a promising future for clinical application.

The research on wound healing has been proceeding rapidly. Numerous efforts are put into the studies of wound pathogenesis and providing insights about the mechanisms of healing process. Moreover, advances in pharmaceutical sciences have resulted in the production of new active molecules that can improve the tissue regeneration rate and accelerate the compromised physiologic processes. Along with these advances, more reliable and applicable drug delivery systems will be designed and tested. The current wound healing preparations still have many limitations, mainly in manufacturing. An ideal wound healing product should be stable and sterile. It is not easy to maintain the activity of ingredients and maintain an aseptic state at the same time, especially for protein drugs. In addition, the manual labor required for synthesis and raw material cost should be considered [80,81]. In the future, drug delivery systems for chronic wound applications will be expected to be commercially developed through different aspects, such as simplifying the preparation methods and establishing quality control of the final products.

Author Contributions: Conceptualization, Y.H. and Z.H.; methodology, P.L.; software, P.L.; validation, P.L. and L.S.; formal analysis, L.S.; investigation, P.L.; resources, L.S.; data curation, L.S.; writing—original draft preparation, P.L.; writing—review and editing, Z.H. and P.H.; visualization, Y.H.; supervision, C.W.; project administration, X.P.; funding acquisition, Z.H. All authors have read and agreed to the published version of the manuscript.

Funding: We are thankful to the funds from China Postdoctoral Science Foundation Special Funded Project (no. 2022T150268) and the Guangzhou Science and Technology Plan Project (no. 202201010589).

Institutional Review Board Statement: Not applicable.

Informed Consent Statement: Not applicable.

Data Availability Statement: Not applicable.

Conflicts of Interest: The authors declare no conflict of interest.

References

1. Powers, J.G.; Higham, C.; Broussard, K.; Phillips, T.J. Wound healing and treating wounds: Chronic wound care and management. *J. Am. Acad. Dermatol.* **2016**, *74*, 607–625. [CrossRef] [PubMed]
2. Maxson, S.; Lopez, E.A.; Yoo, D.; Danilkovitch-Miagkova, A.; LeRoux, M.A. Concise review: Role of mesenchymal stem cells in wound repair. *Stem Cells Transl. Med.* **2012**, *1*, 142–149. [CrossRef]
3. Skrepnek, G.H.; Mills, J.L., Sr.; Lavery, L.A.; Armstrong, D.G. Health care service and outcomes among an estimated 6.7 million ambulatory care diabetic foot cases in the US. *Diabetes Care* **2017**, *40*, 936–942. [CrossRef]
4. Zhao, S.; Li, L.; Wang, H.; Zhang, Y.; Cheng, X.; Zhou, N.; Rahaman, M.N.; Liu, Z.; Huang, W.; Zhang, C. Wound dressings composed of copper-doped borate bioactive glass microfibers stimulate angiogenesis and heal full-thickness skin defects in a rodent model. *Biomaterials* **2015**, *53*, 379–391. [CrossRef]

5. Tottoli, E.M.; Dorati, R.; Genta, I.; Chiesa, E.; Pisani, S.; Conti, B. Skin wound healing process and new emerging technologies for skin wound care and regeneration. *Pharmaceutics* **2020**, *12*, 735. [CrossRef] [PubMed]
6. Ellis, S.; Lin, E.J.; Tartar, D. Immunology of wound healing. *Curr. Dermatol. Rep.* **2018**, *7*, 350–358. [CrossRef]
7. Wang, Z.; Qi, F.; Luo, H.; Xu, G.; Wang, D. Inflammatory microenvironment of skin wounds. *Front. Immunol.* **2022**, *13*, 789274. [CrossRef] [PubMed]
8. Xiang, Y.; Qi, X.; Cai, E.; Zhang, C.; Wang, J.; Lan, Y.; Deng, H.; Shen, J.; Hu, R. Highly efficient bacteria-infected diabetic wound healing employing a melanin-reinforced biopolymer hydrogel. *Chem. Eng. J.* **2023**, *460*, 141852. [CrossRef]
9. Bayat, M.; Sarojini, H.; Chien, S. The role of cluster of differentiation 163-positive macrophages in wound healing: A preliminary study and a systematic review. *Arch. Dermatol. Res.* **2023**, *315*, 359–370. [CrossRef]
10. Li, L.; Ma, Y.; He, G.; Ma, S.; Wang, Y.; Sun, Y. Pilose antler extract restores type I and III collagen to accelerate wound healing. *Biomed. Pharmacother.* **2023**, *161*, 114510. [CrossRef]
11. Belvedere, R.; Novizio, N.; Morello, S.; Petrella, A. The combination of mesoglycan and VEGF promotes skin wound repair by enhancing the activation of endothelial cells and fibroblasts and their cross-talk. *Sci. Rep.* **2022**, *12*, 11041. [CrossRef]
12. Spielman, A.F.; Griffin, M.F.; Parker, J.; Cotterell, A.C.; Wan, D.C.; Longaker, M.T. Beyond the Scar: A Basic Science Review of Wound Remodeling. *Adv. Wound Care* **2023**, *12*, 57–67. [CrossRef] [PubMed]
13. Stahl, H.C.; Ahmadi, F.; Nahzat, S.M.; Dong, H.J.; Stahl, K.W.; Sauerborn, R. Health economic evaluation of moist wound care in chronic cutaneous leishmaniasis ulcers in Afghanistan. *Infect. Dis. Poverty* **2018**, *7*, 63–78. [CrossRef] [PubMed]
14. Liu, Y.S.; Men, J.; Zhang, L.T.; Cheng, L.F.; Yang, W.B.; Zhang, W.H. Microstructural evolution and self-healing mechanism of a 2D C/SiC-BCx composite under constant load in static wet oxygen and dynamic combustion atmosphere. *Mater. Corros.* **2015**, *66*, 128–136. [CrossRef]
15. Steiner, C.A.; Cartwright, I.M.; Taylor, C.T.; Colgan, S.P. Hypoxia-inducible factor as a bridge between healthy barrier function, wound healing, and fibrosis. *Am. J. Physiol.—Cell Physiol.* **2022**, *323*, C866–C878. [CrossRef] [PubMed]
16. Skórkowska-Telichowska, K.; Czemplik, M.; Kulma, A.; Szopa, J. The local treatment and available dressings designed for chronic wounds. *J. Am. Acad. Dermatol.* **2013**, *68*, e117–e126. [CrossRef]
17. Luan, X.; Li, W.; Lou, F. Applied analysis of humanized nursing combined with wet healing therapy to prevent bedsore. *Eur. Rev. Med. Pharmacol. Sci.* **2016**, *20*, 4162–4166.
18. Zahid, M.; Lodhi, M.; Rehan, Z.A.; Tayyab, H.; Javed, T.; Shabbir, R.; Mukhtar, A.; El Sabagh, A.; Adamski, R.; Sakran, M.I. Sustainable development of chitosan/Calotropis procera-based hydrogels to stimulate formation of granulation tissue and angiogenesis in wound healing applications. *Molecules* **2021**, *26*, 3284. [CrossRef]
19. Han, X.; Ju, L.S.; Irudayaraj, J. Oxygenated Wound Dressings for Hypoxia Mitigation and Enhanced Wound Healing. *Mol. Pharm.* **2023**, *20*, 3338–3355. [CrossRef] [PubMed]
20. Han, S.-K. *Innovations and Advances in Wound Healing*; Springer Nature: Berlin/Heidelberg, Germany, 2023.
21. Lei, H.; Zhu, C.; Fan, D. Optimization of human-like collagen composite polysaccharide hydrogel dressing preparation using response surface for burn repair. *Carbohydr. Polym.* **2020**, *239*, 116249. [CrossRef]
22. Wang, X.-F.; Li, M.-L.; Fang, Q.-Q.; Zhao, W.-Y.; Lou, D.; Hu, Y.-Y.; Chen, J.; Wang, X.-Z.; Tan, W.-Q. Flexible electrical stimulation device with Chitosan-Vaseline® dressing accelerates wound healing in diabetes. *Bioact. Mater.* **2021**, *6*, 230–243. [CrossRef]
23. Rodrigues, M.; Govindharajan, T. Study of hydrocellular functional material as microbicidal wound dressing for diabetic wound healing. *J. Appl. Biomater. Funct. Mater.* **2021**, *19*, 22808000211054930. [CrossRef]
24. Agarwal, Y.; Rajinikanth, P.; Ranjan, S.; Tiwari, U.; Balasubramnaiam, J.; Pandey, P.; Arya, D.K.; Anand, S.; Deepak, P. Curcumin loaded polycaprolactone-/polyvinyl alcohol-silk fibroin based electrospun nanofibrous mat for rapid healing of diabetic wound: An in-vitro and in-vivo studies. *Int. J. Biol. Macromol.* **2021**, *176*, 376–386. [CrossRef] [PubMed]
25. Zarei, F.; Soleimaninejad, M. Role of growth factors and biomaterials in wound healing. *Artif. Cells Nanomed. Biotechnol.* **2018**, *46*, 906–911. [CrossRef] [PubMed]
26. Cheng, B.; Yan, Y.; Qi, J.; Deng, L.; Shao, Z.-W.; Zhang, K.-Q.; Li, B.; Sun, Z.; Li, X. Cooperative assembly of a peptide gelator and silk fibroin afford an injectable hydrogel for tissue engineering. *ACS Appl. Mater. Interfaces* **2018**, *10*, 12474–12484. [CrossRef]
27. Gragnani, A.; Tonarelli, E.; Chomiski, V.; Daher, R.P.; Ferreira, L. Fibroblast growth factor in the treatment of burns: A systematic review. *Burns* **2022**, *48*, 104–110. [CrossRef]
28. Kumar, N.; Verma, A.; Mishra, A.; Agrawal, G.; Agrawal, A.; Mishra, S. Platelet Derived Growth Factor in Healing of Large Diabetic Foot Ulcers in Indian Clinical Set-Up: A Protocol-Based Approach. Available online: http://static.webmedcentral.com/article_view/3985 (accessed on 1 February 2013).
29. Landi, F.; Aloe, L.; Russo, A.; Cesari, M.; Onder, G.; Bonini, S.; Carbonin, P.U.; Bernabei, R. Topical treatment of pressure ulcers with nerve growth factor: A randomized clinical trial. *Ann. Intern. Med.* **2003**, *139*, 635–641. [CrossRef] [PubMed]
30. Papanas, D.; Maltezos, E. Benefit-risk assessment of becaplermin in the treatment of diabetic foot ulcers. *Drug Saf.* **2010**, *33*, 455–461. [CrossRef]
31. Ziyadeh, N.; Fife, D.; Walker, A.M.; Wilkinson, G.S.; Seeger, J.D. A matched cohort study of the risk of cancer in users of becaplermin. *Adv. Ski. Wound Care* **2011**, *24*, 31–39. [CrossRef] [PubMed]
32. Uchi, H.; Igarashi, A.; Urabe, K.; Koga, T.; Nakayama, J.; Kawamori, R.; Tamaki, K.; Hirakata, H.; Ohura, T.; Furue, M. Clinical efficacy of basic fibroblast growth factor (bFGF) for diabetic ulcer. *Eur. J. Dermatol.* **2009**, *19*, 461–468. [CrossRef]

33. Goswami, A.G.; Basu, S.; Huda, F.; Pant, J.; Ghosh Kar, A.; Banerjee, T.; Shukla, V.K. An appraisal of vascular endothelial growth factor (VEGF): The dynamic molecule of wound healing and its current clinical applications. *Growth Factors* **2022**, *40*, 73–88. [CrossRef]
34. Lee, K.; Silva, E.A.; Mooney, D.J. Growth factor delivery-based tissue engineering: General approaches and a review of recent developments. *J. R. Soc. Interface* **2011**, *8*, 153–170. [CrossRef]
35. Zhang, X.; Feng, J.; Feng, W.; Xu, B.; Zhang, K.; Ma, G.; Li, Y.; Yang, M.; Xu, F.-J. Glycosaminoglycan-based hydrogel delivery system regulates the wound microenvironment to rescue chronic wound healing. *ACS Appl. Mater. Interfaces* **2022**, *14*, 31737–31750. [CrossRef]
36. Lau, H.-C.; Kim, A. Pharmaceutical perspectives of impaired wound healing in diabetic foot ulcer. *J. Pharm. Investig.* **2016**, *46*, 403–423. [CrossRef]
37. Garcia-Orue, I.; Gainza, G.; Gutierrez, F.B.; Aguirre, J.J.; Evora, C.; Pedraz, J.L.; Hernandez, R.M.; Delgado, A.; Igartua, M. Novel nanofibrous dressings containing rhEGF and Aloe vera for wound healing applications. *Int. J. Pharm.* **2017**, *523*, 556–566. [CrossRef]
38. Xia, G.; Liu, Y.; Tian, M.; Gao, P.; Bao, Z.; Bai, X.; Yu, X.; Lang, X.; Hu, S.; Chen, X. Nanoparticles/thermosensitive hydrogel reinforced with chitin whiskers as a wound dressing for treating chronic wounds. *J. Mater. Chem. B* **2017**, *5*, 3172–3185. [CrossRef] [PubMed]
39. Khan, M.U.A.; Stojanović, G.M.; Hassan, R.; Anand, T.J.S.; Al-Ejji, M.; Hasan, A. Role of Graphene Oxide in Bacterial Cellulose−Gelatin Hydrogels for Wound Dressing Applications. *ACS Omega* **2023**, *8*, 15909–15919. [CrossRef]
40. Al-Arjan, W.S.; Khan, M.U.A.; Almutairi, H.H.; Alharbi, S.M.; Razak, S.I.A. pH-Responsive PVA/BC-f-GO dressing materials for burn and chronic wound healing with curcumin release kinetics. *Polymers* **2022**, *14*, 1949. [CrossRef]
41. Li, J.-Y.; Lin, Y.-T.; Wang, D.K.; Tseng, H.-H.; Wey, M.-Y. Planetary cross-linked structure design of hybrid organosilica membrane by amine-driven polymerization for CO_2 separation. *J. Clean. Prod.* **2023**, *398*, 136568. [CrossRef]
42. Williams, P.A.; Campbell, K.T.; Silva, E.A. Alginate hydrogels of varied molecular weight distribution enable sustained release of sphingosine-1-phosphate and promote angiogenesis. *J. Biomed. Mater. Res. Part A* **2018**, *106*, 138–146. [CrossRef] [PubMed]
43. Park, H.-H.; Ko, S.-C.; Oh, G.-W.; Jang, Y.-M.; Kim, Y.-M.; Park, W.S.; Choi, I.-W.; Jung, W.-K. Characterization and biological activity of PVA hydrogel containing chitooligosaccharides conjugated with gallic acid. *Carbohydr. Polym.* **2018**, *198*, 197–205. [CrossRef]
44. Zepon, K.M.; Marques, M.S.; da Silva Paula, M.M.; Morisso, F.D.P.; Kanis, L.A. Facile, green and scalable method to produce carrageenan-based hydrogel containing in situ synthesized AgNPs for application as wound dressing. *Int. J. Biol. Macromol.* **2018**, *113*, 51–58. [CrossRef]
45. Wang, Z.; An, G.; Zhu, Y.; Liu, X.; Chen, Y.; Wu, H.; Wang, Y.; Shi, X.; Mao, C. 3D-printable self-healing and mechanically reinforced hydrogels with host–guest non-covalent interactions integrated into covalently linked networks. *Mater. Horiz.* **2019**, *6*, 733–742. [CrossRef]
46. Peppas, N.A. Hydrogels and drug delivery. *Curr. Opin. Colloid Interface Sci.* **1997**, *2*, 531–537. [CrossRef]
47. Sennakesavan, G.; Mostakhdemin, M.; Dkhar, L.; Seyfoddin, A.; Fatihhi, S. Acrylic acid/acrylamide based hydrogels and its properties—A review. *Polym. Degrad. Stab.* **2020**, *180*, 109308. [CrossRef]
48. Zhang, X.; Qin, M.; Xu, M.; Miao, F.; Merzougui, C.; Zhang, X.; Wei, Y.; Chen, W.; Huang, D. The fabrication of antibacterial hydrogels for wound healing. *Eur. Polym. J.* **2021**, *146*, 110268. [CrossRef]
49. Zhang, Y.; Dang, Q.; Liu, C.; Yan, J.; Cha, D.; Liang, S.; Li, X.; Fan, B. Synthesis, characterization, and evaluation of poly (aminoethyl) modified chitosan and its hydrogel used as antibacterial wound dressing. *Int. J. Biol. Macromol.* **2017**, *102*, 457–467. [CrossRef]
50. Hoffman, A.S. Hydrogels for biomedical applications. *Adv. Drug Deliv. Rev.* **2012**, *64*, 18–23. [CrossRef]
51. Wang, Y. Swelling behavior of konjac glucomannan/N, n-dimethylene bisacrylamide hydrogel. *Anhui Agron. Bull.* **2015**, *21*, 26–27.
52. Wang, Y.; Zhang, B.; Ma, M.; Lu, W. Preparation of gelma/PEGDA hydrogel by UV copolymerization and crosslinking. *Imaging Sci. Photochem.* **2017**, *35*, 574–580.
53. Wang, Q.; Ren, L.; Wang, Y.; Yao, Y. Preparation and characterization of photocrosslinked n-acryloyl glucosamine/PEGDA hydrogel. *J. South China Univ. Technol.* **2013**, *41*, 62–67.
54. Su, J.; Satchell, S.C.; Wertheim, J.A.; Shah, R.N. Poly (ethylene glycol)-crosslinked gelatin hydrogel substrates with conjugated bioactive peptides influence endothelial cell behavior. *Biomaterials* **2019**, *201*, 99–112. [CrossRef]
55. Wu, X.; He, C.; Wu, Y.; Chen, X. Synergistic therapeutic effects of Schiff's base cross-linked injectable hydrogels for local co-delivery of metformin and 5-fluorouracil in a mouse colon carcinoma model. *Biomaterials* **2016**, *75*, 148–162. [CrossRef]
56. Raia, N.R.; Partlow, B.P.; McGill, M.; Kimmerling, E.P.; Ghezzi, C.E.; Kaplan, D.L. Enzymatically crosslinked silk-hyaluronic acid hydrogels. *Biomaterials* **2017**, *131*, 58–67. [CrossRef]
57. Gyles, D.A.; Castro, L.D.; Silva Jr, J.O.C.; Ribeiro-Costa, R.M. A review of the designs and prominent biomedical advances of natural and synthetic hydrogel formulations. *Eur. Polym. J.* **2017**, *88*, 373–392. [CrossRef]
58. Al-Tabakha, M.M.; Khan, S.A.; Ashames, A.; Ullah, H.; Ullah, K.; Murtaza, G.; Hassan, N. Synthesis, characterization and safety evaluation of sericin-based hydrogels for controlled delivery of acyclovir. *Pharmaceuticals* **2021**, *14*, 234. [CrossRef]
59. Cao, Y.; Wang, P.X.; D'Acierno, F.; Hamad, W.Y.; Michal, C.A.; MacLachlan, M.J. Tunable diffraction gratings from biosourced lyotropic liquid crystals. *Adv. Mater.* **2020**, *32*, 1907376. [CrossRef]

60. Ye, T.-J.; Qian, S.; Gao, Y.; Yu, M.-J.; Wei, Y.-F. Research Status, Problems and Countermeasures of Lyotropic Liquid Crystal in New Drug Delivery Systems of Traditional Chinese Medicine. *Chin. J. Exp. Tradit. Med. Formulae* **2019**, *24*, 229–234. [CrossRef]
61. De Souza, J.F.; Pontes, K.d.S.; Alves, T.F.; Amaral, V.A.; Rebelo, M.d.A.; Hausen, M.A.; Chaud, M.V. Spotlight on biomimetic systems based on lyotropic liquid crystal. *Molecules* **2017**, *22*, 419. [CrossRef]
62. Shan, X.; Li, X.; Luo, Z.; Lin, Q.; Lu, Y.; Jiang, M.; Zhang, J.; Huang, J.; Xie, L.; Guo, X. A Clinically-Achievable Injectable and Sprayable in Situ Lyotropic Liquid Crystalline Platform in Treating Hormone-Sensitive and Castration-Resistant Prostate Cancer. *ACS Nano* **2023**, *17*, 6045–6061. [CrossRef]
63. Mancuso, A.; Cianflone, E.; Cristiano, M.C.; Salerno, N.; Tarsitano, M.; Marino, F.; Molinaro, C.; Fresta, M.; Torella, D.; Paolino, D. Lyotropic liquid crystals: A biocompatible and safe material for local cardiac application. *Pharmaceutics* **2022**, *14*, 452. [CrossRef] [PubMed]
64. Boyd, B.J.; Whittaker, D.V.; Khoo, S.-M.; Davey, G. Lyotropic liquid crystalline phases formed from glycerate surfactants as sustained release drug delivery systems. *Int. J. Pharm.* **2006**, *309*, 218–226. [CrossRef]
65. Silvestrini, A.V.P.; Caron, A.L.; Viegas, J.; Praca, F.G.; Bentley, M.V.L.B. Advances in lyotropic liquid crystal systems for skin drug delivery. *Expert Opin. Drug Deliv.* **2020**, *17*, 1781–1805. [CrossRef]
66. Jain, S.; Yadav, P.; Swami, R.; Swarnakar, N.K.; Kushwah, V.; Katiyar, S.S. Lyotropic liquid crystalline nanoparticles of amphotericin B: Implication of phytantriol and glyceryl monooleate on bioavailability enhancement. *AAPS PharmSciTech* **2018**, *19*, 1699–1711. [CrossRef]
67. Ibrahim, T.M.; El-Megrab, N.A.; El-Nahas, H.M. An overview of PLGA in-situ forming implants based on solvent exchange technique: Effect of formulation components and characterization. *Pharm. Dev. Technol.* **2021**, *26*, 709–728. [CrossRef] [PubMed]
68. Das, K.; Roy, B.; Satpathi, S.; Hazra, P. Impact of topology on the characteristics of water inside cubic lyotropic liquid crystalline systems. *J. Phys. Chem. B* **2019**, *123*, 4118–4128. [CrossRef]
69. Negrini, R.; Mezzenga, R. Diffusion, molecular separation, and drug delivery from lipid mesophases with tunable water channels. *Langmuir* **2012**, *28*, 16455–16462. [CrossRef]
70. Ghanbari, R.; Assenza, S.; Saha, A.; Mezzenga, R. Diffusion of polymers through periodic networks of lipid-based nanochannels. *Langmuir* **2017**, *33*, 3491–3498. [CrossRef] [PubMed]
71. Meikle, T.G.; Yao, S.; Zabara, A.; Conn, C.E.; Drummond, C.J.; Separovic, F. Predicting the release profile of small molecules from within the ordered nanostructured lipidic bicontinuous cubic phase using translational diffusion coefficients determined by PFG-NMR. *Nanoscale* **2017**, *9*, 2471–2478. [CrossRef]
72. Rapalli, V.K.; Waghule, T.; Hans, N.; Mahmood, A.; Gorantla, S.; Dubey, S.K.; Singhvi, G. Insights of lyotropic liquid crystals in topical drug delivery for targeting various skin disorders. *J. Mol. Liq.* **2020**, *315*, 113771. [CrossRef]
73. Smidsrød, O.; Skja, G. Alginate as immobilization matrix for cells. *Trends Biotechnol.* **1990**, *8*, 71–78. [CrossRef] [PubMed]
74. van 't Hag, L.; Li, X.; Meikle, T.G.; Hoffmann, S.V.; Jones, N.C.; Pedersen, J.S.; Hawley, A.M.; Gras, S.L.; Conn, C.E.; Drummond, C.J. How peptide molecular structure and charge influence the nanostructure of lipid bicontinuous cubic mesophases: Model synthetic WALP peptides provide insights. *Langmuir* **2016**, *32*, 6882–6894. [CrossRef]
75. Clapper, J.D.; Guymon, C.A. Nanostructured biodegradable polymer composites generated using lyotropic liquid crystalline media. *Macromolecules* **2007**, *40*, 7951–7959. [CrossRef]
76. Wang, H.; Peng, T.; Wu, H.; Chen, J.; Chen, M.; Mei, L.; Li, F.; Wang, W.; Wu, C.; Pan, X. In situ biomimetic lyotropic liquid crystal gel for full-thickness cartilage defect regeneration. *J. Control. Release* **2021**, *338*, 623–632. [CrossRef] [PubMed]
77. Zhou, C.; Huang, Z.; Huang, Y.; Wang, B.; Yang, P.; Fan, Y.; Hou, A.; Yang, B.; Zhao, Z.; Quan, G.; et al. In situ gelation of rhEGF-containing liquid crystalline precursor with good cargo stability and system mechanical properties: A novel delivery system for chronic wounds treatment. *Biomater. Sci.* **2019**, *7*, 995–1010. [CrossRef]
78. Yue, X.; Zhang, X.; Wang, C.; Huang, Y.; Hu, P.; Wang, G.; Cui, Y.; Xia, X.; Zhou, Z.; Pan, X.; et al. A bacteria-resistant and self-healing spray dressing based on lyotropic liquid crystals to treat infected post-operative wounds. *J. Mater. Chem. B* **2021**, *9*, 8121–8137. [CrossRef]
79. Chen, J.; Wang, H.; Mei, L.; Wang, B.; Huang, Y.; Quan, G.; Lu, C.; Peng, T.; Pan, X.; Wu, C. A pirfenidone loaded spray dressing based on lyotropic liquid crystals for deep partial thickness burn treatment: Healing promotion and scar prophylaxis. *J. Mater. Chem. B* **2020**, *8*, 2573–2588. [CrossRef]
80. Hou, Y.; Li, J.; Guan, S.; Witte, F. The therapeutic potential of MSC-EVs as a bioactive material for wound healing. *Eng. Regen.* **2021**, *2*, 182–194. [CrossRef]
81. Akombaetwa, N.; Bwanga, A.; Makoni, P.A.; Witika, B.A. Applications of Electrospun Drug-Eluting Nanofibers in Wound Healing: Current and Future Perspectives. *Polymers* **2022**, *14*, 2931. [CrossRef]

gels

MDPI

Review

Enhancing Therapeutic Efficacy of Curcumin: Advances in Delivery Systems and Clinical Applications

Hossein Omidian *, Renae L. Wilson and Sumana Dey Chowdhury

Barry and Judy Silverman College of Pharmacy, Nova Southeastern University, Fort Lauderdale, FL 33328, USA; rw1273@mynsu.nova.edu (R.L.W.); sd2236@mynsu.nova.edu (S.D.C.)
* Correspondence: omidian@nova.edu

Abstract: Curcumin, a potent active compound found in turmeric and Curcuma xanthorrhiza oil, possesses a wide range of therapeutic properties, including antibacterial, anti-inflammatory, antioxidant, and wound healing activities. However, its clinical effectiveness is hindered by its low bioavailability and rapid elimination from the body. To overcome these limitations, researchers have explored innovative delivery systems for curcumin. Some promising approaches include solid lipid nanoparticles, nanomicelle gels, and transdermal formulations for topical drug delivery. In the field of dentistry, curcumin gels have shown effectiveness against oral disorders and periodontal diseases. Moreover, Pickering emulsions and floating in situ gelling systems have been developed to target gastrointestinal health. Furthermore, curcumin-based systems have demonstrated potential in wound healing and ocular medicine. In addition to its therapeutic applications, curcumin also finds use as a food dye, contraception aid, corrosion-resistant coating, and environmentally friendly stain. This paper primarily focuses on the development of gel compositions of curcumin to address the challenges associated with its clinical use.

Keywords: curcumin gel formulations; transdermal delivery; periodontal therapy; ocular inserts; gastrointestinal bioavailability

Citation: Omidian, H.; Wilson, R.L.; Chowdhury, S.D. Enhancing Therapeutic Efficacy of Curcumin: Advances in Delivery Systems and Clinical Applications. *Gels* **2023**, *9*, 596. https://doi.org/10.3390/gels9080596

Academic Editors: Ying Huang, Zhengwei Huang and Xuanjuan Zhang

Received: 10 July 2023
Revised: 21 July 2023
Accepted: 24 July 2023
Published: 25 July 2023

1. Introduction

Curcumin, also known as diferuloylmethane (Scheme 1), is an active component in the golden spice turmeric (Curcuma longa) and in Curcuma xanthorrhiza oil [1]. It is a highly pleiotropic molecule that exhibits antibacterial, anti-inflammatory, hypoglycemic, antioxidant, wound healing, and antimicrobial activities. Due to these properties, curcumin has been investigated for the treatment and supportive care of clinical conditions including proteinuria, breast cancer, multiple myeloma, depression, and non-small-cell lung cancer (NSCLC) [2]. Despite proven efficacy against numerous experimental models, poor bioavailability due to poor absorption, rapid metabolism, and rapid systemic elimination have been shown to limit the therapeutic efficacy of curcumin [2,3]. To overcome these challenges, researchers have focused on developing innovative curcumin delivery systems and formulations that enhance stability, bioavailability, and targeted delivery [4,5].

The development of curcumin formulations has been a subject of intense investigation encompassing a wide array of strategies and approaches. Solid lipid nanoparticles (SLNs) loaded with curcumin incorporated into a thermoresponsive gel have shown promise for Alzheimer's disease treatment [6]. Similarly, nanoparticle-incorporated hyaluronic acid gel has demonstrated a preventive effect against peritoneal adhesions [4], and mixed hydrogels of whey protein aggregates and k-carrageenan have shown the ability to protect and deliver curcumin to the colon [5]. A curcumin phytosomal soft gel formulation has also been developed to enhance bioavailability and dissolution characteristics [7]. Surface modification of MIL-100(Fe) with carboxymethyl cellulose enables slow release of curcumin [8]. Starch-based emulsion gel beads and curcumin–pectin calcium gel beads have shown potential for targeted delivery and functional food applications [9,10].

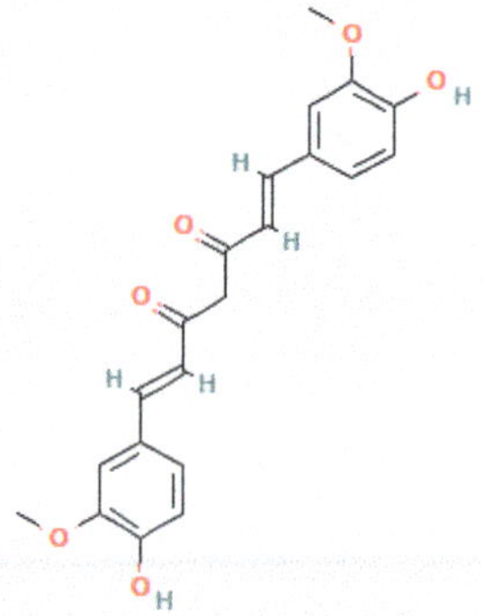

Scheme 1. Chemical structure of curcumin, diferuloylmethane [1].

Liposomal solid gels and chitosan-based hydrogels have been investigated to address curcumin's solubility and bioavailability issues [11,12]. Ophthalmic in situ gels, optical pH sensors, cryogel encapsulation, and film-forming hydrogels have been developed as efficient drug delivery systems [13–16]. Curcumin has shown promise in malarial parasite management systems [17] and demonstrated improved anti-hepatitis-C-virus delivery in liposomal solid gels [18].

In the field of topical applications, various curcumin delivery systems have been developed to overcome its poor bioavailability and limited skin permeability. Proniosomal gels, nanomicelle gels, and transdermal gel formulations have been explored for topical drug delivery [19–21]. Nanostructured lipid carriers, organogels, mucoadhesive gels, and nanosponge-based gels offer alternative approaches [22–25]. Transdermal delivery systems such as ethosomal gels, molecular inclusion complexes, transfersomes, and gel microemulsions have shown promise in enhancing curcumin's skin permeation [26–29].

In dentistry, curcumin gel formulations have demonstrated therapeutic benefits for various oral conditions. Studies have explored curcumin gel for treating minor recurrent aphthous stomatitis (RAS) and premalignant oral disorders (oral leukoplakia) [30,31]. Curcumin's antimicrobial capacity has been investigated in root canals [32], and it has shown antiplaque and anti-inflammatory effects when used in gingivitis management [33]. Furthermore, curcumin gel has been studied in the context of periodontal disease management, including management of chronic periodontitis [34], periodontal pockets [35], and azole-resistant oral candidiasis [36].

Curcumin-based delivery systems have also shown potential in managing conditions in oral health. In the case of oral submucous fibrosis (OSMF), curcumin gel and buccal mucoadhesive patches have demonstrated efficacy [37]. Combination therapies involving curcumin gel, triamcinolone–hyaluronidase gel, aloe vera gel, and oral physiotherapy have been explored to enhance OSMF management [38,39].

Gastrointestinal health has also been a focus of curcumin delivery system research. Curcumin-loaded Pickering emulsions, emulsion gels, and oil-filled aerogels offer potential solutions for enhancing curcumin's bioavailability and stability during digestion [40,41]. Gastroretentive floating in situ gelling systems have been formulated for gastric ulcer treatment, targeting Helicobacter-pylori-associated ulcers [42,43]. Moreover, ultra-high-pressure technology has been utilized to fabricate emulsion gels with unique physical characteristics and controlled release properties [44].

In wound healing, curcumin-based delivery systems have shown promise in overcoming solubility and bioavailability challenges. Nanogel formulations, such as silver–curcumin nanoparticles, have been evaluated for burn healing efficacy and safety [45]. Self-assembled nanogels, gelatin-based gels, and self-microemulsifying gels offer innovative solutions for curcumin delivery and wound healing [46–48]. Furthermore, nanocrystals and sol-gel coatings have been developed to enhance curcumin's solubility, stability, and wound healing potential [49,50]. Comparisons with conventional treatments, such as triamcinolone oral paste, have further highlighted the potential of curcumin gel in wound healing [51].

In ocular medicine, innovative delivery systems have been developed based on curcumin's therapeutic potential. Ocular inserts and in situ gelling systems have been developed for long-acting ocular delivery of curcumin, addressing its poor solubility and short ocular residence time [52,53]. These advancements offer promise for the effective treatment of ocular diseases using curcumin-based therapies.

Moreover, other than therapeutic applications, curcumin has been investigated for its use as a natural dye alternative in food products [54], a component in vaginal gels for contraception [55], a corrosion-resistant coating for mild steel surfaces [56], and an environmentally friendly protein stain [57]. Curcumin-based delivery systems have also been explored for controlled curcumin delivery in food matrices [58,59], efficient drug delivery [60,61], and intranasal neurological applications [62].

This study explores the wide-ranging potential of curcumin as a therapeutic agent in various fields. It aims to review and analyze innovative curcumin delivery systems that address challenges such as poor solubility and limited bioavailability. By enhancing our understanding of these delivery methods, the study seeks to pave the way for more effective treatments and improved patient outcomes in different medical areas.

2. Curcumin Delivery Systems

To overcome the poor pharmacokinetic properties of curcumin, researchers have delved into developing various delivery systems. One approach investigated the development of a microemulsion-based in situ ion-sensitive gelling system for the intranasal administration of curcumin. This novel system was designed to improve the absorption and brain targeting of curcumin through nasal delivery. The microemulsion was optimized using a simple lattice design, and its physicochemical properties were thoroughly investigated. Notably, histological section studies demonstrated the safety of this system for nasal mucosa. The pharmacokinetic results showed a significant improvement in the absolute bioavailability of curcumin, reaching 55.82% by intranasal administration, and a remarkable brain-targeting index (BTI) of 6.50. This enhanced brain targeting was attributed to direct nose-to-brain drug transport, making the microemulsion-based in situ ion-sensitive gelling system an effective and safe vehicle for the intranasal delivery of curcumin [62].

Another approach was the development of a thermoresponsive ophthalmic in situ gel containing curcumin-loaded albumin nanoparticles (Cur-BSA-NPs-Gel) for ocular drug delivery. Albumin nanoparticles were prepared via a desolvation method, and the gels were prepared using a cold method. The formulation was optimized to undergo a sol–gel transition at a specific temperature, allowing for easy administration and improved bioavailability. In vitro and in vivo studies in rabbits demonstrated the safety and increased bioavailability of curcumin in the aqueous humor with this optimized formulation. The Cur-BSA-NPs-Gel achieved superior sustained release effects, and the incorporation of albumin nanoparticles had minimal impact on the gel structure, highlighting its potential for ocular application [13].

Furthermore, researchers explored a nanogel combining cationic nanostructured lipid carriers (CNLC) and a thermosensitive gelling agent to enhance the preocular retention and ocular permeation capacity of curcumin. The thermosensitive ophthalmic in situ nanogel of CUR-CNLC (CUR-CNLC-GEL) demonstrated a solution–gel transition temperature suitable for ocular administration. The CUR-CNLC-GEL exhibited zero-order release kinetics and significantly improved curcumin permeation compared to a curcumin solution. The formulation's controlled release properties indicate its potential as a promising option for enhancing bioavailability in the aqueous humor and improving corneal permeation and retention capacity [63].

Moreover, researchers explored the use of copolymeric micelles to modify the pharmacokinetics and tissue distribution of curcumin. A synthesized poly(D,L-lactide-co-glycolide)-b-poly(ethylene glycol)-b-poly(D,L-lactide-co-glycolide) (PLGA-PEG-PLGA) copolymer was used to prepare CUR-loaded micelles. These micelles demonstrated improved pharmacokinetic parameters, including increased plasma AUC(0–∞), t1/2α, t1/2β, and MRT compared to a CUR solution. In vivo biodistribution studies in mice showed reduced drug uptake by the liver and spleen, while drug distribution was enhanced in the lung and brain. These results highlight the potential of PLGA-PEG-PLGA micelles as an effective carrier for curcumin, offering improved tissue distribution and bioavailability [64].

Other innovative strategies have been explored to enhance the properties of curcumin and its drug delivery systems. Amalia et al. modified MIL-100(Fe) with carboxymethyl cellulose (CMC) to achieve pH-responsive drug delivery potential and suppressed release of curcumin [8]. Furthermore, Bu et al. developed starch-based emulsion gel beads capable of dual loading proanthocyanidin and curcumin which showed tolerance to gastric conditions and released the loaded compounds in simulated intestinal fluid [9]. Cai et al. compared different de-esterification methods of low-methoxyl citrus pectin (LMP) and studied curcumin–pectin calcium gel beads, which exhibited controlled release and colon-targeted delivery potential [10]. Gugleva et al. developed a hybrid drug delivery platform with niosomal in situ gel for intravesical co-delivery of curcumin and gentamicin sulfate. This thermosensitive gel exhibited synergistic effect and controlled drug release (Figure 1) [65].

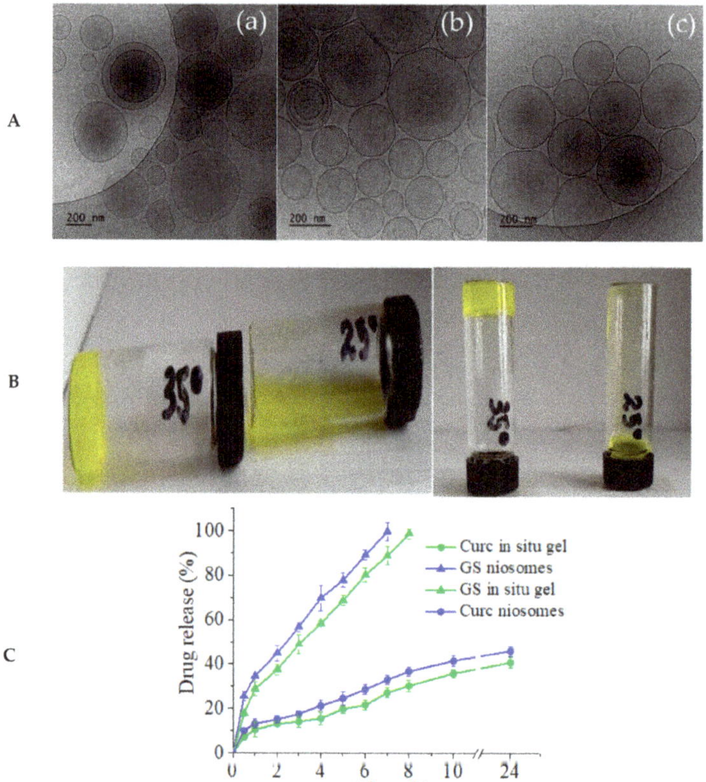

Figure 1. (**A**) Cryo-TEM images of niosomes (a) empty, (b) loaded with curcumin, and (c) loaded with curcumin and gentamicin (Curc/GS co-loaded); (**B**) Curc-GS niosomal thermosensitive in situ gel at 35 °C and at ambient temperature; (**C**) in vitro release profiles of curcumin and gentamicin sulfate from selected Curc/GS co-loaded niosomes and from hybrid niosomal in situ gel [65].

The development of curcumin delivery systems also involves analytical techniques for characterization and analysis. Ghodake et al. developed a stability-indicating HPTLC method for simultaneous determination of donepezil hydrochloride and curcumin in nasal gel, providing accurate results suitable for stability-indicating analysis [66]. Iwunze and McEwan explored the use of sol–gel-encapsulated curcumin as a sensor for small biomolecules, suggesting its potential for characterizing biomolecules [67]. Mohammad et al. developed an optical pH sensor using curcumin immobilized in sol–gel/chitosan–melamine hybrid matrices, demonstrating a linear response in the pH range of 9–13 with good reproducibility and fast response time [14].

2.1. *Bioavailability Enhancement and Sustained Release of Curcumin in Food and Pharmaceutical Systems*

Researchers have explored the use of protein-based gels as delivery vehicles for curcumin. Aliabbasi et al. developed an acid-induced cold-set gel using pinto bean protein isolate (PBI) and successfully embedded curcumin within the gel matrix. The gel exhibited sustained release of curcumin in simulated intestinal conditions, indicating its potential as a delivery system [68]. Du et al. utilized soybean protein isolate (SPI) to develop a low-oil-phase emulsion gel for curcumin delivery. The gel demonstrated excellent wettability and prevented oxidative deterioration of curcumin, making it suitable for protecting easily oxidized lipid-soluble nutrients in low-fat diets [69].

Gelation agents such as gellan gum (GG) and xanthan gum (XG) have also been investigated for their ability to improve the properties of curcumin-loaded gels. Du et al. studied the effect of GG on the gelation of yellow croaker roe protein isolate (pcRPI) and found that GG enhanced the water-holding capacity and microstructure of the gels. Moreover, GG effectively protected curcumin and slowed down its release in the gastrointestinal environment [70]. Geremias-Andrade et al. employed whey protein isolate and xanthan gum to produce curcumin-loaded emulsion-filled gels with improved color stability. The encapsulation of curcumin in solid lipid microparticles within the gels enhanced its stability [71].

In addition to protein-based gels, researchers have explored the use of Pickering emulsion gels as carriers for curcumin. Lei et al. developed Pickering emulsion gels stabilized by zein hydrolysate–chitin nanocrystal (CNC) coacervates. These emulsion gels exhibited improved rheological properties, emulsion stability, and antioxidant activity. The zein hydrolysate–chitin nanocrystals effectively enhanced the stability and controlled release of curcumin, increasing its bioaccessibility during digestion [72]. Lv et al. fabricated whey protein isolate (WPI) gel particles using high hydrostatic pressure (HHP) treatment and successfully formed Pickering emulsion gels. The resulting gels showed high loading efficiency of curcumin and excellent stability against light degradation [73]. Qiao et al. prepared Pickering emulsion gels stabilized by pea protein nanoparticles (PNPs) using heat-assisted pH shifting. The resulting emulsion gels exhibited improved curcumin bioaccessibility and stability. This research highlights the potential of this delivery system for enhancing the bioavailability of hydrophobic nutraceuticals such as curcumin [60].

Polysaccharide-based high-internal-phase emulsion (HIPE) gels have also been explored for curcumin delivery. Miao et al. formulated HIPE gels using sugar beet pectin, tannic acid, and chitosan complexes. These HIPEs exhibited a gel-like texture, improved stability, and enhanced curcumin bioaccessibility [74].

Additionally, Qazi et al. investigated the influence of gel structures on the gastric digestion and bioaccessibility of curcumin in dairy gels. The type of gel affected the release of curcumin-enriched oil and solubilized curcumin during gastric and intestinal digestion. This study provided insights for designing therapeutic dairy products that optimize the bioavailability of curcumin [59]. Pan et al. delved into the formation mechanism and curcumin bioaccessibility of emulsion gels based on sugar beet pectin (SBP). Emulsion gels stabilized by modified whey protein hydrolysates and SBP showed uniform networks and the highest curcumin bioaccessibility. These findings highlight the potential of emulsion gels for delivering curcumin and improving its bioavailability [58].

Furthermore, researchers have focused on developing oral delivery systems for curcumin. Tan et al. developed a food-grade oral delivery system using curcumin-loaded medium-chain triglycerides (MCTs)-encapsulated kappa-carrageenan (MCT-KC) gel beads. The MCT oil and kappa-carrageenan improved curcumin's solubility, gastric resistance, and encapsulation efficiency. In vitro studies demonstrated the performance of the MCT-KC system, indicating its potential as a delivery carrier for hydrophobic compounds [75].

To maximize stability and minimize lipid oxidation, Vellido-Perez et al. designed an oil gelled-in-water emulsified curcumin delivery system. Through optimization of critical operating parameters, an optimal formulation was identified that achieved high curcumin load and minimal lipid oxidation [76].

2.2. *Topical Delivery Systems for Inflammatory Conditions*

Several studies have focused on developing optimized curcumin-based topical formulations. Abd El-Halim et al. developed a curcumin proniosomal gel with high entrapment efficiency and controlled release, showing promise for herpes simplex virus type 1 (HSV-1) infections [19]. Araujo et al. evaluated a curcumin transdermal gel for inflammatory bowel diseases, demonstrating a protective effect against oxidative stress [77]. Bakhshi et al. conducted a clinical trial comparing 1% curcumin nanomicelle gel and 2% curcumin gel for recurrent aphthous stomatitis (RAS), both showing efficacy in reducing pain and lesion size [20].

Transdermal gel formulations, such as those developed by Chaudhary et al., have been optimized using statistical design, emphasizing the importance of formulation optimization [21]. Innovative dermal delivery systems such as nanostructured lipid carriers (NLCs) and in situ gels have also been explored. Chen et al. developed curcumin-loaded NLCs and a thermosensitive in situ gel, demonstrating high entrapment efficiency and enhanced anti-inflammatory effects [22]. The use of two topical vehicle organogels and nanogels for curcumin delivery has shown potential, as seen in the study by Gonzalez-Ortega et al. [23].

Mucoadhesive gels loaded with curcumin solid lipid nanoparticles (CurSLNs) have shown promise in treating oral precancerous lesions. Hazzah et al. formulated a curcumin loaded CurSLN gel, demonstrating good mucoadhesion and complete healing in patients with oral erythroplakia [24]. For psoriasis treatment, various gel formulations have been developed. Iriventi et al. developed a nanosponge-based topical gel of curcumin and caffeine, showing sustained release and anti-psoriatic activity [25]. Jain et al. developed a liposphere gel formulation loaded with tacrolimus and curcumin, improving skin penetration and ameliorating psoriatic features [78]. Jain et al. investigated the synergistic potential of a nanostructured lipid carrier (NLC) gel of ibrutinib with curcumin, demonstrating enhanced drug flux and reductions in psoriasis severity score [79]. Combination formulations, such as the nanoemulsion gel containing curcumin, thymoquinone, and resveratrol developed by Khatoon et al., have also shown promise for psoriasis therapy [80].

To address challenges related to solubility and permeability, researchers have developed various delivery systems. Kumar et al. formulated a curcumin-loaded ethosomal gel for transdermal delivery, addressing solubility and permeability challenges [26]. Innovative gel formulations incorporating curcumin, such as the topical gel delivery system developed by Patel et al., have demonstrated enhanced skin permeation and anti-inflammatory effects [81]. Novel approaches involving complexation with beta-cyclodextrin have also been investigated, enhancing the solubility and permeability of curcumin [27].

Targeted delivery systems have been explored for rheumatoid arthritis treatment. Sana et al. developed curcumin-loaded transfersomes, exhibiting improved therapeutic efficacy and reduced pro-inflammatory cytokines [28]. Gel microemulsions have been investigated for topical cosmetic applications, showing potential for incorporating hydrophobic active ingredients such as curcumin [29]. Shehata et al. formulated curcumin into proniosome gels that can be hydrated into niosomal formulations, demonstrating enhanced skin permeability and significant anti-inflammatory activity [82].

Solid lipid nanoparticles (SLNs) have been studied for curcumin delivery in the treatment of pigmentation and irritant contact dermatitis (ICD), exhibiting controlled drug release and efficacy in reducing ICD-related symptoms [83]. Singh and Dabre formulated a controlled-release gel containing curcumin microspheres and diclofenac diethylamine for rheumatoid-arthritis-associated inflammation, showing improved anti-inflammatory efficacy [84]. To enhance solubility and transdermal delivery, Tripathi et al. utilized a hydrotropic solid-dispersion (HSD) approach with sodium salicylate, resulting in enhanced solubility and improved drug release [85]. Vila et al. developed a stable emulsion with curcumin and locust bean gum, demonstrating enhanced permeation of curcumin suitable for cutaneous applications [86].

Clinical studies have evaluated the efficacy of curcumin gel in the treatment of specific inflammatory conditions. Yaghoobi et al. conducted a study showing significant reduction in psoriasis symptoms with 1% curcumin gel compared to placebo [87].

2.3. Oral Health and Dental Applications of Curcumin

Curcumin has shown promising results in the treatment of recurrent aphthous stomatitis (RAS), a common oral condition characterized by painful ulcers. Deshmukh and Bagewadi compared the efficacy of curcumin gel with triamcinolone acetonide gel in treating RAS. Both gels demonstrated significant improvements in pain, size, and number of ulcers. Curcumin gel in particular exhibited strong antioxidant and anti-inflammatory properties, suggesting it as an effective alternative to steroids in RAS treatment [30].

In the context of oral leukoplakia, Fathima and Manoharan investigated the efficacy of curcumin oral gel compared to bleomycin. While bleomycin showed greater resolution of leukoplakic lesions, curcumin exhibited histopathological improvement of dysplasia [31]. These findings highlight the potential of curcumin as a therapeutic agent in managing precancerous oral lesions.

The development of effective drug delivery systems is crucial for enhancing the bioavailability and therapeutic efficacy of curcumin in oral applications. Fonseca-Santos et al. designed a mucoadhesive vehicle for buccal administration of curcumin to treat oral candidiasis. The formulation demonstrated mucoadhesive properties and ex vivo study showed enhanced curcumin retention in porcine esophageal mucosa. In vitro studies further revealed enhanced antifungal activity against Candida albicans [36]. This research highlights the potential of curcumin-based formulations in the management of oral fungal infections.

Furthermore, curcumin has been evaluated as a potential therapeutic agent in managing orthodontic-treatment-related inflammation. Samita et al. studied the effect of locally applied curcumin gel on myeloperoxidase (MPO) enzymatic activity in gingival crevicular fluid during orthodontic tooth movement. The gel significantly decreased MPO activity, indicating its potential anti-inflammatory effect in orthodontic treatment [88]. These findings suggest that curcumin may offer benefits in reducing inflammation associated with orthodontic procedures.

The antimicrobial properties of curcumin have also been explored in the context of root canal disinfection. Oda et al. evaluated the antimicrobial action of curcumin photoactivated by LED curing light in the presence of carbopol gel. The effectiveness of curcumin + LED curing light was comparable to methylene blue + diode laser, demonstrating its potential as an alternative disinfection method [32].

Moreover, curcumin has been investigated as an adjunctive therapy in the treatment of gingivitis. Pandey et al. evaluated the effects of oral curcumin gel as an adjunct to scaling and root planing (SRP) for treating gingivitis. The curcumin gel exhibited significant antiplaque and anti-inflammatory effects when used alongside SRP, improving clinical parameters [33]. These findings suggest the potential of curcumin as an adjunctive therapy for periodontal management.

2.4. Periodontal Therapy

Periodontal disease, a prevalent oral health condition characterized by inflammation and destruction of the supporting structures of the teeth, necessitates effective treatment strategies. While scaling and root planing (SRP) remain the conventional approach, researchers have been exploring adjunctive therapies to enhance treatment outcomes.

Anuradha et al. conducted a clinical study on curcumin gel as an adjunct to periodontal treatment, reporting significant reductions in plaque index, gingival index, and probing depth and a gain in clinical attachment level, signifying its potential effectiveness in periodontal therapy [89]. Moreover, Dave et al. found that topical curcumin gel combined with SRP resulted in greater reductions in plaque accumulation, sulcular bleeding, and pocket probing depth compared to SRP alone, thus emphasizing its effectiveness as an adjunct to periodontal treatment [90].

In experimental periodontitis, Hosadurga et al. formulated a 2% curcumin gel and observed significant inhibition of edema and improvements in gingival index and probing pocket depth compared to in the control group, further highlighting its effectiveness in treating periodontitis [91]. Sha et al. demonstrated the antibacterial activity of curcumin against *P. gingivalis*, suggesting its utility in managing periodontal infections and promoting oral health (Figure 2) [92].

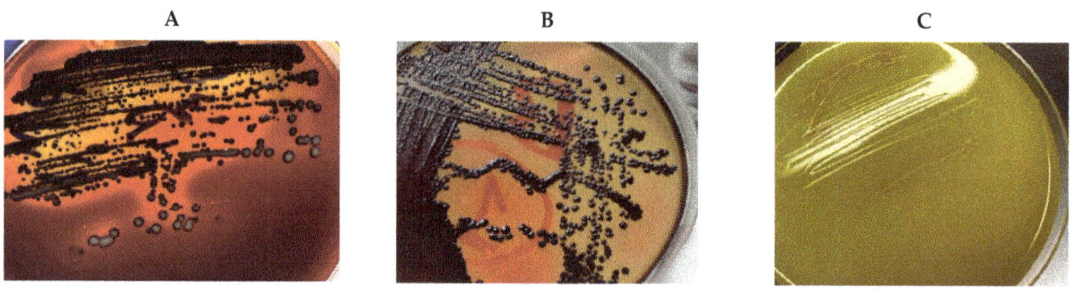

Figure 2. Pure isolated bacterial colonies of *P. gingivalis* after (**A**) 7 days and (**B**) 10 days on Columbia agar plate; (**C**) agar-formed media cultured from the tube with the MIC of curcumin showing no *P. gingivalis* growth [92].

Comparative studies have been conducted to evaluate the efficacy of curcumin gel compared to other adjunctive treatments. Hugar et al. compared the efficacy of curcumin gel and chlorhexidine gel as adjuncts to SRP in periodontal therapy and found that curcumin gel demonstrated greater reduction in periodontal parameters, establishing its superiority as an adjunct to SRP [93]. Similarly, Ravishankar et al. evaluated the effect of curcumin gel compared to ornidazole gel in treating chronic periodontitis and reported significant improvements in pocket probing depth, plaque index, and clinical attachment loss with curcumin gel, indicating its potential as an adjunct to nonsurgical periodontal therapy [94].

Furthermore, curcumin gel has shown promising results in managing specific conditions. It was evaluated as an adjunct to scaling and root planing (SRP) in the treatment of chronic periodontitis by Kaur et al., who reported a mild adjunctive benefit in reducing gingival inflammation [95]. Additionally, Meghana et al. compared the efficacy of curcumin gel and noneugenol periodontal dressing following periodontal flap surgery, and both treatments effectively reduced tissue edema, enhanced wound healing, and reduced pain perception [96].

Curcumin gel has also been investigated for its impact on systemic factors and its potential synergistic effects with existing therapies. Mohammad evaluated the effect of curcumin gel on serum micronutrients and pro-inflammatory cytokines in chronic periodontitis patients and found significant reductions in clinical parameters, inflammatory mediators, and copper levels, while zinc and magnesium levels increased [97]. Moreover, in experimental periodontitis, Mohammad et al. demonstrated that curcumin gel effectively reduced inflammation, bone resorption, and osteoclast numbers, indicating its potential osteogenesis and healing effects [98]. In a study by Mohammad et al., the antioxidant effects of curcumin gel in diabetes-induced periodontitis were evaluated, revealing a significant reduction in oxidative stress and increased antioxidant enzyme levels [99].

2.5. Oral Submucous Fibrosis (OSMF) Management

Oral submucous fibrosis (OSMF) is a chronic, progressive, and potentially malignant disorder characterized by fibrosis of the oral submucosal tissues. It is associated with various symptoms such as burning sensation and restricted mouth opening, leading to significant morbidity and impaired quality of life for affected individuals. Current treatment options for OSMF are limited and often associated with side effects. Therefore, there is a growing interest in exploring natural compounds with therapeutic potential for managing OSMF.

Chandrashekar et al. conducted a study to compare the efficacy of topical curcumin gel and buccal mucoadhesive patches for OSMF. Both formulations demonstrated significant improvements in reducing burning sensation, increasing mouth opening, and reducing serum lactate dehydrogenase levels. These findings indicate the potential of curcumin-based formulations for the treatment of OSMF [37].

In a study by Lanjekar et al., the efficacy of curcumin gel, triamcinolone–hyaluronidase gel, and their combination was evaluated for OSMF treatment. The combination therapy showed the greatest improvement in mouth opening, while the triamcinolone–hyaluronidase group reported reduced burning sensation. This study highlights the therapeutic effects of curcumin on OSMF and demonstrated that combination therapy enhances its utilization and drug delivery [38].

Furthermore, Nerkar Rajbhoj et al. compared the efficacy of curcumin gel and aloe vera gel in managing OSMF. Both gels exhibited improvements in mouth opening; however, aloe vera gel provided better relief from burning sensation, making it a more effective option without notable side effects [39].

2.6. Gastrointestinal Applications of Curcumin

Researchers have been exploring various delivery systems and formulations to optimize the bioavailability and therapeutic efficacy of curcumin. In this section, we discuss several studies that investigated different approaches for curcumin delivery in the gastrointestinal tract.

Araiza-Calahorra et al. conducted a study on Pickering emulsions stabilized by colloidal gel particles for curcumin delivery. Their modified emulsions demonstrated improved stability against coalescence and enhanced cellular uptake of curcumin. This research suggests the potential of utilizing Pickering emulsions as a promising delivery system for lipophilic bioactive compounds, including curcumin [40].

Another study by Fontes-Candia et al. focused on agar and kappa-carrageenan emulsion gels and oil-filled aerogels as curcumin carriers. The researchers observed distinct behavior and structural changes of the gels during in vitro digestion influenced by the presence of curcumin. This investigation provides valuable insights into the interaction between curcumin and emulsion gels, shedding light on their potential application in gastrointestinal delivery systems [41].

Kathpalia et al. formulated a gastroretentive floating in situ gelling system of solubility-enhanced curcumin specifically designed for gastric ulcers. This system exhibited extended release of curcumin for up to 12 h, indicating its potential as a reliable delivery platform for curcumin in treating gastric ulcers [42]. Similarly, Padhan et al. developed a gastroretentive in situ gelling system of curcumin for gastric ulcers associated with Helicobacter pylori. Their curcumin in situ gel demonstrated suitable properties such as floating lag time, duration, and controlled drug release, which could be regulated by the concentration of sodium alginate [43].

In another approach, Su et al. fabricated whey protein isolate (WPI)/kappa-carrageenan (kappa-CG) composite emulsion gels using ultra-high-pressure (UHP) technology for curcumin delivery. This study investigated the physical properties and release behavior of the emulsion gels, examining different formulations and processing conditions. It also evaluated the release of curcumin during simulated gastrointestinal digestion, providing valuable insights into the potential of WPI/kappa-CG emulsion gels as an effective curcumin delivery system [44].

2.7. Wound Healing

One approach to harnessing the wound healing properties of curcumin is through the development of novel delivery systems. Aghamoosa et al. developed a silver–curcumin nanogel that exhibited desirable properties and demonstrated efficacy in promoting wound healing in a burn model [45]. Similarly, Das et al. prepared a composite gel containing gelatin, F127, and lecithin for enhanced wound healing which facilitated the controlled release of curcumin and improved wound healing in a murine model [46]. These studies highlight the potential of curcumin-based formulations as effective tools for wound management.

The use of nanogels as delivery vehicles for curcumin has garnered attention in wound healing research. El-Refaie et al. evaluated curcumin-loaded gel–core hyaluosomes (GC-HS) and demonstrated their high entrapment efficiency, bilayer structure, and improved healing rate with reduced scar formation in vivo (Figure 3) [47]. Guo et al. investigated the therapeutic effect of curcumin-loaded self-microemulsifying gel and found that it exhibited higher skin flux and enhanced wound healing compared to commercial gels [48]. These studies underline the potential of nanogel-based systems in facilitating efficient delivery of curcumin to promote wound healing.

Furthermore, the development of three-dimensional (3D) scaffolds incorporating curcumin has shown promise in wound healing applications. Karahaliloglu demonstrated the effectiveness of an electroresponsive silk fibroin (SF) hydrogel loaded with curcumin as a 3D scaffold for wound healing which exhibited reduced bacteria and no toxicity to healthy cells [100]. This approach offers a versatile platform for delivering curcumin to wound sites and promoting tissue regeneration.

In addition to its individual effects, curcumin has been investigated in combination with other compounds to enhance its wound healing properties. Khan et al. formulated a gel containing allicin and curcumin which exhibited superior wound healing activity compared to conventional treatments [101]. These findings highlight the potential synergistic effects of curcumin in combination with other bioactive compounds for improved wound management.

The wound healing potential of curcumin has also been explored in various contexts. Kim et al. investigated the wound healing effect of transdermal curcumin gel, which exhibited antioxidant activity, inhibition of nitric oxide production, and promotion of wound healing in vivo [102]. Raman and Pitty conducted a study comparing the effectiveness of topical curcumin gel and triamcinolone acetonide oral paste in reducing pain and size of recurrent minor oral aphthous ulcers, demonstrating the efficacy of curcumin gel in wound healing applications [51]. These studies emphasize the diverse applications of curcumin in wound management.

Moreover, advancements in formulation techniques have facilitated the development of curcumin-based delivery systems with improved efficacy. Kotian et al. prepared curcumin nanocrystals and incorporated them into a gel, resulting in enhanced drug release, skin permeation, and wound healing activity compared to unprocessed curcumin without causing skin irritation [49]. Pisitsak and Ruktanonchai embedded curcumin in a sol–gel coating on cotton textiles, which exhibited controlled release and antibacterial properties, suggesting their potential as wound dressing materials [50]. These studies highlight the importance of formulation strategies in maximizing the therapeutic potential of curcumin for wound healing applications.

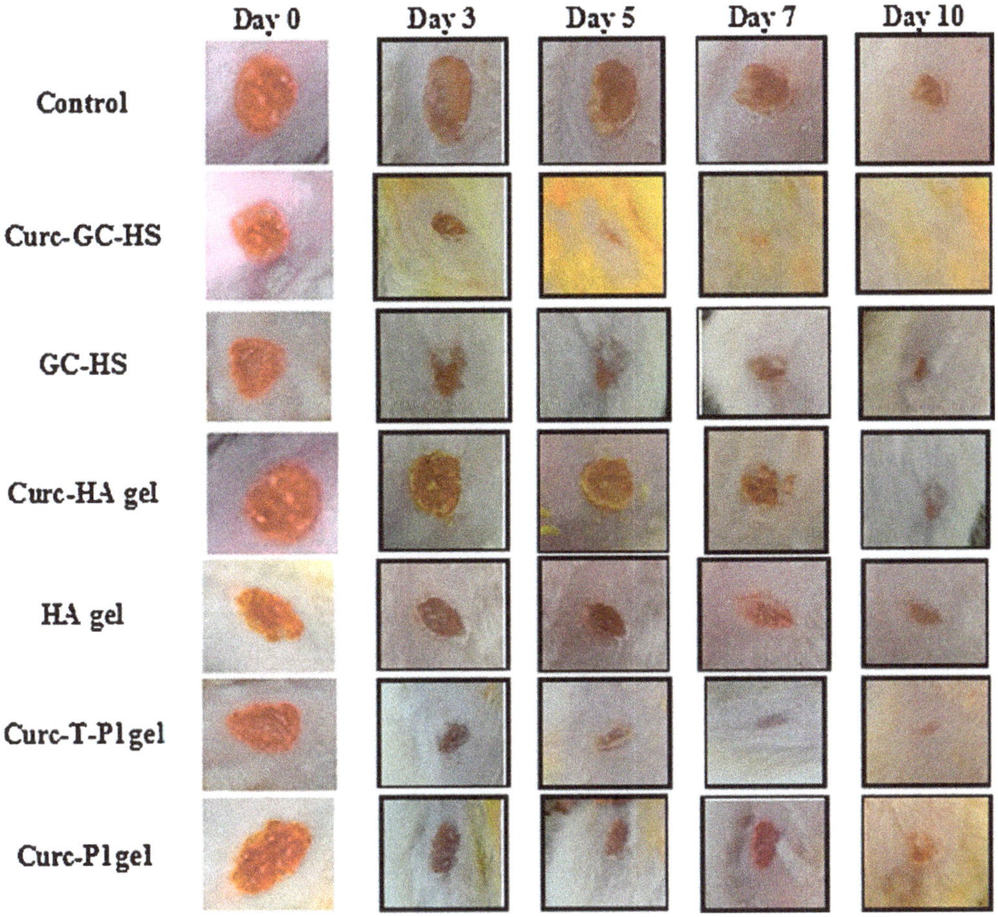

Figure 3. Stages of burn wound healing potential study of curcumin-loaded novel self-assembled nanogels gel–core hyaluosome (Curc-GC-HS) compared to conventional transfersomal gel (Curc-T-Pl gel) and other conventional gels Curc–HA gel, Curc in pluronic gel (Curc–Pl gel), gel-core hyaluosome (GC-HS), and HA gel. Adapted with permission [47].

2.8. Ophthalmic Delivery

The use of curcumin in ophthalmic drug delivery has emerged as a promising strategy to overcome the limitations of conventional eye drops and enhance the therapeutic efficacy of the drug. Several innovative formulations have been developed to improve the delivery and release of curcumin to the ocular surface and its permeation.

Abdelkader et al. investigated the use of in situ gelling polymeric inserts for curcumin delivery to the ocular surface. Their study demonstrated that these inserts exhibited improved characteristics, release, and permeation compared to a traditional suspension. Furthermore, the inserts were well tolerated and provided sustained release of curcumin. This research suggests that in situ gelling polymeric inserts could serve as a viable alternative to conventional eye drops for efficient curcumin delivery (Figure 4) [52].

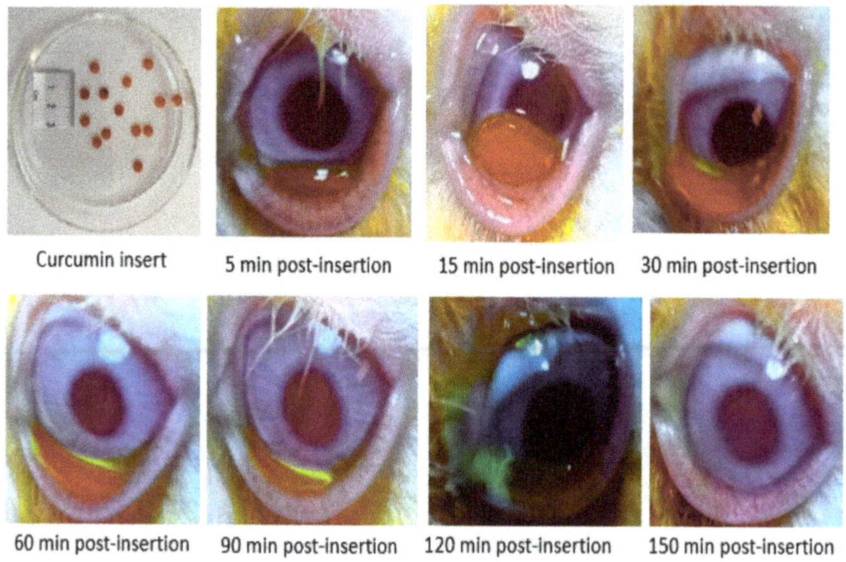

Figure 4. Image of curcumin in situ gelling polymeric solid insert into the inferior conjunctival cul de sac of rabbit eye, transformed from solid to gel for enhancing ocular performance [52].

Duan et al. focused on the development of an ion-sensitive mixed micelle in situ gel system for ophthalmic delivery of curcumin. Their system demonstrated sustained release properties and higher corneal permeation when compared to a curcumin solution. These findings highlight the potential of the ion-sensitive mixed micelle in situ gel system as an effective carrier for curcumin delivery to the eye [53].

In a similar vein, Sai et al. developed a mixed micelle in situ gelling system of curcumin for ophthalmic drug delivery. This innovative system not only improved the solubility and corneal penetration of curcumin but also exhibited a longer retention time on the corneal surface. The prolonged retention time is crucial for enhancing drug bioavailability and therapeutic efficacy. The study concluded that the mixed micelle in situ gelling system holds promise as an effective option for ophthalmic drug therapy utilizing curcumin [103].

2.9. Other Applications of Curcumin

In addition to ophthalmic drug delivery, curcumin gels have shown promise in several other therapeutic applications, ranging from food products to corrosion protection, protein staining, contraception, and nutraceutical delivery.

Brito-Oliveira et al. focused on developing emulsion-filled gels using soy protein isolate and xanthan gum as the gel matrix. The incorporation of curcumin-loaded solid lipid microparticles in these gels enhanced their mechanical properties and stability. This research suggests that curcumin encapsulation in gels could serve as an alternative to artificial dyes in gelled food products, providing a natural and functional ingredient [54].

Asefi et al. investigated the effect of curcumin on orthodontic tooth movement (OTM) and found that it inhibited root and bone resorption, osteoclastic recruitment, and angiogenesis, suggesting its potential benefits in preventing side effects associated with orthodontic treatment [104].

Gaurav et al. aimed to develop a safe and effective vaginal contraceptive using a copper–curcumin complex incorporated into a cyclodextrin inclusion complex. The vaginal gel demonstrated excellent spermicidal activity and safety in preclinical studies. This novel contraceptive platform utilizing curcumin offers a promising approach for contraception [52].

Ishak et al. investigated the incorporation of curcumin extracts into sol–gel coatings for corrosion protection of mild steel. The resulting coatings exhibited high inhibition efficiency and hydrophobic properties, indicating their potential for corrosion mitigation. This research highlights the potential of curcumin-based sol–gel coatings in the field of corrosion protection [56].

Kurien et al. explored the utility of curcumin as a protein stain gel. They demonstrated that heat-solubilized curcumin could effectively stain proteins in a manner comparable to Coomassie brilliant blue (CBB) at a significantly lower cost and without requiring destaining. This cost-effective and environmentally friendly alternative offers curcumin as an ideal protein stain [57]. Furthermore, Kurien et al. solubilized curcumin in Tween 80 or Triton X-100, creating an efficient protein stain gel. The solubilized curcumin exhibited high solubility, efficient staining, and stability without the need for destaining, providing a nontoxic and environmentally friendly alternative to traditional protein stains [105].

Raduly et al. focused on coating cotton fabrics with multifunctional nanosols generated through sol–gel reactions. The resulting coatings exhibited hydrophobicity, fluorescent and antimicrobial properties, as well as pH-dependent color change. These multifunctional coatings retained their properties even after washing cycles, offering potential applications in textiles requiring signaling, self-cleaning, or antibacterial properties [106].

Scomoroscenco et al. investigated the stability and antioxidant properties of curcumin encapsulated in microemulsion and gel microemulsion systems. These systems effectively protected curcumin from degradation and demonstrated a synergistic effect between curcumin and vegetable oil in terms of antioxidant capacity. This research highlights the potential of curcumin-based microemulsion and gel microemulsion systems as stable and effective antioxidant formulations [61].

Wang et al. developed an ion-sensitive in situ gelling system for the intranasal administration of curcumin. The system, based on a microemulsion formulation, demonstrated desirable properties and enhanced the bioavailability of curcumin through intranasal administration. Additionally, the system exhibited potential brain-targeting properties, making it a promising approach for delivering curcumin to the central nervous system [62].

Table 1 summarizes recent research on the preparation of therapeutic curcumin gel formulations and delivery systems.

Table 1. A summary of curcumin gel compositions and their use in various areas.

Curcumin Delivery Systems		
Materials and Methods	**Applications and Advancements**	**Ref.**
Curcumin-loaded solid lipid nanoparticles (SLN) in thermoresponsive gel	Enhanced physicochemical properties, sustained release, potential for Alzheimer's disease treatment	[6]
Curcumin–soybean phosphatidylcholine (CUR-SPC)-complex-loaded NPs in HA gel	Prevention of postoperative peritoneal adhesion, harnessing advantages of CUR, SPC, and HA	[4]
Curcumin loaded whey protein aggregates/k-carrageenan hydrogels	Potential delivery system, protecting curcumin during digestion, enabling targeted delivery to the colon	[5]
Curcumin phytosomal soft gel formulations using different vehicles	Enhancing bioavailability, overcoming solubility limitations, improved delivery of curcumin	[7]
Curcumin encapsulated in surface-modified MIL-100(Fe) with CMC	pH-responsive drug delivery, controlling release profile of hydrophobic drugs, biomedical applications	[8]
Starch-based emulsion gel beads prepared by inverse emulsion gelation	Ideal carriers for loading hydrophilic and hydrophobic ingredients, controlled release systems	[9]
Low-methoxyl-citrus-pectin-based curcumin–pectin calcium gel beads	Potential in functional foods and drug delivery, colon-targeted delivery, novel delivery systems	[10]
Stability-indicating HPTLC method for simultaneous determination of donepezil hydrochloride and curcumin in nasal gel	Reliable analytical tool for quality control and stability testing in nasal gel formulation	[66]
Study on dissolution, gel properties, and characterization of amorphous curcumin	Understanding gelation potential of amorphous materials, improving drug delivery systems	[107]

Table 1. *Cont.*

Curcumin Delivery Systems		
Materials and Methods	**Applications and Advancements**	**Ref.**
Sol–gel-encapsulated curcumin	Potential sensor for biomolecules, contributes to sensor technologies	[67]
Curcumin in kappa-carrageenan gel with metal chlorides	Prolonging antioxidant and anti-inflammatory effects, benefiting healthcare	[108]
Polyphenon 60 (P60) + curcumin (CUR) nanoemulsion gel	Combined antibacterial treatment for E. coli infections, enhanced drug availability	[109]
Nanostructured lipid carriers (NLCs)-loaded chitosan/carbopol hybrid gel	Promising delivery system for malarial parasite management	[17]
Curcumin-loaded poly(2-hydroxyethyl methacrylate) nanoparticles	Therapeutic potential in ovarian cancer treatment	[11]
pH-sensitive curcumin-loaded microemulsion gel system	Smart drug delivery system with enhanced controlled release	[110]
Hydrogel carrier of oligo-conjugated linoleic acid vesicles	Injectable/3D printable drug delivery system with antioxidant property	[12]
Albumin nanoparticles loaded with curcumin and thermoresponsive gels	Potential for ocular drug delivery with enhanced bioavailability	[13]
Curcumin reagent immobilized in sol–gel/chitosan–melamine hybrid matrices	Optical pH sensor for pH measurement	[14]
Cryogel-based encapsulation of curcumin in chitosan, kappa-carrageenan, and NaCMC	Advancements in controlled release technology	[15]
Film-forming nanogels containing poly(lactide-co-glycolide) and gelatine nanoparticles	Improved permeability and sustained diffusion	[16]
Preparation of a hydrotropic solid dispersion (HSD) loaded with curcumin	Enhancing topical delivery and potential for transdermal drug delivery	[85]
Development of a microemulsion-based in situ ion-sensitive gelling system	Enhanced curcumin absorption and brain targeting, intranasal drug delivery	[62]
Preparation of Ca2+-induced emulsion gels using HHP–pectin and erythritol	Low-calorie food options and delivery of fat-soluble nutrients	[111]
Curcumin-loaded liposomal solid gels prepared with freeze-drying technique	Enhanced therapeutic efficacy against hepatitis C virus	[18]
Niosomal in situ gel system prepared by thin-film hydration method	Treating urinary bladder-related conditions with sustained release	[65]
Bioavailability Enhancement and Sustained Release of Curcumin in Food and Pharmaceutical Systems		
Acid-induced cold-set gels based on pinto bean protein isolate (PBI) with curcumin	Sustained and controlled release of curcumin, enhancing bioavailability	[68]
Low-oil-phase emulsion gel prepared via high-pressure homogenization of SPI-Cur-NPs	Protection of oxidized lipid-soluble nutrients, advancements in antioxidant-rich food	[69]
Binary gels loaded with curcumin formed from large yellow croaker roe protein isolate and gellan gum	Improved water-holding ability, controlled release properties, versatile applications	[70]
Curcumin-loaded solid lipid microparticles incorporated in whey protein isolate and xanthan gum gels	Reducing lipid content in food, improving color stability while maintaining texture	[71]
Pickering emulsion gels stabilized by zein hydrolysate–chitin nanocrystals coacervates	Enhanced stability, delayed lipolysis, sustained nutrient release, applications in food/pharma	[72]
Whey protein isolate (WPI) gel particles with curcumin	Food-grade stabilizers for Pickering emulsions, enhanced stability against degradation	[73]
Polysaccharide-based high-internal-phase emulsion gels (HIPEs) with SBP, TA, and CS + curcumin	Soft solids with natural ingredients, stability enhancement, delayed lipid digestion	[74]
Emulsion gels with modified whey protein hydrolysates and curcumin	Delivery systems enhancing curcumin bioavailability	[58]
Curcumin nanoemulsion-containing dairy gels	Understanding gel structure impact on curcumin bioavailability	[59]
Pea protein nanoparticles (PNPs) and curcumin	Stabilizing Pickering emulsion gels for enhancing hydrophobic nutraceutical bioavailability	[60]
Curcumin-loaded MCT-KC gel bead formulation	Natural and food-grade carrier for enhancing hydrophobic compound delivery	[75]
Designing an oil gelled-in-water curcumin-loaded emulsion	Rational design for stabilizing and delivering bioactive compounds	[76]

Table 1. *Cont.*

Curcumin Delivery Systems		
Materials and Methods	**Applications and Advancements**	**Ref.**
Topical Delivery Systems for Inflammatory Conditions		
Curcumin proniosomal gel	Effective topical delivery system for HSV-1 treatment, insights into molecular interactions	[19]
Curcumin transdermal gel with Pentravan®	Reducing oxidative stress and inflammation in inflammatory bowel diseases	[77]
Efficacy of 1% curcumin nanomicelle gel and 2% curcumin gel	Enhanced healing of recurrent aphthous stomatitis, improved treatment outcomes	[20]
Gel formulations of diclofenac diethylamine (DDEA) and curcumin (CRM)	Optimized gel formulations for transdermal delivery, advancements in formulation	[21]
Curcumin-loaded nanostructured lipid-carriers-based thermosensitive in situ gel	Efficient dermal delivery system, improved permeation, anti-inflammatory effects	[22]
Curcumin-loaded gel carriers (organogels and O/W-type nanogels)	Improved topical formulations for inflammatory conditions	[23]
Curcumin-loaded solid lipid nanoparticles (CurSLNs) in mucoadhesive gel matrix	Local treatment of precancerous lesions with reduced side effects	[24]
Nanosponge with beta-CD and curcumin-loaded topical gel	More effective anti-psoriatic therapy, advancing psoriasis treatment	[25]
Liposphere gel formulation of tacrolimus and curcumin	Treatment strategy for psoriasis, addressing solubility and skin penetration	[78]
Ibrutinib and curcumin-loaded nanostructured lipid carrier (NLC)	Management of psoriasis, integrated benefits in reducing inflammation	[79]
Silica-based formulations with glycyrrhizic acid and curcumin	Superior treatment option for psoriasis, enhancing anti-psoriatic efficacy	[112]
Nanoemulsion gel formulation containing curcumin, resveratrol, and thymoquinone	Promising approach for psoriasis therapy, overcoming solubility and permeation issues	[80]
Curcumin-loaded ethosomal gel prepared by ethanol and soya lecithin	Potential solution for transdermal delivery, overcoming solubility and permeability challenges	[26]
Curcumin gel with Carbopol 934P and hydroxypropyl cellulose	Topical application with anti-inflammatory effects, alternative treatment option	[81]
Tailored water-in-oil emulsion systems with polyglycerol polyricinoleate (PGPR) and curcumin	Effective delivery of curcumin for various applications, controlled release platform	[113]
Curcumin–beta-cyclodextrin nanoparticle complex (BCD-CUR-N)	Improved solubility and skin penetration, potential for transdermal delivery	[27]
Curcumin-loaded transferosomes (Cur-TF) embedded in Carbopol 934 gel	Targeted topical treatment of rheumatoid arthritis, improved therapeutic efficacy	[28]
Gel microemulsions with grape seed oil and curcumin	Effective systems for curcumin delivery in dermatocosmetics	[29]
Curcumin-loaded niosomal emulgel with nonionic surfactants	Efficient carrier for transdermal drug delivery, enhanced anti-inflammatory effect	[82]
Curcumin solid lipid nanoparticles (CUR-SLNs) in carbopol gel	Safe and effective alternative for irritant contact dermatitis treatment	[83]
Curcumin microspheres in gel formulation	Potential treatment option for arthritis inflammation	[84]
Hydrotropic solid dispersion (HSD) in curcumin gel	Greener approach for enhancing topical delivery of poorly soluble compounds	[85]
Curcumin with choline and geranic acid ionic liquid (CAGE-IL)	Potential for transdermal drug delivery and cutaneous drug delivery systems	[86]
1% curcumin gel	Modest but continuous effect on plaque psoriasis	[87]
Curcumin-loaded water-responsive gel (CUR-WRG)	Enhanced efficacy of topical drug delivery against psoriasis	[114]
Oral Health and Dental Applications of Curcumin		
Curcumin gel	Safer and accessible therapeutic option for minor recurrent aphthous stomatitis (RAS)	[30]
Topical bleomycin and curcumin oral gel with antioxidants	Potential in treating oral leukoplakia	[31]
Mucoadhesive vehicle formulation of PPG-5-CETETH-20 and oleic acid	Effective treatment for oral candidiasis	[36]

Table 1. *Cont.*

Curcumin Delivery Systems		
Materials and Methods	**Applications and Advancements**	**Ref.**
Clinical trial comparing amlexanox paste and curcumin oral gel	Safe alternative for managing RAS minor	[115]
Curcumin + carbopol gel + LED curing light	Alternative to conventional photodynamic therapy for disinfecting root canals	[32]
Randomized clinical trial evaluating the effectiveness of curcumin gel	Potential alternative for managing gingival diseases	[33]
Clinical study evaluating locally applied 1% curcumin gel during orthodontic treatment	Implications for managing inflammation during orthodontic tooth movement	[88]
Periodontal Therapy		
Adjunctive use of curcumin gel with scaling and root planning (SRP)	Enhancing outcomes of periodontal therapy, anti-inflammatory properties	[89]
Adjunctive use of curcumin gel with SRP	Improved outcomes in chronic periodontitis, potential complementary treatment	[90]
Adjunctive use of curcumin oral gel	Effective improvement in periodontal pocket parameters	[34]
2% curcumin gel	Effective treatment of experimental periodontitis	[91]
Comparison of chlorhexidine gel and curcumin gel	Both gels as adjuncts to scaling and root planing, curcumin as an alternative	[93]
Curcumin gel as an adjunct to scaling and root planing	Limited additional benefits in reducing gingival inflammation	[95]
Curcumin gel and noneugenol periodontal dressing for post-surgery treatment	Effectiveness in reducing tissue edema and promoting wound healing	[96]
Curcumin gel injection with scaling and root planing	Reduction of clinical parameters and inflammatory mediators	[97]
Scaling and root planing combined with curcumin	Comparable results to tetracycline in reducing inflammation and bone resorption	[98]
Curcumin gel as an adjunct to SRP in diabetes-induced periodontitis	Reducing oxidative stress and improving antioxidant enzyme levels	[99]
In situ gel formulation of curcumin	Effective treatment of periodontal pockets	[35]
Curcumin + carbopol gel + LED curing light	Alternative to conventional photodynamic therapy for root canal disinfection	[32]
Split-mouth design study with curcumin gel	Potential adjunct to nonsurgical periodontal therapy	[94]
Evaluation of curcumin's antibacterial activity against *P. gingivalis*	Effective antibacterial action at low concentrations	[92]
Curcumin gel in experimental periodontitis using Wistar rats	Effective in treating experimental periodontitis and preventing bone destruction	[116]
Clinical trial evaluating subgingival delivery of curcumin gel as an adjunct to SRP	Effective anti-inflammatory and antibacterial properties, alternative to chlorhexidine	[117]
Clinical trial comparing scaling and root planing alone, curcumin application, and curcumin photodynamic therapy	Curcumin photodynamic therapy as an adjunct to scaling and root planing improves treatment outcomes	[118]
Oral Submucous Fibrosis (OSMF) Management		
Curcumin gel and buccal mucoadhesive patches	Noninvasive and effective treatment for oral submucous fibrosis (OSMF)	[37]
Curcumin mucoadhesive semisolid gel and combination gels	Therapeutic effects in treating OSMF, potential role in treating oral cancers and precancerous lesions	[38]
Randomized clinical trial comparing curcumin gel and aloe vera gel	Efficacy in managing OSMF symptoms, aloe vera gel as adjuvant treatment	[39]
Gastrointestinal Applications of Curcumin		
Pickering emulsions with whey protein nanogel and whey protein isolate + dextran conjugate microgel	Enhanced delivery of curcumin, potential for functional foods	[40]
Emulsion gels with agar, kappa-carrageenan, and oil-filled aerogels + curcumin	Effective carriers with controlled release for gastrointestinal applications	[41]
Dry suspension floating in situ gelling system of curcumin	Locally treating gastric ulcers, controlled drug release in gastric environment	[42]
Gastroretentive floating in situ gelling system with curcumin	Extended gastric retention for improved therapeutic effectiveness	[43]
Composite emulsion gels with whey protein isolate/kappa-carrageenan	Controlled release and improved stability, gastrointestinal fate study	[44]

Table 1. *Cont.*

Curcumin Delivery Systems		
Materials and Methods	**Applications and Advancements**	**Ref.**
Wound Healing		
Silver–curcumin nanoparticles prepared by high-pressure homogenization	Enhanced wound healing and potential for burn treatment	[45]
Composite gel prepared from gelatin, F127, and lecithin	Promising wound healing approach with improved properties and bioavailable curcumin release	[46]
Curcumin gel-core hyaluosome (Curc-GC-HS) nanogels	Enhanced curcumin skin penetration, wound healing promotion, potential for skin cancer therapy	[47]
Curcumin-loaded self-microemulsifying gel (olive oil:glycerol = 1:2, PBS, and surfactant RH40)	Advanced tool for wound healing, controlled and sustained drug delivery	[48]
Curcumin-loaded silk fibroin e-gel scaffolds	Improving wound healing activity and serving as a wound dressing material	[100]
Allicin and curcumin gel using Carbopol 940	Excellent wound healing activity, alternative for wound care	[101]
Curcumin gel prepared with Carbopol 934 and propylene glycol	Promotes wound healing through transdermal delivery of curcumin	[102]
Curcumin nanocrystal-based gel prepared by sonoprecipitation	Enhances wound healing efficacy of curcumin	[49]
Coating cotton textiles with acid-catalyzed silica xerogel	Potential as wound dressing materials with prolonged curcumin release	[50]
Single-blinded randomized study on topical 2% curcumin gel	Effective reduction of pain and ulcer size in oral aphthous ulcers	[51]
Curcumin-loaded chitosan-based nanoemulsion gel	Enhanced skin deposition and retention for effective wound healing	[119]
Ophthalmic Delivery		
Polymeric inserts for ocular delivery of curcumin	Improved mechanical properties, sustained release for ophthalmic drug delivery	[52]
Curcumin-loaded ion-sensitive mixed micelle in situ gel (Cur-MM-ISGs)	Prolonged ocular retention, improved permeability, sustained release	[53]
Curcumin mixed micelles (Cur-MMs) in gellan gum gels	Advances ocular drug delivery, addressing curcumin's solubility challenges	[103]
Other Therapeutic Applications of Curcumin		
Synthesis of solid-state fluorescent selenium quantum dots by solvothermal-assisted sol–gel method	Curcumin sensing applications, advancements in fluorescence-based sensing techniques	[120]
Copper–curcumin complex with beta-cyclodextrin	Novel metal–herbal approach for vaginal delivery, safe and effective topical contraception	[55]
Curcumin-incorporated hybrid sol–gel	Potential applications in corrosion protection for mild steel in acidic environments	[56]
Curcumin solubilization in water with heat/pressure treatment and detergents	Efficient, nontoxic, and environmentally friendly protein stain, increased bioavailability with Tween 80	[105]
Fabrication of multifunctional coatings on cotton fabrics using nanosols and curcumin derivatives	Coated cotton fabrics exhibit signaling, self-cleaning, and antibacterial properties	[106]
Encapsulation of curcumin in O/W microemulsion systems and gel microemulsions	Stable products with enhanced antioxidant activity for pharmaceuticals and cosmetics	[61]
Non-aqueous gel and curcumin analogue-loaded nanoparticle gel formulations	Topical formulations with potential applications and antioxidant activity	[121]
Curcumin gel prepared in gelatin hydrogel and chitosan hydrogel	Adjunctive treatment to preserve tooth and bone integrity during orthodontic treatment	[104]
Genotoxicity of curcumin in human lymphocytes	Highlights potential genotoxicity in human lymphocytes, warrants further investigation	[122]

3. Concluding Remarks

The field of research on curcumin gel compositions for therapeutic purposes has seen significant advancements and promising developments. Various innovative approaches have been explored to enhance curcumin's delivery, stability, dissolution characteristics, and targeted release [5–11,15,16,75,76,108,109]. These formulations offer controlled release properties, improved bioavailability, enhanced therapeutic benefits, and targeted delivery for specific applications such as Alzheimer's treatment, gastrointestinal stability, controlled drug release, and intravaginal delivery.

In food and pharmaceutical systems, the focus has been on enhancing bioavailability and sustaining curcumin release [68–76]. These advancements have contributed to the development of functional food ingredients and delivery systems, offering improved oxidative stability, enhanced antioxidant properties, and reduced lipid content.

For inflammatory conditions, researchers have explored topical delivery systems [19–21,23,25,26,28,77–79,81–87]. These approaches provide potential anti-inflammatory activity, improved efficacy, permeation, stability, and targeted delivery, showing promise for managing inflammatory conditions such as rheumatoid arthritis and psoriasis.

Curcumin has demonstrated its potential in various oral health applications [30–33,36,88,115]. It has been investigated for the treatment of minor recurrent aphthous stomatitis (RAS) and oral leukoplakia and as an adjunct in endodontic therapy, gingivitis treatment, and oral wound healing. Additionally, curcumin gel shows potential as an adjunctive therapy in periodontal treatment [92–94,96,97,99,104,116–118], reducing bone and root resorption during orthodontic treatment and enhancing the efficacy of scaling and root planing (SRP) in treating chronic periodontitis.

Moreover, curcumin gel and mucoadhesive patches show potential as noninvasive management options for oral submucous fibrosis (OSMF) [37–39], offering symptom relief, improved oral function, and enhanced drug delivery. Complementary treatments using curcumin alongside other compounds have shown effectiveness in managing OSMF symptoms.

In gastrointestinal applications, curcumin's potential lies in enhancing bioavailability and cellular uptake [40–44]. Various formulations offer extended-release properties, gastric floatability, and sustained drug release, contributing to improved therapeutic outcomes.

The research has also highlighted curcumin's promise in wound healing applications [45–51,100–102,119]. Various formulations have shown positive effects in wound healing, making curcumin gel a valuable option for managing oral aphthous ulcers.

Furthermore, curcumin shows potential in ophthalmic drug delivery [52,53,103], with polymeric in situ gelling inserts and ion-sensitive mixed micelle in situ gel systems offering improved properties for ocular therapy.

Additionally, curcumin has demonstrated its versatility beyond traditional therapeutic uses [54–57,61,62,106], finding applications as a natural food dye alternative, a topical contraception option, a corrosion inhibitor, a reversible protein stain, and more. These findings highlight the diverse applications and potential benefits of curcumin in various fields, including food, contraception, corrosion protection, and drug delivery.

Despite its numerous therapeutic benefits, researchers have also noted potential safety concerns related to curcumin's genotoxic properties [122], warranting further investigation in this area.

4. Conclusions

Curcumin holds great promise in advancing various fields of research and therapeutic applications. Significant advancements have been made in terms of its physicochemical properties and drug delivery systems, enhancing curcumin's delivery, stability, and targeted release for specific applications. Bioavailability enhancement and sustained release strategies have been developed for food and pharmaceutical systems, improving curcumin's stability, antioxidant properties, and controlled release properties. Topical delivery systems show potential for managing inflammatory conditions, while curcumin has shown promise in various oral health applications, periodontal therapy, and the management of oral submucous fibrosis. Gastrointestinal applications, wound healing interventions, ophthalmic drug delivery, and diverse therapeutic applications have further demonstrated the potential of curcumin. However, it is important to consider the safety concerns associated with curcumin, as reports suggest its genotoxic potential and DNA-damaging effects. Continued research and development in these areas are crucial to fully understand curcumin's therapeutic potential, optimize its delivery systems, and ensure its safe and effective use in various biomedical and pharmaceutical applications.

Author Contributions: The authors confirm contribution to the paper as follows: Conceptualization, writing, review and editing, H.O.; Investigation, review and editing, R.L.W. and S.D.C. All authors have read and agreed to the published version of the manuscript.

Funding: This review article received no external funding.

Institutional Review Board Statement: Not applicable.

Informed Consent Statement: Not applicable.

Data Availability Statement: Not applicable.

Conflicts of Interest: The authors partly used OpenAI's large-scale language-generation model. The authors reviewed, revised, and edited the document for accuracy and take full responsibility for the content of this publication. The authors used Bing AI image creator to draw the graphical abstract.

References

1. Curcumin. Available online: https://pubchem.ncbi.nlm.nih.gov/compound/969516#section=2D-Structure (accessed on 20 July 2023).
2. Curcumin. Available online: https://pubchem.ncbi.nlm.nih.gov/compound/969516 (accessed on 20 July 2023).
3. Amalraj, A.; Pius, A.; Gopi, S.; Gopi, S. Biological activities of curcuminoids, other biomolecules from turmeric and their derivatives. *J. Tradit. Complement. Med.* **2016**, *7*, 205–233. [CrossRef] [PubMed]
4. Akhlaghi, S.; Rabbani, S.; Karimi, H.; Haeri, A. Hyaluronic Acid Gel Incorporating Curcumin-Phospholipid Complex Nanoparticles Prevents Postoperative Peritoneal Adhesion. *J. Pharm. Sci.* **2023**, *112*, 587–598. [CrossRef]
5. Alavi, F.; Emam-Djomeh, Z.; Yarmand, M.S.; Salami, M.; Momen, S.; Moosavi-Movahedi, A.A. Cold gelation of curcumin loaded whey protein aggregates mixed with k-carrageenan: Impact of gel microstructure on the gastrointestinal fate of curcumin. *Food Hydrocoll.* **2018**, *85*, 267–280. [CrossRef]
6. Agrawal, M.; Pradhan, M.; Singhvi, G.; Patel, R.; Ajazuddin; Alexander, A. Thermoresponsive in situ gel of curcumin loaded solid lipid nanoparticle: Design, optimization and in vitro characterization. *J. Drug Deliv. Sci. Technol.* **2022**, *71*, 13. [CrossRef]
7. Allam, A.N.; Komeil, I.A.; Abdallah, O.Y. Curcumin phytosomal soft gel formulation: Development, optimization and physicochemical characterization. *Acta Pharm.* **2015**, *65*, 285–297. [CrossRef]
8. Amalia, A.; Lestari, W.W.; Pratama, J.H.; Wibowo, F.R.; Larasati, L.; Saraswati, T.E. Modification of dry-gel synthesized MIL-100(Fe) with carboxymethyl cellulose for curcumin slow-release. *J. Polym. Res.* **2022**, *29*, 13. [CrossRef]
9. Bu, X.T.; Guan, M.H.; Dai, L.; Ji, N.; Qin, Y.; Xu, X.F.; Xiong, L.; Shi, R.; Sun, Q.J. Fabrication of starch-based emulsion gel beads by an inverse gelation technique for loading proanthocyanidin and curcumin. *Food Hydrocoll.* **2023**, *137*, 13. [CrossRef]
10. Cai, R.; Pan, S.; Li, R.; Xu, X.; Pan, S.; Liu, F. Curcumin loading and colon release of pectin gel beads: Effect of different de-esterification method. *Food Chem.* **2022**, *389*, 133130. [CrossRef]
11. Kumar, S.S.D.; Mahadevan, S.; Vijayaraghavan, R.; Mandal, A.B.; MacFarlane, D.R. Curcumin loaded poly(2-hydroxyethyl methacrylate) nanoparticles from gelled ionic liquid—In vitro cytotoxicity and anti-cancer activity in SKOV-3 cells. *Eur. J. Pharm. Sci.* **2014**, *51*, 34–44. [CrossRef]
12. Liu, H.; Meng, X.Y.; Li, L.; Xia, Y.M.; Hu, X.Y.; Fang, Y. The incorporated hydrogel of chitosan-oligoconjugated linoleic acid vesicles and the protective sustained release for curcumin in the gel. *Int. J. Biol. Macromol.* **2023**, *227*, 17–26. [CrossRef]
13. Lou, J.; Hu, W.J.; Tian, R.; Zhang, H.; Jia, Y.T.; Zhang, J.Q.; Zhang, L.K. Optimization and evaluation of a thermoresponsive ophthalmic in situ gel containing curcumin-loaded albumin nanoparticles. *Int. J. Nanomed.* **2014**, *9*, 2517–2525. [CrossRef]
14. Mohammad, R.; Ahmad, M.; Daud, J.M. Development of Optical pH Sensor Based on Curcumin Reagent Immobilized in Sol-gel/Chitosan Melamine Hybrid Matrices). *Sains Malays.* **2009**, *38*, 413–418.
15. Nakagawa, K.; Sowasod, N.; Charinpanitkul, T.; Soottitantawat, A.; Tanthapanichakoon, W. Encapsulation of curcumin loaded oil droplets by cryotropic gel formation from O/W emulsion. In Proceedings of the 11th International Congress on Engineering and Food (ICEF), Athens, Greece, 22–26 May 2011; pp. 1973–1979.
16. Nguyen, K.T.; Tran, P.H.L.; Ngo, H.V.; Tran, T.T.D. Film-Forming Nanogels: Effects of Nanocarriers and Film-Forming Gel on the Sustained Release of Curcumin. *Anticancer Agents Med. Chem.* **2021**, *21*, 658–666. [CrossRef]
17. Kumar, A.; Behl, T.; Uniyal, T.; Chadha, S. Synthesis of Nanostructured Lipid Carriers Loaded Chitosan/Carbopol Hybrid Nanocomposite Gel for Oral Delivery of Artemether and Curcumin. *Pharm. Nanotechnol.* **2020**, *8*, 418–432. [CrossRef]
18. Yusuf, H.; Novitasari, E.; Purnami, N.L.W.; Mahbub, A.W.; Sari, R.; Setyawan, D. Formulation Design and Cell Cytotoxicity of Curcumin-Loaded Liposomal Solid Gels for Anti-Hepatitis C Virus. *Adv. Pharm. Pharm. Sci.* **2022**, *2022*, 8. [CrossRef]
19. Abd El-Halim, S.M.; Mamdouh, M.A.; El-Haddad, A.E.; Soliman, S.M. Fabrication of Anti-HSV-1 Curcumin Stabilized Nanostructured Proniosomal Gel: Molecular Docking Studies on Thymidine Kinase Proteins. *Sci. Pharm.* **2020**, *88*, 16. [CrossRef]

20. Bakhshi, M.; Mahboubi, A.; Jaafari, M.R.; Ebrahimi, F.; Tofangchiha, M.; Alizadeh, A. Comparative Efficacy of 1% Curcumin Nanomicelle Gel and 2% Curcumin Gel for Treatment of Recurrent Aphthous Stomatitis: A Double-Blind Randomized Clinical Trial. *J. Evid. Based Dent. Pract.* **2022**, *22*, 101708. [CrossRef]
21. Chaudhary, H.; Kohli, K.; Amin, S.; Rathee, P.; Kumar, V. Optimization and Formulation Design of Gels of Diclofenac and Curcumin for Transdermal Drug Delivery by Box-Behnken Statistical Design. *J. Pharm. Sci.* **2011**, *100*, 580–593. [CrossRef]
22. Chen, P.; Zhang, H.; Cheng, S.C.; Zhai, G.X.; Shen, C.W. Development of curcumin loaded nanostructured lipid carrier based thermosensitive in situ gel for dermal delivery. *Colloid Surf. A-Physicochem. Eng. Asp.* **2016**, *506*, 356–362. [CrossRef]
23. Gonzalez-Ortega, L.A.; Acosta-Osorio, A.A.; Grube-Pagola, P.; Palmeros-Exsome, C.; Cano-Sarmiento, C.; Garcia-Varela, R.; Garcia, H.S. Anti-inflammatory Activity of Curcumin in Gel Carriers on Mice with Atrial Edema. *J. Oleo Sci.* **2020**, *69*, 123–131. [CrossRef]
24. Hazzah, H.A.; Farid, R.M.; Nasra, M.M.A.; Zakaria, M.; Gawish, Y.; El-Massik, M.A.; Abdallah, O.Y. A new approach for treatment of precancerous lesions with curcumin solid-lipid nanoparticle-loaded gels: In vitro and clinical evaluation. *Drug Deliv.* **2016**, *23*, 1409–1419. [CrossRef]
25. Iriventi, P.; Gupta, N.V.; Osmani, R.A.M.; Balamuralidhara, V. Design & development of nanosponge loaded topical gel of curcumin and caffeine mixture for augmented treatment of psoriasis. *Daru* **2020**, *28*, 489–506. [CrossRef] [PubMed]
26. Kumar, B.; Sahoo, P.K.; Manchanda, S. Curcumin Loaded Ethosomal Gel for Improved Topical Delivery: Formulation, Characterization and Ex-vivo Studies. *Pharm. Nanotechnol.* **2021**, *9*, 281–287. [CrossRef] [PubMed]
27. Rachmawati, H.; Edityaningrum, C.A.; Mauludin, R. Molecular inclusion complex of curcumin-beta-cyclodextrin nanoparticle to enhance curcumin skin permeability from hydrophilic matrix gel. *AAPS PharmSciTech* **2013**, *14*, 1303–1312. [CrossRef] [PubMed]
28. Sana, E.; Zeeshan, M.; Ain, Q.U.; Khan, A.U.; Hussain, I.; Khan, S.; Lepeltier, E.; Ali, H. Topical delivery of curcumin-loaded transfersomes gel ameliorated rheumatoid arthritis by inhibiting NF-kappabeta pathway. *Nanomedicine* **2021**, *16*, 819–837. [CrossRef] [PubMed]
29. Scomoroscenco, C.; Teodorescu, M.; Raducan, A.; Stan, M.; Voicu, S.N.; Trica, B.; Ninciuleanu, C.M.; Nistor, C.L.; Mihaescu, C.I.; Petcu, C.; et al. Novel Gel Microemulsion as Topical Drug Delivery System for Curcumin in Dermatocosmetics. *Pharmaceutics* **2021**, *13*, 505. [CrossRef]
30. Deshmukh, R.A.; Bagewadi, A.S. Comparison of effectiveness of curcumin with triamcinolone acetonide in the gel form in treatment of minor recurrent aphthous stomatitis: A randomized clinical trial. *Int. J. Pharm. Investig.* **2014**, *4*, 138–141. [CrossRef]
31. Fathima, A.I.S.; Manoharan, G. Comparison of Efficacy of Topical Application of Bleomycin with Adjuvant Antioxidants vs. Topical Curcumin Oral Gel with Adjuvant Antioxidants in the Treatment of Oral Leukoplakia. *J. Indian Acad. Oral Med. Radiol.* **2022**, *34*, 409–413. [CrossRef]
32. Oda, D.F.; Duarte, M.A.H.; Andrade, F.B.; Moriyama, L.T.; Bagnato, V.S.; de Moraes, I.G. Antimicrobial action of photodynamic therapy in root canals using LED curing light, curcumin and carbopol gel. *Int. Endod. J.* **2019**, *52*, 1010–1019. [CrossRef]
33. Pandey, V.; Kumar, D.; Nisha, S.; Gupta, A.K.; Verma, T.; Kumari, A. Evaluation of Anti-plaque and Anti-inflammatory Effects of Oral Curcumin Gel as Adjunct to Scaling and Root Planing: A Clinical Study. *Int. J. Appl. Basic Med. Res.* **2021**, *11*, 90–94. [CrossRef]
34. Farhood, H.T.; Ali, B.G. Clinical and Anti-Inflammatory Effect of Curcumin Oral Gel as Adjuncts in Treatment of Periodontal Pocket. *J. Res. Med. Dent. Sci.* **2020**, *8*, 83–90.
35. Nasra, M.M.A.; Khiri, H.M.; Hazzah, H.A.; Abdallah, O.Y. Formulation, in-vitro characterization and clinical evaluation of curcumin in-situ gel for treatment of periodontitis. *Drug Deliv.* **2017**, *24*, 133–142. [CrossRef]
36. Fonseca-Santos, B.; Bonifacio, B.V.; Baub, T.M.; Gremiao, M.P.D.; Chorilli, M. In-Situ Gelling Liquid Crystal Mucoadhesive Vehicle for Curcumin Buccal Administration and Its Potential Application in the Treatment of Oral Candidiasis. *J. Biomed. Nanotechnol.* **2019**, *15*, 1334–1344. [CrossRef]
37. Chandrashekar, A.; Annigeri, R.G.; Va, U.; Thimmasetty, J. A clinicobiochemical evaluation of curcumin as gel and as buccal mucoadhesive patches in the management of oral submucous fibrosis. *Oral Surg. Oral Med. Oral Pathol. Oral Radiol.* **2021**, *131*, 428–434. [CrossRef]
38. Lanjekar, A.B.; Bhowate, R.R.; Bakhle, S.; Narayane, A.; Pawar, V.; Gandagule, R. Comparison of Efficacy of Topical Curcumin Gel with Triamcinolone-hyaluronidase Gel Individually and in Combination in the Treatment of Oral Submucous Fibrosis. *J. Contemp. Dent. Pract.* **2020**, *21*, 83–90.
39. Nerkar Rajbhoj, A.; Kulkarni, T.M.; Shete, A.; Shete, M.; Gore, R.; Sapkal, R. A Comparative Study to Evaluate Efficacy of Curcumin and Aloe Vera Gel along with Oral Physiotherapy in the Management of Oral Submucous Fibrosis: A Randomized Clinical Trial. *Asian Pac. J. Cancer Prev.* **2021**, *22*, 107–112. [CrossRef]
40. Araiza-Calahorra, A.; Wang, Y.Q.; Boesch, C.; Zhao, Y.S.; Sarkar, A. Pickering emulsions stabilized by colloidal gel particles complexed or conjugated with biopolymers to enhance bioaccessibility and cellular uptake of curcumin. *Curr. Res. Food Sci.* **2020**, *3*, 178–188. [CrossRef]
41. Fontes-Candia, C.; Martinez, J.C.; Lopez-Rubio, A.; Salvia-Trujillo, L.; Martin-Belloso, O.; Martinez-Sanz, M. Emulsion gels and oil-filled aerogels as curcumin carriers: Nanostructural characterization of gastrointestinal digestion products. *Food Chem.* **2022**, *387*, 11. [CrossRef]
42. Kathpalia, H.; Salunkhe, S.; Juvekar, S. Formulation of gastroretentive sustained release floating in situ gelling drug delivery system of solubility enhanced curcumin-soy lecithin complex. *J. Drug Deliv. Sci. Technol.* **2019**, *53*, 6. [CrossRef]

43. Padhan, A.; Nanda, B.K.; Behera, B.C. Formulation and Evaluation of Gastro Retentive Novel Floating In-Situ Gelling System of Curcumin. *Int. J. Pharm. Sci. Res.* **2019**, *10*, 4574–4586. [CrossRef]
44. Su, J.Q.; Wang, L.L.; Dong, W.X.; Wei, J.; Liu, X.; Yan, J.X.; Ren, F.Z.; Yuan, F.; Wang, P.J. Fabrication and Characterization of Ultra-High-Pressure (UHP)-Induced Whey Protein Isolate/kappa-Carrageenan Composite Emulsion Gels for the Delivery of Curcumin. *Front. Nutr.* **2022**, *9*, 18. [CrossRef]
45. Aghamoosa, M.; Bidgoli, S.A.; Ghafari, S.; Sabokbar, A.; Harzandi, N. Preclinical Evaluation of Silver-Curcumin Nano- gel: A complete assessment on a new topical antimicrobial product for burn. *J. Nanoanal.* **2020**, *7*, 62–72.
46. Das, R.P.; Gandhi, V.V.; Verma, G.; Ajish, J.K.; Singh, B.G.; Kunwar, A. Gelatin-lecithin-F127 gel mediated self-assembly of curcumin vesicles for enhanced wound healing. *Int. J. Biol. Macromol.* **2022**, *210*, 403–414. [CrossRef]
47. El-Refaie, W.M.; Elnaggar, Y.S.R.; El-Massik, M.A.; Abdallah, O.Y. Novel curcumin-loaded gel-core hyaluosomes with promising burn-wound healing potential: Development, in-vitro appraisal and in-vivo studies. *Int. J. Pharm.* **2015**, *486*, 88–98. [CrossRef] [PubMed]
48. Guo, J.W.; Pu, C.M.; Liu, C.Y.; Lo, S.L.; Yen, Y.H. Curcumin-Loaded Self-Microemulsifying Gel for Enhancing Wound Closure. *Skin Pharmacol. Physiol.* **2020**, *33*, 300–308. [CrossRef] [PubMed]
49. Kotian, V.; Koland, M.; Mutalik, S. Nanocrystal-Based Topical Gels for Improving Wound Healing Efficacy of Curcumin. *Crystals* **2022**, *12*, 20. [CrossRef]
50. Pisitsak, P.; Ruktanonchai, U. Preparation, characterization, and in vitro evaluation of antibacterial sol-gel coated cotton textiles with prolonged release of curcumin. *Text. Res. J.* **2015**, *85*, 949–959. [CrossRef]
51. Raman, P.; Pitty, H.R. Correlation of pain score with ulcer size in oral aphthous ulcers using 2% curcumin gel and 0.1% triamcinolone oral paste—A parallel comparison study. *J. Indian Acad. Oral Med. Radiol.* **2021**, *33*, 53–59. [CrossRef]
52. Abdelkader, H.; Wertheim, D.; Pierscionek, B.; Alany, R.G. Curcumin In Situ Gelling Polymeric Insert with Enhanced Ocular Performance. *Pharmaceutics* **2020**, *12*, 17. [CrossRef]
53. Duan, Y.W.; Cai, X.Q.; Du, H.L.; Zhai, G.X. Novel in situ gel systems based on P123/TPGS mixed micelles and gellan gum for ophthalmic delivery of curcumin. *Colloid Surf. B-Biointerfaces* **2015**, *128*, 322–330. [CrossRef]
54. Brito-Oliveira, T.C.; Bispo, M.; Moraes, I.C.F.; Campanella, O.H.; Pinho, S.C. Stability of curcumin encapsulated in solid lipid microparticles incorporated in cold-set emulsion filled gels of soy protein isolate and xanthan gum. *Food Res. Int.* **2017**, *102*, 759–767. [CrossRef]
55. Gaurav, C.; Goutam, R.; Rohan, K.N.; Sweta, K.T.; Abhay, C.S.; Amit, G.K. (Copper-curcumin) beta-cyclodextrin vaginal gel: Delivering a novel metal-herbal approach for the development of topical contraception prophylaxis. *Eur. J. Pharm. Sci.* **2014**, *65*, 183–191. [CrossRef]
56. Ishak, N.A.; Hamidon, T.S.; Tan, Z.H.; Hussin, M.H. Extracts of curcumin-incorporated hybrid sol-gel coatings for the corrosion mitigation of mild steel in 0.5 M HCl. *J. Coat. Technol. Res.* **2020**, *17*, 1515–1535. [CrossRef]
57. Kurien, B.T.; Dorri, Y.; Scofield, R.H. Curcumin/Turmeric as an Environment-Friendly Protein Gel Stain. *Methods Mol. Biol.* **2018**, *1853*, 121–131. [CrossRef]
58. Pan, Y.; Li, X.M.; Meng, R.; Xu, B.C.; Zhang, B. Investigation of the Formation Mechanism and Curcumin Bioaccessibility of Emulsion Gels Based on Sugar Beet Pectin and Laccase Catalysis. *J. Agric. Food Chem.* **2021**, *69*, 2557–2563. [CrossRef]
59. Qazi, H.J.; Ye, A.Q.; Acevedo-Fani, A.; Singh, H. In vitro digestion of curcumin-nanoemulsion-enriched dairy protein matrices: Impact of the type of gel structure on the bioaccessibility of curcumin. *Food Hydrocoll.* **2021**, *117*, 12. [CrossRef]
60. Qiao, X.; Liu, F.D.; Kong, Z.H.; Yang, Z.Y.; Dai, L.; Wang, Y.F.; Sun, Q.J.; McClements, D.J.; Xu, X.F. Pickering emulsion gel stabilized by pea protein nanoparticle induced by heat-assisted pH-shifting for curcumin delivery. *J. Food Eng.* **2023**, *350*, 13. [CrossRef]
61. Scomoroscenco, C.; Teodorescu, M.; Burlacu, S.G.; Gifu, I.C.; Mihaescu, C.I.; Petcu, C.; Raducan, A.; Oancea, P.; Cinteza, L.O. Synergistic Antioxidant Activity and Enhanced Stability of Curcumin Encapsulated in Vegetal Oil-Based Microemulsion and Gel Microemulsions. *Antioxidants* **2022**, *11*, 854. [CrossRef]
62. Wang, S.; Chen, P.; Zhang, L.; Yang, C.F.; Zhai, G.X. Formulation and evaluation of microemulsion-based in situ ion-sensitive gelling systems for intranasal administration of curcumin. *J. Drug Target.* **2012**, *20*, 831–840. [CrossRef]
63. Liu, R.; Sun, L.; Fang, S.; Wang, S.; Chen, J.; Xia, X.; Liu, C. Thermosensitive in situ nanogel as ophthalmic delivery system of curcumin: Development, characterization, in vitro permeation and in vivo pharmacokinetic studies. *Pharm. Dev. Technol.* **2016**, *21*, 576–582. [CrossRef]
64. Song, Z.; Feng, R.; Sun, M.; Guo, C.; Gao, Y.; Li, L.; Zhai, G. Curcumin-loaded PLGA-PEG-PLGA triblock copolymeric micelles: Preparation, pharmacokinetics and distribution in vivo. *J. Colloid Interface Sci.* **2011**, *354*, 116–123. [CrossRef] [PubMed]
65. Gugleva, V.; Michailova, V.; Mihaylova, R.; Momekov, G.; Zaharieva, M.M.; Najdenski, H.; Petrov, P.; Rangelov, S.; Forys, A.; Trzebicka, B.; et al. Formulation and Evaluation of Hybrid Niosomal In Situ Gel for Intravesical Co-Delivery of Curcumin and Gentamicin Sulfate. *Pharmaceutics* **2022**, *14*, 747. [CrossRef] [PubMed]
66. Ghodake, V.A.; Agarwal, B.A.; Tekade, A.R. Stability-indicating densitometric method for simultaneous determination of donepezil hydrochloride and curcumin in insitu nasal gel. *Indo Am. J. Pharm. Sci.* **2018**, *5*, 4097–4106. [CrossRef]
67. Iwunze, M.O.; McEwan, D. The characterization of the sol-gel encapsulated curcumin as a possible sensor for small biologically important molecules. *Cell. Mol. Biol.* **2007**, *53*, 81–87. [PubMed]

68. Aliabbasi, N.; Emam-Djomeh, Z.; Askari, G.; Salami, M. Pinto bean protein-based acid-induced cold-set gels as carriers for curcumin delivery: Fabrication and characterization. *Food Hydrocoll. Health* **2021**, *1*, 15. [CrossRef]
69. Du, C.X.; Xu, J.J.; Luo, S.Z.; Li, X.J.; Mu, D.D.; Jiang, S.T.; Zheng, Z. Low-oil-phase emulsion gel with antioxidant properties prepared by soybean protein isolate and curcumin composite nanoparticles. *LWT-Food Sci. Technol.* **2022**, *161*, 10. [CrossRef]
70. Du, Y.N.; Yan, J.N.; Xu, S.Q.; Wang, Y.Q.; Wang, X.C.; Wu, H.T. Formation and characteristics of curcumin-loaded binary gels formed from large yellow croaker (*Pseudosciaena crocea*) roe protein isolate and gellan gum. *Food Chem.* **2023**, *405*, 10. [CrossRef]
71. Geremias-Andrade, I.M.; Souki, N.; Moraes, I.C.F.; Pinho, S.C. Rheological and mechanical characterization of curcumin-loaded emulsion-filled gels produced with whey protein isolate and xanthan gum. *LWT-Food Sci. Technol.* **2017**, *86*, 166–173. [CrossRef]
72. Lei, L.; Chen, Y.L.; Zhu, C.H.; Wu, H.F.; Wan, Z.L.; Yang, X.Q.; Yuan, Y. The novel pickering emulsion gels stabilized by zein hydrolysate-chitin nanocrystals coacervates: Improvement on stability and bioaccessibility for curcumin. *Food Res. Int.* **2022**, *161*, 14. [CrossRef]
73. Lv, P.F.; Wang, D.; Dai, L.; Wu, X.J.; Gao, Y.X.; Yuan, F. Pickering emulsion gels stabilized by high hydrostatic pressure-induced whey protein isolate gel particles: Characterization and encapsulation of curcumin. *Food Res. Int.* **2020**, *132*, 11. [CrossRef]
74. Miao, J.Y.; Xu, N.; Cheng, C.; Zou, L.Q.; Chen, J.; Wang, Y.; Liang, R.H.; McClements, D.J.; Liu, W. Fabrication of polysaccharide-based high internal phase emulsion gels: Enhancement of curcumin stability and bioaccessibility. *Food Hydrocoll.* **2021**, *117*, 13. [CrossRef]
75. Tan, K.X.; Ng, L.L.E.; Loo, S.C.J. Formulation Development of a Food-Graded Curcumin-Loaded Medium Chain Triglycerides-Encapsulated Kappa Carrageenan (CUR-MCT-KC) Gel Bead Based Oral Delivery Formulation. *Materials* **2021**, *14*, 22. [CrossRef]
76. Vellido-Perez, J.A.; Ochando-Pulido, J.M.; Brito-de la Fuente, E.; Martinez-Ferez, A. Effect of operating parameters on the physical and chemical stability of an oil gelled-in-water emulsified curcumin delivery system. *J. Sci. Food Agric.* **2021**, *101*, 6395–6406. [CrossRef]
77. Araujo, F.O.; Felicio, M.B.; Lima, C.F.; Piccolo, M.S.; Pizziolo, V.R.; Diaz-Munoz, G.; Bastos, D.S.S.; Oliveira, L.L.; Peluzio, M.; Diaz, M.A.N. Antioxidant and anti-inflammatory activity of curcumin transdermal gel in an IL-10 knockout mouse model of inflammatory bowel disease. *An. Acad. Bras. Cienc.* **2022**, *94*, e20201378. [CrossRef]
78. Jain, A.; Doppalapudi, S.; Domb, A.J.; Khan, W. Tacrolimus and curcumin co-loaded liposphere gel: Synergistic combination towards management of psoriasis. *J. Control. Release* **2016**, *243*, 132–145. [CrossRef]
79. Jain, H.; Devabattula, G.; Bhat, A.; Dalvi, H.; Rangaraj, N.; Godugu, C.; Srivastava, S. Topical delivery of Bruton's tyrosine kinase inhibitor and curcumin-loaded nanostructured lipid carrier gel: Repurposing strategy for the psoriasis management. *Pharm. Dev. Technol.* **2022**, *27*, 975–988. [CrossRef]
80. Khatoon, K.; Ali, A.; Ahmad, F.J.; Hafeez, Z.; Rizvi, M.M.A.; Akhter, S.; Beg, S. Novel nanoemulsion gel containing triple natural bio-actives combination of curcumin, thymoquinone, and resveratrol improves psoriasis therapy: In vitro and in vivo studies. *Drug Deliv. Transl. Res.* **2021**, *11*, 1245–1260. [CrossRef]
81. Patel, N.A.; Patel, N.J.; Patel, R.P. Formulation and evaluation of curcumin gel for topical application. *Pharm. Dev. Technol.* **2009**, *14*, 80–89. [CrossRef]
82. Shehata, T.M.; Ibrahim, M.M.; Elsewedy, H.S. Curcumin Niosomes Prepared from Proniosomal Gels: In Vitro Skin Permeability, Kinetic and In Vivo Studies. *Polymers* **2021**, *13*, 14. [CrossRef]
83. Shrotriya, S.; Ranpise, N.; Satpute, P.; Vidhate, B. Skin targeting of curcumin solid lipid nanoparticles-engrossed topical gel for the treatment of pigmentation and irritant contact dermatitis. *Artif. Cells Nanomed. Biotechnol.* **2018**, *46*, 1471–1482. [CrossRef]
84. Singh, P.; Dabre, S. Controlled Release Gel Encompassing Curcumin Microspheres and Diclofenac Diethylamine for Feat Against Arthritis Inflammation. *Curr. Rheumatol. Rev.* **2020**, *16*, 110–119. [CrossRef] [PubMed]
85. Tripathi, D.; Mishra, S.; Rai, A.K.; Sahoo, J.; Sharma, D.K.; Singh, Y. Curcumin-loaded Hydrotropic Solid Dispersion Topical Gel Development and Evaluation: A Greener Approach Towards Transdermal Delivery of Drugs. *Curr. Green Chem.* **2022**, *9*, 26–39. [CrossRef]
86. Vila, M.; Boscariol, R.; Balcao, V.M. Nanostructured curcumin with choline and geranic acid ionic liquid (cage-il): Potential for incorporation into pharmaceutical gel formulations. *Quím. Nova* **2023**, *46*, 250–256. [CrossRef]
87. Yaghoobi, R.; Soghrati, M.; Pazyar, N.; Parvin, N. INVESTIGATING THE EFFECT OF 1% CURCUMIN GEL AND PLACEBO ON PATIENTS WITH PLAQUE PSORIASIS. *Int. J. Life Sci. Pharma Res.* **2018**, *8*, L48–L57. [CrossRef]
88. Samita; Verma, S.K.; Sharma, V.K.; Moinuddin; Ahad, A. Effect of 1% curcumin gel on myeloperoxidase activity in GCF and periodontal status in the initial phase of orthodontic tooth movement. *J. Orthod. Sci.* **2022**, *11*, 55. [CrossRef]
89. Anuradha, B.R.; Bai, Y.D.; Sailaja, S.; Sudhakar, J.; Priyanka, M.; Deepika, V. Evaluation of Anti-Inflammatory Effects of Curcumin Gel as an Adjunct to Scaling and Root Planing: A Clinical Study. *J. Int. Oral Health* **2015**, *7*, 90–93.
90. Dave, D.H.; Patel, P.; Shah, M.; Dadawala, S.M.; Saraiya, K.; Sant, A.V. Comparative Evaluation of Efficacy of Oral Curcumin Gel as an Adjunct to Scaling and Root Planing in the Treatment of Chronic Periodontitis. *Adv. Hum. Biol.* **2018**, *8*, 79–82. [CrossRef]
91. Hosadurga, R.R.; Rao, S.N.; Jose, J.; Rompicharla, N.C.; Shakil, M.; Shashidhara, R. Evaluation of the efficacy of 2% curcumin gel in the treatment of experimental periodontitis. *Pharmacogn. Res.* **2014**, *6*, 326–333. [CrossRef]
92. Sha, A.M.; Garib, B.T. Antibacterial Effect of Curcumin against Clinically Isolated Porphyromonas gingivalis and Connective Tissue Reactions to Curcumin Gel in the Subcutaneous Tissue of Rats. *Biomed. Res. Int.* **2019**, *2019*, 6810936. [CrossRef]
93. Hugar, S.S.; Patil, S.; Metgud, R.; Nanjwade, B.; Hugar, S.M. Influence of application of chlorhexidine gel and curcumin gel as an adjunct to scaling and root planing: A interventional study. *J. Nat. Sci. Biol. Med.* **2016**, *7*, 149–154. [CrossRef]

94. Ravishankar, P.L.; Kumar, Y.P.; Anila, E.N.; Chakraborty, P.; Malakar, M.; Mahalakshmi, R. Effect of local application of curcumin and ornidazole gel in chronic periodontitis patients. *Int. J. Pharm. Investig.* **2017**, *7*, 188–192. [CrossRef]
95. Kaur, H.; Grover, V.; Malhotra, R.; Gupta, M. Evaluation of Curcumin Gel as Adjunct to Scaling & Root Planing in Management of Periodontitis- Randomized Clinical & Biochemical Investigation. *Infect. Disord. Drug Targets* **2019**, *19*, 171–178. [CrossRef]
96. Meghana, M.V.S.; Deshmukh, J.; Devarathanamma, M.V.; Asif, K.; Jyothi, L.; Sindhura, H. Comparison of effect of curcumin gel and noneugenol periodontal dressing in tissue response, early wound healing, and pain assessment following periodontal flap surgery in chronic periodontitis patients. *J. Indian Soc. Periodontol.* **2020**, *24*, 54–59. [CrossRef]
97. Mohammad, C.A. Efficacy of Curcumin Gel on Zinc, Magnesium, Copper, IL-1beta, and TNF-alpha in Chronic Periodontitis Patients. *Biomed. Res. Int.* **2020**, *2020*, 8850926. [CrossRef]
98. Mohammad, C.A.; Ali, K.M.; Al-Rawi, R.A.; Gul, S.S. Effects of Curcumin and Tetracycline Gel on Experimental Induced Periodontitis as an Anti-Inflammatory, Osteogenesis Promoter and Enhanced Bone Density through Altered Iron Levels: Histopathological Study. *Antibiotics* **2022**, *11*, 521. [CrossRef]
99. Mohammad, C.A.; Ali, K.M.; Sha, A.M.; Gul, S.S. Antioxidant Effects of Curcumin Gel in Experimental Induced Diabetes and Periodontitis in Rats. *Biomed. Res. Int.* **2022**, *2022*, 7278064. [CrossRef]
100. Karahaliloglu, Z. Curcumin-loaded silk fibroin e-gel scaffolds for wound healing applications. *Mater. Technol.* **2018**, *33*, 276–287. [CrossRef]
101. Khan, I.A.; Lodhi, A.H.; Munawar, S.H.; Manzoor, A.; Raza, M.A. Formulation and Evaluation of Allicin and Curcumin Gel Improves Normal and Diabetic Ulcers in Rabbits. *Lat. Am. J. Pharm.* **2018**, *37*, 1602–1607.
102. Kim, J.; Kim, M.J.; Lee, C.Y. Wound Healing Effect of Curcumin Gel for Transdermal Delivery. *Polymer* **2013**, *37*, 387–392. [CrossRef]
103. Sai, N.; Dong, X.; Huang, P.Q.; You, L.T.; Yang, C.J.; Liu, Y.; Wang, W.P.; Wu, H.M.; Yu, Y.C.; Du, Y.Y.; et al. A Novel Gel-Forming Solution Based on PEG-DSPE/Solutol HS 15 Mixed Micelles and Gellan Gum for Ophthalmic Delivery of Curcumin. *Molecules* **2020**, *25*, 15. [CrossRef]
104. Asefi, S.; Seifi, M.; Fard, G.H.; Lotfi, A. Innovative evaluation of local injective gel of curcumin on the orthodontic tooth movement in rats. *Dent. Res. J.* **2018**, *15*, 40–49. [CrossRef]
105. Kurien, B.T.; Thomas, R.; Payne, A.; Scofield, R.H. Heat/Pressure Treatment with Detergents Significantly Increases Curcumin Solubility and Stability: Its Use as an Environment-Friendly Protein Gel Stain. *Methods Mol. Biol.* **2018**, *1853*, 237–246. [CrossRef] [PubMed]
106. Raduly, F.M.; Raditoiu, V.; Raditoiu, A.; Frone, A.N.; Nicolae, C.A.; Raut, I.; Constantin, M.; Grapin, M. Multifunctional Finishing of Cotton Fabric with Curcumin Derivatives Coatings Obtained by Sol-Gel Method. *Gels* **2023**, *9*, 369. [CrossRef] [PubMed]
107. Han, J.W.; Li, L.Y.; Pang, Z.T.; Su, M.L.; He, X.S.; Qian, S.; Zhang, J.J.; Gao, Y.; Wei, Y.F. Mechanistic insight into gel-induced aggregation of amorphous curcumin during dissolution process. *Eur. J. Pharm. Sci.* **2022**, *170*, 14. [CrossRef] [PubMed]
108. Kadota, K.; Nogami, S.; Uchiyama, H.; Tozuka, Y. Controlled release behavior of curcumin from kappa-carrageenan gels with flexible texture by the addition of metal chlorides. *Food Hydrocoll.* **2020**, *101*, 8. [CrossRef]
109. Kaur, A.; Saxena, Y.; Bansal, R.; Gupta, S.; Tyagi, A.; Sharma, R.K.; Ali, J.; Panda, A.K.; Gabrani, R.; Dang, S. Intravaginal Delivery of Polyphenon 60 and Curcumin Nanoemulsion Gel. *AAPS PharmSciTech* **2017**, *18*, 2188–2202. [CrossRef]
110. Li, Y.H.; Wang, Y.S.; Zhao, J.S.; Li, Z.Y.; Chen, H.H. A pH-sensitive curcumin loaded microemulsion-filled alginate and porous starch composite gels: Characterization, in vitro release kinetics and biological activity. *Int. J. Biol. Macromol.* **2021**, *182*, 1863–1873. [CrossRef]
111. Yang, J.Y.; Wan, L.; Duan, X.K.; Wang, H.D.; Yang, Z.X.; Liu, F.X.; Xu, X.Y.; Pan, S.Y. Potential low-calorie model that inhibits free fatty acid release and helps curcumin deliver in vitro: Ca2+-induced emulsion gels from low methyl-esterified pectin with the presence of erythritol. *Int. J. Biol. Macromol.* **2022**, *200*, 449–457. [CrossRef]
112. Jin, N.; He, J.W.; Wu, C.Y.; Chen, Z.J.; Li, Y.L.; Chen, J.M.; Lin, J.H. Glycyrrhizic acid assists anti-psoriatic efficacy of a self-deformable curcumin loaded transdermal gel. *Pharm. Dev. Technol.* **2022**, *27*, 282–289. [CrossRef]
113. Perez-Salas, J.L.; Medina-Torres, L.; Rocha-Guzman, N.E.; Calderas, F.; Gonzalez-Laredo, R.F.; Bernad-Bernad, M.J.; Moreno-Jimenez, M.R.; Gallegos-Infante, J.A. A Water in Oil Gelled Emulsion as a Topical Release Vehicle for Curcumin. *Starch-Starke* **2022**, *74*, 9. [CrossRef]
114. Yao, Q.; Zhai, Y.Y.; He, Z.M.; Wang, Q.; Sun, L.N.; Sun, T.Y.; Lv, L.Y.; Li, Y.T.; Yang, J.Y.; Lv, D.H.; et al. Water-responsive gel extends drug retention and facilitates skin penetration for curcumin topical delivery against psoriasis. *Asian J. Pharm. Sci.* **2023**, *18*, 15. [CrossRef]
115. Gauthaman, J.; Ganesan, A. Therapeutic Evaluation of 5% Topical Amlexanox Paste and 2% Curcumin Oral Gel in the Management of Recurrent Aphthous Stomatitis-A Randomized Clinical Trial. *J. Indian Acad. Oral Med. Radiol.* **2022**, *34*, 17–21. [CrossRef]
116. Sha, A.M.; Garib, B.T.; Azeez, S.H.; Gul, S.S. Effects of curcumin gel on osteoclastogenic bone markers in experimental periodontitis and alveolar bone loss in wistar rats. *J. Dent. Sci.* **2021**, *16*, 905–914. [CrossRef]
117. Siddharth, M.; Singh, P.; Gupta, R.; Sinha, A.; Shree, S.; Sharma, K. A Comparative Evaluation of Subgingivally Delivered 2% Curcumin and 0.2% Chlorhexidine Gel Adjunctive to Scaling and Root Planing in Chronic Periodontitis. *J. Contemp. Dent. Pract.* **2020**, *21*, 494–499.

118. Sreedhar, A.; Sarkar, I.; Rajan, P.; Pai, J.; Malagi, S.; Kamath, V.; Barmappa, R. Comparative evaluation of the efficacy of curcumin gel with and without photo activation as an adjunct to scaling and root planing in the treatment of chronic periodontitis: A split mouth clinical and microbiological study. *J. Nat. Sci. Biol. Med.* **2015**, *6*, S102–S109. [CrossRef]
119. Thomas, L.; Zakir, F.; Mirza, M.A.; Anwer, M.K.; Ahmad, F.J.; Iqbal, Z. Development of Curcumin loaded chitosan polymer based nanoemulsion gel: In vitro, ex vivo evaluation and in vivo wound healing studies. *Int. J. Biol. Macromol.* **2017**, *101*, 569–579. [CrossRef]
120. Anupama, K.; Paul, T.; Mary, K.A.A. Solid-State Fluorescent Selenium Quantum Dots by a Solvothermal-Assisted Sol-Gel Route for Curcumin Sensing. *ACS Omega* **2021**, *6*, 21525–21533. [CrossRef]
121. Gokhale, K.M.; Pandey, D.S. Formulation Development of Novel Curcumin Analogue Loaded Non-aqueous Gel and Curcumin Analogue Loaded Nanoparticle (CA-NP) Gel for Topical Use and In-vitro Antioxidant Study. *J. Pharm. Res. Int.* **2021**, *33*, 284–302. [CrossRef]
122. Kowalik, J.; Trzeciak, A.; Wojewodzka, M.; Blasiak, J. In vitro genotoxic effect of curcumin assessed by alkaline single cell gel/comet assay. *Neoplasma* **1999**, *46*, 64–65.

gels

MDPI

Review

Pectin Based Hydrogels for Drug Delivery Applications: A Mini Review

Sung Soo Han [1,2], Seong Min Ji [1], Min Jung Park [1], Maduru Suneetha [1] and Uluvangada Thammaiah Uthappa [1,*]

1 School of Chemical Engineering, Yeungnam University, 280 Daehak-Ro, Gyeongsan 38541, Republic of Korea
2 Institute of Cell Culture, Yeungnam University, 280 Daehak-ro, Gyeongsan 38541, Republic of Korea
* Correspondence: sanjuuthappa@gmail.com

Abstract: Over the past few decades, hydrogel systems using natural polymers have been expansively employed in drug delivery applications. Among the various reported biopolymer-based hydrogel drug delivery systems, pectin (Pec) is an exceptional natural polymer due to its unique functionalities and excellent properties such as biocompatibility, biodegradability, low-cost, and simple gelling capability, which has received considerable interest in the drug delivery fields. Since there is an increasing need for biomaterials with unique properties for drug delivery applications, in this review, hydrogels fabricated from natural pectin polymers were thoroughly investigated. Additionally, the present mini review aims to bring collectively more concise ways such as sources, extraction, properties, and various forms of Pec based hydrogel drug delivery systems and their toxicity concerns are summarized. Finally, the potential objectives and challenges based on pectin-based hydrogel drug delivery systems are also discussed.

Keywords: hydrogels; pectins; controlled release; biopolymers; targeted drug delivery

Citation: Han, S.S.; Ji, S.M.; Park, M.J.; Suneetha, M.; Uthappa, U.T. Pectin Based Hydrogels for Drug Delivery Applications: A Mini Review. *Gels* **2022**, *8*, 834. https://doi.org/10.3390/gels8120834

Academic Editor: Lifeng Kang

Received: 28 November 2022
Accepted: 15 December 2022
Published: 17 December 2022

1. Introduction

Biomaterials with exceptional properties have gained a lot of study interest, specifically in drug delivery applications. Polymers, both synthetic and natural, are regarded as better candidates in the fabrication of biomaterials [1–4]. Hydrogels, films, nanoparticles, and nanocomposites are just a few of the drug formulations that have been designed and advanced in drug delivery fields [5–8]. Among the formulations above-mentioned, hydrogels have grown in popularity due to their intriguing properties such as biocompatibility, biodegradability, and exclusive "soft-wet" nature in correlation to biological tissue [9,10]. It is worthwhile mentioning that hydrogels have a high-water content, which could swell and adsorb liquid due to their porous nature, and an injectable hydrogel is highly efficient for clinical use. In terms of tumor application, hydrogels possess excellent biocompatibility and controllability, and some of these hydrogel systems are used in various other applications such as additives, the chemical industry, energy, and water treatment [11–16]. Despite these exceptional benefits, hydrogels in bio-related applications face some challenges due to limitations such as mechanical stiffness, water sensitivity, and instability in physiological conditions [17].

Polymeric materials are classified into two types: synthetic and natural. Because of their biodegradability and biocompatibility, natural polymers have distinct merits over synthetic polymer systems. The main disadvantages of using natural polymers are their low mechanical properties over synthetic polymers, which makes them unsuitable for a variety of biomedical applications. Several reports have specified that the most synthetic polymers have drawbacks such as high cytotoxicity and low biocompatibility [18,19]. Generally, biopolymers are comprised of monomeric units covalently attached to form bigger biomolecules. Usually, pectic substances are differentiated into four different types: protopectin, pectic acid, pectinic acid, and Pec [20]. Among the various biopolymers, Pec is a kind of water-soluble anionic heteropolysaccharide found from the primary cell walls of terrestrial plants extracted using chemical or enzymatic process [21]. Pec possesses a higher range of heterogeneity in their structure due to their sources and methods of different extraction process [22]. Pec is considered as a promising candidate in the drug delivery field due to its excellent features such as non-toxicity, biocompatibility, biodegradability, low-cost, antibacterial, and anti-inflammatory properties [23].

To our best of knowledge, recently there have been no reviews published specifically focusing on "Pec based hydrogel for drug delivery systems". However, there are two general reviews on Pec based biomaterials for biomedical applications [23,24]. Thus, this is a sole mini review where we tried to compile the latest progress and advances specifically on Pec based hydrogels for drug delivery applications. Thus, this present mini review could target a wide audience/researchers who exclusively work on Pec and Pec based hydrogel systems. This mini review details the fabrication of hydrogels from natural Pec polymers and aims to collectively bring more concise ways such as sources, extraction, properties, and various forms of Pec based hydrogels in drug delivery applications and their toxicity concerns. Finally, the possible purposes and challenges based on Pec based hydrogel drug delivery systems are also discussed.

2. Sources

Some of the sources for Pec are from apple pomace, citrus peels, and, more recently, sugar beet pulp [25]. Certainly, tropical and subtropical fruit by-products are primarily a significant source of Pec. Nevertheless, it is important to mention that the Pec yield and physicochemical properties of Pec are affected by the extraction technique as well as additional variables such as the extraction time, type of acid, pH, temperature, and the liquid–solid ratios [26].

3. Pectin Extraction

The yield of extracted Pec as well as the quality can be used to evaluate the suitability of the extraction method because mass transfer into the extraction solvents governs Pec extraction. To extract Pec from natural sources, several methods have been used including traditional hot extraction and advanced procedures such as ultrasound, microwave, and enzymatic processes. Huge efforts are being made to promote "green" chemistry and technology. In terms of Pec extraction, hot conventional extraction necessitates a lengthy protocol, more energy, and the use of strong acids, which is contrary to "green" chemistry principles. Thus, an outline is depicted in the below section on conventional and non-conventional extraction methods.

3.1. Conventional Extraction

Extraction temperature, solid–liquid ratio, pH, particle size, and extraction time are all factors that influence the yield and quality of the extracted Pec. The utilization of mineral acids for Pec extraction is linked to environmental concerns as well as higher costs. Concerning the emergent concept of "green" chemistry and the drawbacks associated with the practice of mineral acids, the emphasis is now shifting to "food" compatible acids [27].

3.2. Ultrasound Mediated Extraction

Ultrasonic waves with frequencies from 20 to 100 kHz are commonly used. It is important to note that ultrasound frequency influences the extraction process because it influences the size of the microbubbles and their resistance to mass transfer. Furthermore, an upsurge in ultrasound frequency results is a decrease in the production and intensity of cavitation in liquid [28,29]. Several studies backing up the substantial assistance of the ultrasound-assisted extraction has several merits including low energy, less extraction time, minimal solvent, and enhanced extraction yield in support of using ultrasound as a "green" extraction method.

3.3. Microwave Mediated Extraction

This technique needs less processing time and solvent, and produces a higher extraction yield and generate superior qualities [30]. Microwave extraction is the process of applying a microwave field to a dielectric material. Ionic conduction and dipole rotation heat the solvent–sample matrix. Microwave energy initiates the electrophoretic transfer of ions and electrons, resulting in an electric field that drives particle movement, whereas dipole rotation is instigated by the substitute movement of polar molecules. Microwave power, measured in Watts (W), is a key factor in Pec extraction. Increased microwave power was found to be positively related to extraction efficiency [31].

3.4. Enzyme Aided Extraction

For enzyme-aided Pec extractions, enzymes must be able to display reactions with precise specificity and selectivity. Enzymes used in Pec extraction disturb features of the plant cell wall, enabling pectin release and reducing the complete extraction period [32]. There are more additional benefits of using enzyme aided extraction such as avoiding the corrosion of equipment by acids, reduced energy consumption, and the specificity of enzymes yield an improved quality of Pec [33].

3.5. Combination of Non-Conventional Technologies

Researchers have looked at how non-traditional extraction techniques combine to effectively extract Pec from tropical and subtropical fruit waste. Ultrasound-microwave-assisted extraction, which combines ultrasonic and microwave extraction approaches, is viewed as an efficient process [34,35]. Ultrasound-microwave-mediated extraction involves rapid yield and competent Pec extraction at low temperatures at ambient conditions, saving energy, time, and is economically viable [36].

4. Structure of Pec

Pec is widely present in the cell walls of terrestrial/earthy plants [37]. Pec was made and explored in the powder form, which is very simple to use and handle [38]. Pec is recognized as a significant component of the middle lamella, which helps to keep cells organized. Every part of the plant contains different amounts of Pec and chemical assemblies. In terms of the chemical composition and molecular density, Pec in fruits and vegetables exists in poly-molecular and poly-disperse forms [39]. The monomeric units of Pec may vary depending on the sources, procedure used for separation, and successive examinations. Depending on the origin and method of isolation, diverse properties of Pec can be used to prepare its innumerable forms [40]. The Pec is comprised of chemical moieties such as the carboxylic (-COOH) group, ester and amide (-NH_2) groups [41], as shown in (Figure 1). Figure 2 displays the representative overview of rhamnose addition, which leads to the existence of the galacturonic acid (GA) chain, where S indicates the presence of neutral sugars [42].

5. Physical-Chemical Properties of Pec

Pec is a class of substances that when it is dissolved in water under certain environments, it can form gels. It is obtained from protopectin, which is found in the plant cell middle lamellae [43]. All of Pec's physical properties are due to its bi-linear poly-anion configuration (poly-carboxylate) [44]. When it comes to chemical features, the depolymerization of dissolved Pec occurs in aquatic classifications, and Pec has the highest stability at pH 4. The Pec de-esterifies below and above this pH, resulting in decreased stability. Depolymerization occurs at low pH levels via the acid catalyst hydrolysis of glycosidic bonds [45].

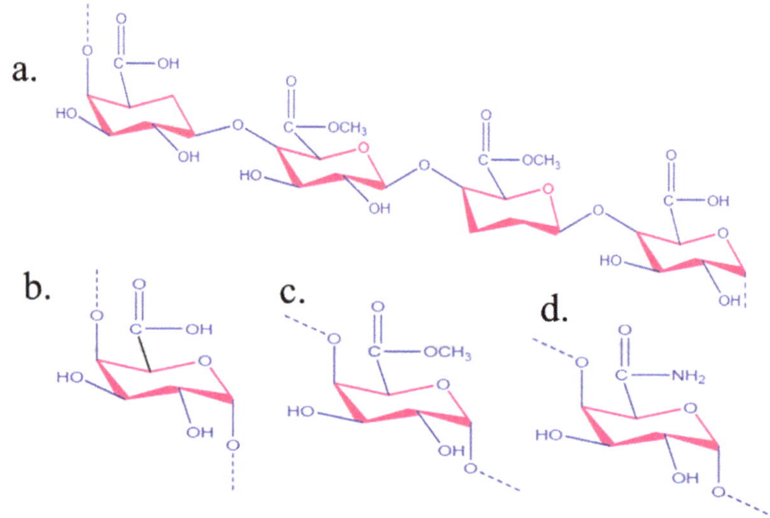

Figure 1. (**a**) The frequent unit of chemical moieties in the Pec chemical structure, (**b**) carboxylic, (**c**) ester, and (**d**) amide groups. Reproduced with permission [46].

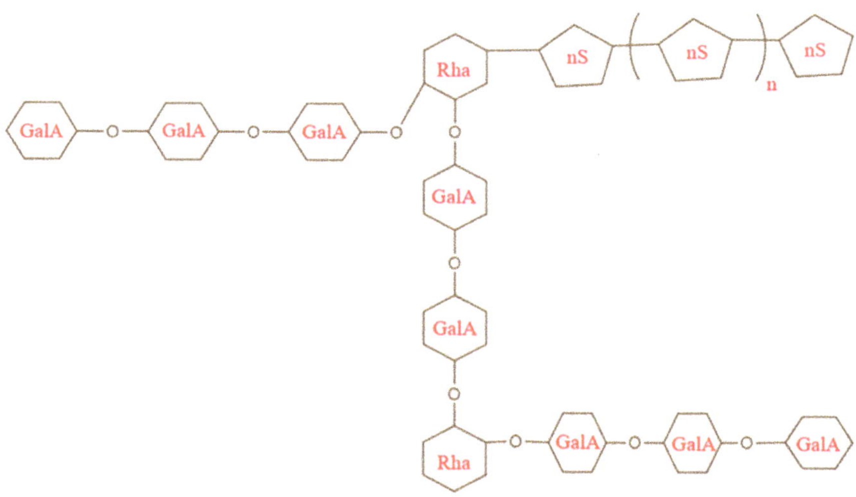

Figure 2. Rhamnose (Rha) insertion occurrence of galacturonic acid (Ga) and nS (neutral sugar). Reproduced with permission [46].

6. Application of Pec-Based Hydrogels in Drug Delivery

Over the last 50 years, hydrogels based on biodegradable natural polymers have been widely used in drug delivery systems [47]. Hydrogels have expanded in the drug delivery field as their three-dimensional structures have exclusive properties such as being hydrophilic in nature, biocompatibility, biodegradability, moist environments in surrounding tissues, and low cost [23,48]. Thus, in this section, various forms of Pec-based hydrogels in drug delivery applications are showcased.

In 2018, a low-density lipoprotein (LDL)-Pec nanogels in the presence of alginate hydrogel beads was successfully fabricated via the ionotropic gelation process. The designed LDL/pectin nanogels were confined in the core of the alginate-based hydrogel beads without affecting the properties of the hydrogel beads. Furthermore, curcumin (CUR) was encapsulated into the LDL-Pec nanogels to assess their role as pH dependent studies. The release of CUR was greatly prolonged by adding nanogels to alginate hydrogel beads, where the release profile of curcumin proved to have a slightly slower rate in the simulated GI conditions, signifying a role as an oral drug delivery system [49].

In another work, hydrogels were fabricated by graft polymerization and magnetic nanoparticles (MNPs) comprised in Pec-based hydrogels through the in situ method for the controlled release of the diclofenac sodium (DS) drug. Due to the presence of MNPs, there was a substantial enhancement in the mechanical properties, swelling capacity, drug loading efficiency, and drug release performances. Due to the porous network and high surface of the MNPs, the MNP-based pectin hydrogels displayed 68.84% of drug loading efficiency, however, without MNPs, the Pec-based hydrogels showed only 44.84% of loading efficiency. It was identified that around 95% of the DS drug was released from the MNP based Pec hydrogels, suggesting that swelling controlled the diffusion mode of the drug release profiles [50].

Other research investigations have been reported to improve the therapeutic effects and lessen the toxic effects of the systematic drug administration of vancomycin hydrochloride (vanco HCl), a hydrogel scaffold of silk fibroin/oxidized pectin (SF/OP) designed based on the Schiff-base reaction. To obtain a sustained release profile of the drug, electrospun fibers of poly(L-lactide) (PLLA) were incorporated into the hydrogel and unveiled high 97% of drug loading efficiency, followed by a 61% decrease in drug release content. For the first 24 h, the drug release profile from hydrogel was 39.97% and in the hydrogel/fiber system, a 13.83% decrease in the drug release content was observed, followed by promising sustained release up to 192 h. In addition, the designed hydrogel drug delivery systems proved to be non-toxic against human adipose derived mesenchymal stem cells (hAD-MSCs) [51].

It is important to mention that most often, the reagents utilized are generally expensive, toxic, and could lead to allergic effects. Meanwhile, the preparation of hydrogels using a chemical process with high energy beams (ex: gamma rays/electron beams) has gained significant attention due to the formation of pure hydrogel products. Thus, when compared to the gamma ray technique, the electron beam irradiation showed immense potential due to its rapid speed and a particular beam direction during the production of hydrogels. However, one of the complications in preparing hydrogels of natural polymers in an aqueous solution by irradiation is the higher possibility of chain breakage reactions than crosslinking reactions. Lately, phenolic compounds have been conjugated to natural polymers trailed by crosslinking with chemical reagents or irradiation. Against this background, using electron beam irradiation, porous and non-porous Pec-based hydrogel systems were fabricated using a combination of Pec and 5-hydroxytryptophan (5-HTP) with or without a surfactant for the delivery of tetracycline (TET). The drug loading for the porous hydrogel and non-porous hydrogel systems was 103 and 77 mg/g, respectively. The drug release from the non-porous hydrogel was comparatively less when compared with the porous hydrogel and showed higher amounts and a faster rate. Consequently, the biocompatibility and non-toxicity of both hydrogels were within the acceptable limits [52].

Abbasi et al. reported an efficient hydrogel-based controlled drug delivery system where Pec was grafted with polyethylene glycol (PEG) and methacrylic acid (MAA) via free radical polymerization (Figure 3) for the treatment of ulcerative colitis. Sulfasalazine (SZ) was used as a model drug in order to load the designed hydrogel systems. The swelling and release studies revealed that the hydrogels could release drugs explicitly at colonic pH. The toxicological studies revealed that they were safe in mouse animal models [53].

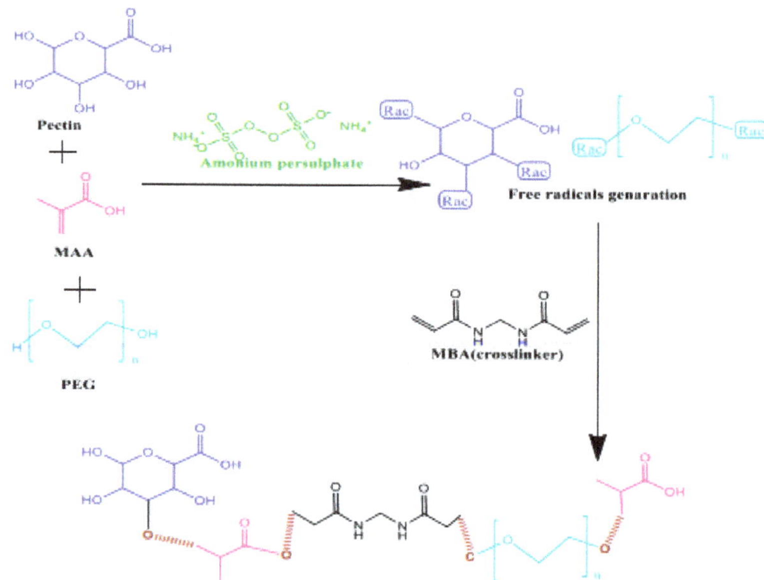

Figure 3. Scheme showing the formation of Pec-g-PEG-MAA hydrogels. Reproduced with permission [53].

Generally, protein administration via the oral route has continued as an appealing strategy due to less pain, superior suitability, and improved patient compliance. Herein, calcium carbonate microparticles ($CaCO_3$) were mineralized in situ in a Pec/poly(ethylene glycol) (PEG) hydrogel blend to shield and release the bovine serum albumin (BSA) protein drug (Figure 4) at the specific colon site. The BSA encapsulation efficiency for the blended hydrogels was around 98%. In vitro swelling and protein drug release studies of a-based hydrogel revealed the drug carrier's ability to release protein for around 9 h at the colon site [54].

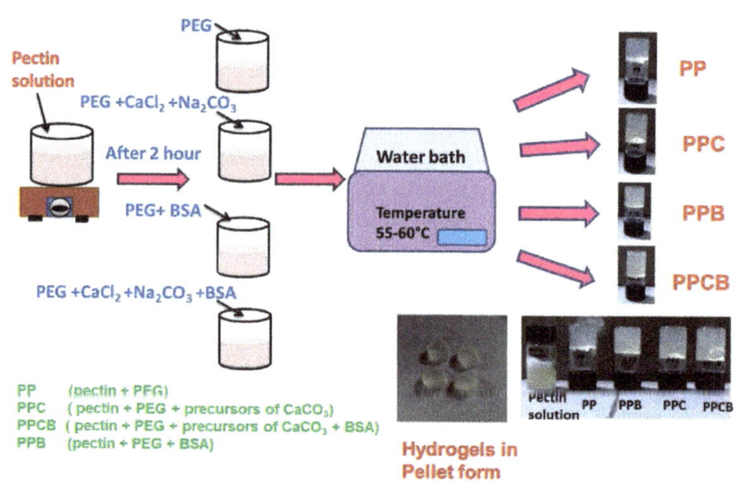

Figure 4. Schematic representation involved in the preparation of Pec-based hydrogels. Reproduced with permission [54].

In another work, Pec based lactic acid (LA) hydrogels were fabricated using free radical polymerizations. As the concentrations of Pec and lactic acid increased swelling, drug loading and drug release were observed, whereas methacrylic acid (MAA) showed the opposite effect. The loaded oxaliplatin (OL) to the developed hydrogel systems displayed around 49% of entrapment efficiency and substantiated with 18, 41, and 47% of drug release at pH 1.2, 6.8, and 7.4, respectively. Based on the MTT assay, the drug loaded hydrogels confirmed controlled inhibition against HCT-116 and MCF-7 cells [55].

Other research studies involved amine grafted high methoxy Pec-arabic gum (AG) incorporated montmorillonite (MMT) composite hydrogels for ziprasidone HCl (ZIP) delivery (Figure 5). The hydrogels showed around 39–64% of entrapment efficiency for ZIP and proved that slow ZIP release up to 8 h indicated excellent gastroretention ability and biodegradable properties. Overall, the designed hydrogels, fabricated using green synthesis, have the potential to be used as an effective intragastric drug carrier for the treatment of schizophrenia [56].

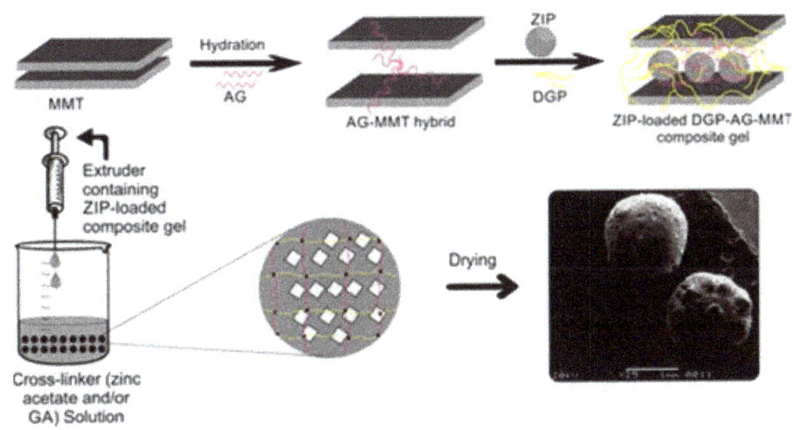

Figure 5. Scheme for the formation of Pec-based MMT hydrogels and ZIP loading. Reproduced with permission [56].

Usually, thiolated polymers are used in drug delivery applications due to their better permeation features, which provide superior bioavailability. However, the thiolation process is time intense due to a series of chemical reactions. To address such drawbacks, 2-thiobarbituric acid (TBA) was integrated onto the Pec hydrogels via noncovalent interactions. Importantly, due to the incorporation of TBA, the Pec hydrogels showed an additional flexible nature and their fragmentation delayed from 4 h to 4 days. Furthermore, the loaded theophylline (THP) drug to the hydrogel systems showed a loading capacity of 30 mg/g and showed a controlled drug release of up to 400 min [57].

It is significant to mention that injectable hydrogels with self-healing properties are important for drug delivery. One such work was reported by Li et al. by varying the ratios of oxidized Pec/chitosan (CS) to nano iron oxides (n-IO) (Figure 6), which showed remarkable injectable, self-healing, biocompatible, and anticancer features for the loaded 5-flurouracil (5-FU) drug with a drug release profile of more than 12 h [58].

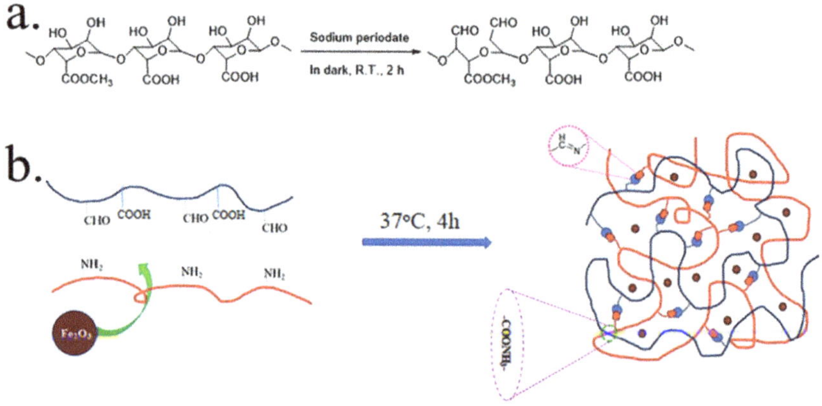

Figure 6. Proposed scheme (**a**) oxidation reaction and (**b**) hydrogel formation process. Reproduced with permission [58].

Other work have reported a combination of hydroxypropyl methylcellulose (HPMC) and Pec-based hydrogels fabricated by the free radical polymerization process. Furthermore, loaded galantamine hydrobromide (GHBr) showed varied drug entrapment efficiency in the range from 63 to 95% based on the swelling capacity of the designed hydrogel systems. The formulated hydrogels exhibited pH dependent behavior and intelligent response to environmental conditions by controlling the drug release for around 37 h. Additionally, the toxicity studies conducted on albino male rabbits were proven to have safe efficacy for hydrogel systems [59].

Gazzi et al. developed a Pec hydrogel comprising imiquimod loaded polymer nanocapsules for melanoma treatment. The loaded imiquimod (IQ) drug content was around 0.52 mg/mL and in vitro release study disclosed 63% of imiquimod release from the hydrogel systems for 2 h, while 60% of the drug was released after 8 h, followed by controlled release up to 24 h. In addition, the designed hydrogel systems displayed superior adhesiveness and a higher penetration of the drug inside the skin was observed [60].

Another work reported pH responsive hydrogels containing zein protein nanoparticles (ZPN) and Pec biopolymer for the encapsulation of doxorubicin (DOX) and release studies. It is important to note that the nanoparticles aided in the formation of complete gel networks for the loading of DOX. Interestingly, DOX loaded hydrogels showed better cytotoxicity effects against cervical cancer cell lines. In addition, the designed hydrogels were responsible for the pH dependent release of DOX to the cytosolic acid environment of HeLa cells. Altogether, this unique combination of zein and Pec-based hydrogels was favored with controlled release, improved shell life of the drug, and are capable of creating an intrinsic environment for the drug [61].

Keeping our focus on ulcerative colitis (UC), pH sensitive and enzymatically triggered hydrogels containing Pec and polyacrylamide (PA) were used to load budesonide (BUD). Based on the gel fraction and swelling behavior of the optimized hydrogel formulations revealed 80% of encapsulation efficiency and 8.8% of drug loading capacity. In vitro release of BUD from the hydrogel unveiled a sustained release behavior with non fickian diffusion mechanism over a time period of 1400 min. However, future studies such as stability, cytotoxicity and in vivo studies should be of focus to prove the potent drug delivery systems [62].

In another work, cellulose nanofiber (CNF)-alginate (Alg)-Pec-based hydrogel systems was developed for breast cancer treatments. Furthermore, the 5-FU drug was loaded to the hydrogel system and for the different hydrogel formulations, the encapsulation efficiency varied from 62 to 76%. The developed hydrogel formulations were facilitated by the initial burst release, followed by controlled drug release up to 24 h. Remarkably, the developed hydrogel systems enabled modulating the viability of breast tumor cells and inflammasome activities [63].

Similarly, in another work using nanocellulose fibers (NFs), low methoxy Pec and sodium alginate-based biocomposite hydrogels were synthesized. For the designed biocomposite-based hydrogel, clindamycin hydrochloride (CH) was loaded. By varying the ratios of hydrogels and drug, it could be possible to attain 82–94% of drug loading content. In vitro drug release profile showed 30–38% of drug release for the first 3 h and 100% of drug release was observed for 48 h. In addition, a cell viability study revealed superior cytocompatibility for human keratinocyte cells. It is worthwhile mentioning that these kinds of formulations are highly required in transdermal drug delivery. However, additional studies need to be carried out for the complete evaluation of the potential pharmaceutical applications toward the designed hydrogel systems [64].

One of the significant properties of self-healing hydrogels is pH stimuli and aid to protect the drug from being destroyed until it reaches the target site. Thus, in this report, through the Diels–Alder reaction, Pec/chitosan (CS) hydrogel systems were fabricated. The 5-FU drug was loaded to the designed hydrogel systems, which proved to have superior loading efficiency and sustained drug release profiles. As expected for the series of the designed hydrogel formulations, varied drug loading efficiency was observed from 53–65%. In vitro drug release displayed 30% of drug release for the first 4 h, followed by sustained drug release up to 12 h. Significantly, the developed hydrogel systems were cytocompatible for fibroblast L929 cells [65].

Lemos et al. developed magnetic (Mag) hydrogel microspheres using Pec coated chitosan for smart drug release. Herein, magnetic Pec microspheres attained by ionotropic gelation trailed by polyelectrolyte complexation with chitosan (Figure 7). In this study, metamizole (MTZ) was loaded and showed an encapsulation efficiency of 85%. In vitro drug release was performed at pH 1.2 and 6.8, which suggested a pH dependent drug release profile. Evidently, at pH 6.8, the drug release was favored by attaining 75%, even after 12 h. The utilization of the magnetic field amplified the drug release to 91% at pH 6.8, suggesting the role as magnetic reliant [66].

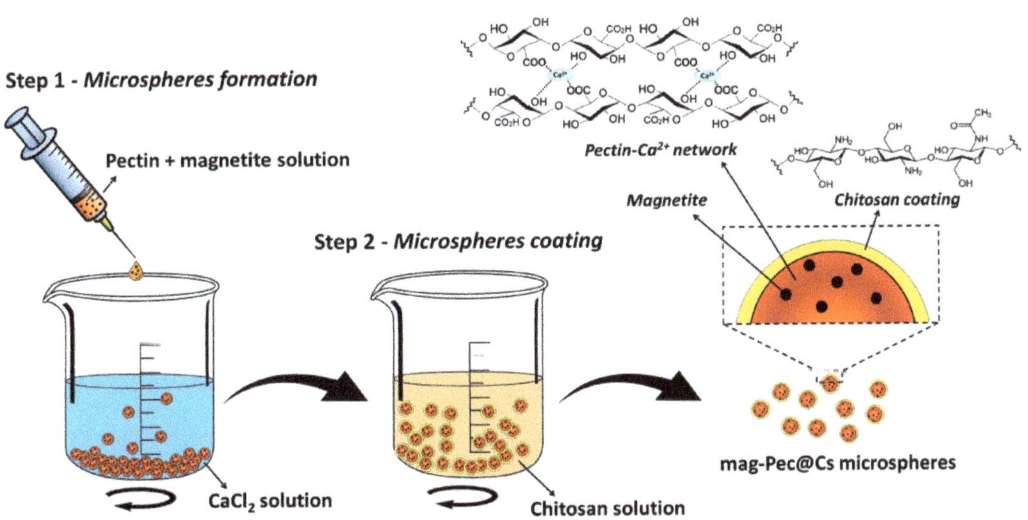

Figure 7. Scheme for the illustration and formation of magnetic Pec microspheres coated with chitosan. Reproduced with permission [66].

Another motivating work report involves new injectable and self-healing hydrogels fabricated using aldehyde terminated Pec with poly(N-isopropylacrylamide-stat-acylhydrazide) for an improved anticancer, DOX drug release property. In vitro and in vivo studies showed that the hydrogel had good biocompatibility, biodegradability, reduced drug toxicity in living bodies, and displayed controlled drug release behavior as synergetic anti-tumor drug delivery carriers [67].

It is well-known that amphiphilic (Amp) polymer systems have gained more interest due to their benefit in increasing drug penetration over the skin. Thus, focusing on this special aspect, an amphiphilic alkylated Pec through glycidyl tert-butyl ether functionalization was carried out to obtain hydrogels for fusidic acid (FSA) diffusion for topical treatment. The hydrogels were constructed via ionic interactions of negatively charged Pec and positively charged crosslinkers with varied 93–95% of FSA drug loading capacity. The swelling percentage of alkylated Pec hydrogels was lower than that of native Pec, resulting in a slower fusidic acid release up to 185 min. The effect of pH on the swelling rate and drug release was also studied, with outcomes showing that higher pH increased the swelling percentage and drug release. Interestingly, in vitro co-related with HaCaT cells displayed significantly less cytotoxicity, however, further extensive investigation is required [68].

Cai et al. reported low methoxyl citrus Pec (LMP) hydrogels as an actual drug carrier to load curcumin (CUR) for colon targeted delivery systems. The encapsulation efficiency (EE) for the designed hydrogels varied from 37 to 40% and loading capacity (LC) of 2.5–3.0%, respectively. Due to the better EE, the designed hydrogels displayed improved texture properties, inhibited premature release in the gastrointestinal (GI) tract, and were able to release the drug in the colon area at a faster rate, which attained a drug release rate at around 20 h [69].

In another work, Pec-Alg-based zinc (Zn) alginate hydrogel particles obtained based on callus culture Pec with varied structures were formed. The development of a Pec-alginate interpenetrating network was confirmed by the increase in hydrogen bonds between Pec and alginate. Usually, grape seed (GS) extract displays an anti-inflammatory outcome for inflammatory bowel disease (IBD) and thus grape seed extract was loaded to the designed hydrogel systems and revealed 95% of encapsulation efficiency. The drug release studies were carried out at different pH conditions and showed drug release specifically in colon conditions, which might be alternative candidates for colon targeted drug delivery systems [70].

Other findings reported functionalized kappa-carrageenan (K-CRG)-Pec hydrogel patches for the treatment of buccal fungal treatment. Herein, kappa-carrageenan-g-acrylic acid was surface functionalized with thiolated agents. Furthermore, in an ex vivo mucoadhesion study, a swelling test was extensively carried out to prove the applicability of the hydrogel patches. The triamcinolone acetonide (TA) was encapsulated within the poly (lactic-co-glycolic acid) nanoparticles. The EE and drug loading were 79% and 10%, respectively. The in vitro drug release parameter showed that the amount of drug release was around 3.28 mg/g polymer after 7 h. The cell culture studies on the hydrogel patches revealed that none of the patch formulations were toxic. All of these interesting findings suggest that novel thiolated grafted hydrogel patches could be utilized for buccal drug delivery systems [71].

Another work showed that to enhance the gel properties, Pec was additionally modified with phenylalanine (Phe) using an ultra-low temperature supported enzymatic process (Figure 8). Thus, the designed hydrogels exhibited better mechanical properties and superior water holding capacity. Due to the good gel features, matrine (MT) was loaded and showed sustained release properties with swelling properties. In cases of poor drug release, the ultra-low temperature enzymatic process might be a viable approach [72].

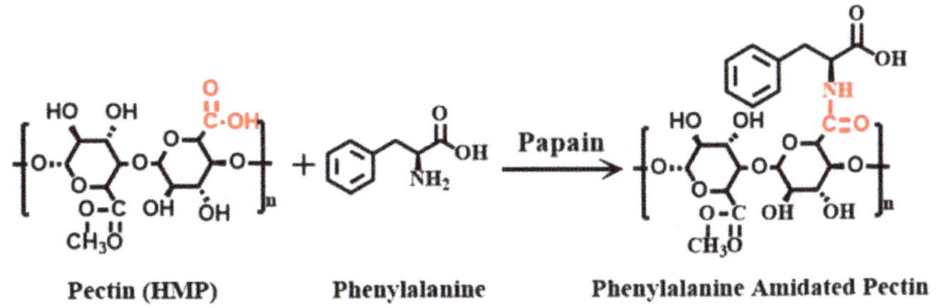

Figure 8. Scheme for the formation of phenylalanine amidated Pec. Reproduced with permission [72].

Regulating the optimal drug concentration and controlling drug release from hydrogels necessitate a large number of experiments and is overall expensive. To address such issues, in this work molecular dynamics (MD) stimulations were used to envisage the actual drug concentration to load on the LMP-based hydrogels, which allow for structural integrity and controlled drug release. When compared to other samples, Pec hydrogels loaded with 30 mg procaine (PRO)/g had a low hydrogel degradation rate of 0.001 g/min and a controlled in vitro drug release, releasing all 30 mg of the loaded PRO from the 670 mg hydrogel in 24 h [73].

Another study examined the fabrication of the Pec/chitosan nanoparticle (PEC/CSNP) beads as nanocarriers by encapsulating quercetin (QR) to overcome the solubility and sensitivity issues. The fabricated hydrogel beads exhibited 34–56% of EE and 12–24% of loading capacities. The prepared beads were able to release quercetin in a sustained release pattern up to 480 mins, as demonstrated in an in vitro drug release study. Furthermore, an in vitro cytotoxicity study revealed that the designed beads displayed a cell viability above 80% on the L929 cell line [74].

In another part of the work, a self-healing hydrogel was created by cross-linking Pec acylhydrazide (Pec-AH) with polyethylene glycol dialdehyde (PEG-DA) (Figure 9), and its use as a doxorubicin (DOX) delivery carrier for operative tumor treatment was examined. Significantly, the hydrogels demonstrated excellent in vitro and in vivo biocompatibility and biodegradability, with controlled drug release for around 50 h at varied pH conditions due to their microporous structure. The xenograft CT-26 tumor model in the mice trial exposed that the DOX-loaded hydrogel could prevent tumor growth associated with the direct injection of DOX and eliminate the associated drawbacks [75].

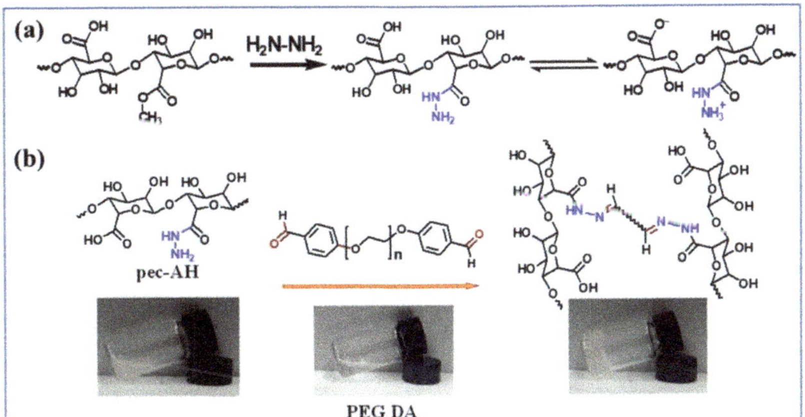

Figure 9. (**a**) Scheme for the synthesis of Pec-AH and (**b**) formation of the Pec-AH and PEG-DA hydrogel. Reproduced with permission [75].

Another study attempted to develop Pec-based layered zinc hydroxide (LZH) hydrogels comprised of baclofen. Through in vitro studies, they revealed that Pec-LZH containing baclofen (BFN) displayed a lower release rate when compared with BFN loaded with LZH. An addition, the MTT study suggests that for the HFFF2 cells, the prepared hydrogel system was biocompatible at 1.564–25μg/mL doses. The developed hydrogel beads appear to be promising as efficient carriers for targeted delivery to the colon [76].

Due to the lack of accurate drug targets, lung cancer is considered as the most common malignant tumor. This study involves limonin (LM), which prevents proliferation and encourages apoptosis in lung adenocarcinoma cells by directing a specific high expressed TMEM16A ion channel. Furthermore, a new class of self-healing hydrogels was created using acylhydrazide functionalized carboxymethyl cellulose (CMC-AH) and oxidized (Pec-CHO) (Figure 10) to decrease limonin's adverse effects on the body. The hydrogels demonstrated rapid gelation, better biocompatibility, and long-term limonin release up to 12 h. The limonin-loaded hydrogel expressively inhibited the development of lung adenocarcinoma in xenograft mice [77].

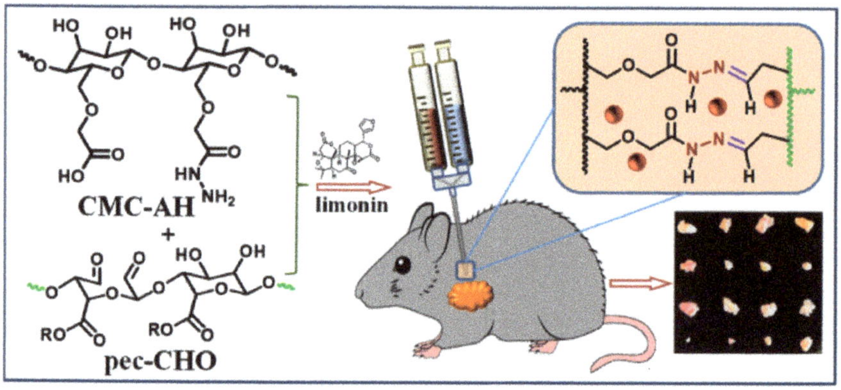

Figure 10. Scheme for the limonin loaded hydrogels for lung cancer treatment. Reproduced with permission [77].

Other work reported with photo-crosslinkable PC-CN formed by functionalizing the polysaccharide Pec (PC) with the photo-responsive cinnamic acid hydrazide (CNH). The photo-crosslinked hydrogel was then assessed as a carrier for the encapsulation of aspirin (AS). The developed hydrogels demonstrated better potential as a drug carrier, allowing for controlled drug release up to 60 h at different pH conditions by improving both the degree of cinnamic functionality and the photo-curing time [78]. Overall, Table 1 summarizes the various types of Pec-based hydrogel systems, the drugs used, and their key features involved in drug delivery applications.

Table 1. Summary of various types of Pec-based hydrogel systems, drugs used and their key features.

Serial no.	Hydrogel Systems	Drugs Used	Key Features	Reference
1	LDL-Pec-Alg	CUR	• Release profile of drug demonstrated with little slower rate in the simulated GI conditions, signifying role as oral drug delivery systems	[49]
2	MNP-Pec	DS	• Suggesting swelling controlled diffusion mode of drug release profiles	[50]
3	SF-OP-PLLA	Vanco	• Significant sustained release up to 192 h and non-toxic against hAD-MSCs	[51]
4	Pec-5-HTP	TET	• Improved biocompatibility and non-toxicity are in tolerable limits	[52]
5	Pec-PEG-MAA	SZ	• Drug release explicitly at colonic pH and toxicological studies revealed safe efficacy	[53]
6	Pec-$CaCO_3$-PEG	BSA	• Encapsulation efficiency was around 98% and drug release profile was around 9 h at the colon site	[54]
7	Pec-LA-MAA	OL	• Controlled inhibition against HCT-116 and MCF-7 cells	[55]
8	Pec-AG-MMT	ZIP	• Slow release and effective carrier for the treatment of schizophrenia	[56]
9	Pec-TBA	THP	• Additional flexible nature and controlled release of drug	[57]
10	Pec-CS-n-IO	5-FU	• Remarkable injectable, self-healing and biocompatible	[58]
11	Pec-HPMC	GHBr	• Intelligent response to environmental conditions and toxicity studies conducted proved with safe efficacy	[59]
12	Pec	IQ	• Displayed superior adhesiveness and higher penetration of the drug inside the skin	[53]
13	Pec-ZPN	DOX	• pH dependent release cytotoxicity effects against cervical cancer cell lines	[61]
14	Pec-PA	BUD	• Sustained release behavior and sustained release of 1400 min	[62]
15	CNF-Alg-Pec	5-FU	• Enabled in modulating breast tumor cells	[63]
16	NF-Pec-Alg	(CH)	• Superior cytocompatibility	[64]
17	Pec-CS	5-FU	• Cytocompatible for fibroblast L929 cells	[65]
18	Pec-Mag-CS	MTZ	• Improved drug release and pH dependent	[66]
19	Pec-PNIPAAm	DOX	• Good biocompatibility and biodegradability	[67]
20	Amp-Pec-FSA	FSA	• Proven with less cytotoxicity	[68]
21	LMP	CUR	• Inhibit premature release in GI and able to release drug in colon area	[69]
22	Pec-Alg-Zn	GS	• 95% of encapsulation efficiency and drug release, specifically in colon conditions	[70]
23	Prc-K-CRG	TA	• Cell culture studies revealed that none of the patch formulations were toxic	[71]
24	Pec-Phe	MT	• Sustained release with swelling properties	[72]
25	LMP	PRO	• Improved structural integrity	[73]
26	Pec-CSNP	QR	• Sustained release and cell viability above 80% on the L929 cell line	[74]
27	Pec-AH-DA	DOX	• Xenograft CT-26 tumor model in mice trial aided in preventing tumor growth	[75]
28	Pec-LZH	BFN	• Low release rate and biocompatible	[76]
30	Pec-CMC-CHO	LM	• Inhibited the development of lung adenocarcinoma in xenograft mice	[77]
31	Pec-CNH	AS	• Controlled drug release up to 60 h at different pH	[78]

7. Toxicity Concern of Pec-Based Hydrogels

Due to the excellent properties of Pec such as its biocompatibility, biodegradability, and non-toxic nature, researchers have been attracted to utilize Pec-based hydrogel systems in drug delivery fields. Prominently, for drug delivery applications, the developed hydrogel systems must be nontoxic while being able to perform their functions in response to the host's action. Based on some research findings, toxic materials can show negative impacts on the immune system. Some of the preliminary extensive research investigations have reported that Pec-based hydrogel systems have no such toxicity effects [52,75]. However, additional pre-clinical and clinical trials remain unresolved and must be prioritized as future prospects.

8. Conclusions and Future Perspectives

The present review mirrors based on the advancement of Pec-based hydrogels for drug delivery applications. The significance of Pec-based hydrogels is augmented by their unique functional groups, biocompatibility, biodegradability, easy gelling capability, low-cost, and simple modifications, which allow these systems to be astonishing candidates for the design and advance of potent drug delivery systems. It is significant to mention that by using a combination of other polymers and nanomaterials, the overall structural properties of Pec-based hydrogel systems can be considerably enhanced. Furthermore, more precise chemical modifications of Pec as well as their combination with other polymers or integration with other nanobiomaterials will enhance the overall structural behavior of Pec-based hydrogel systems and aid in tuning the interaction with the drug molecules at the molecular and nanoscale levels. With these substantial key features and the growing task of research groups on Pec-based hydrogel formulations, it could be projected that Pec applications in the drug delivery domain will expand in the near future. To this end, efforts should be increased to advance Pec-based hydrogel systems into clinical use, with the goal of dealing with regulatory problems, which are currently regarded as the main impediment.

Author Contributions: U.T.U.: Writing-original draft, conceptualization, preparation, reviewing, validation, investigation, and editing; S.M.J.: Conceptualization and investigation; M.J.P.: Validation and editing; M.S.: Formal analysis and validation; S.S.H.: Supervision, editing, resource, and funding acquisition. All authors have read and agreed to the published version of the manuscript.

Funding: This work was supported by the Basic Science Research Program through the National Research Foundation of Korea (NRF) funded by the Ministry of Education, Science and Technology (2020R1A2C1012586) and by the Ministry of Education (2020R1A6A1A03044512)

Data Availability Statement: Not applicable.

Conflicts of Interest: The authors declare no conflict of interest.

References

1. Eivazzadeh-Keihan, R.; Zare-Bakheir, E.; Aliabadi, H.A.M.; Gorab, M.G.; Ghafuri, H.; Maleki, A.; Madanchi, H.; Mahdavi, M. A novel, bioactive and antibacterial scaffold based on functionalized graphene oxide with lignin, silk fibroin and ZnO nanoparticles. *Sci. Rep.* **2022**, *12*, 1–12.
2. Eivazzadeh-Keihan, R.; Ganjali, F.; Aliabadi, H.A.M.; Maleki, A.; Pouri, S.; Mahdavi, M.; Shalan, A.E.; Lanceros-Méndez, S. Synthesis and characterization of cellulose, β-cyclodextrin, silk fibroin-based hydrogel containing copper-doped cobalt ferrite nanospheres and exploration of its biocompatibility. *J. Nanostruct. Chem.* **2022**, 1–11. [CrossRef]
3. Eivazzadeh-Keihan, R.; Ahmadpour, F.; Aliabadi, H.A.; Radinekiyan, F.; Maleki, A.; Madanchi, H.; Mahdavi, M.; Shalan, A.E.; Lanceros-Méndez, S. Pectin-cellulose hydrogel, silk fibroin and magnesium hydroxide nanoparticles hybrid nanocomposites for biomedical applications. *Int. J. Biol. Macromol.* **2021**, *192*, 7–15. [CrossRef] [PubMed]
4. Eivazzadeh-Keihan, R.; Khalili, F.; Khosropour, N.; Aliabadi, H.A.M.; Radinekiyan, F.; Sukhtezari, S.; Maleki, A.; Hamid Madanchi, H.; Hamblin, M.R.; Mahadvi, M. Hybrid bionanocomposite containing magnesium hydroxide nanoparticles embedded in a carboxymethyl cellulose hydrogel plus silk fibroin as a scaffold for wound dressing applications. *ACS Appl. Mater. Interfaces* **2021**, *13*, 33840–33849. [CrossRef] [PubMed]
5. Eivazzadeh-Keihan, R.; Noruzi, E.B.; Chenab, K.K.; Jafari, A.; Radinekiyan, F.; Hashemi, S.M.; Ahmadpour, F.; Behboudi, A.; Mosafer, J.; Mokhtarzadeh, A. Metal-based nanoparticles for bone tissue engineering. *J. Tissue Eng. Regen. Med.* **2020**, *14*, 1687–1714. [CrossRef]
6. B Mendes, B.B.; Daly, A.C.; Reis, R.L.; Domingues, R.M.A.; Gomes, M.E.; Burdick, J.A. Injectable hyaluronic acid and platelet lysate-derived granular hydrogels for biomedical applications. *Acta Biomater.* **2021**, *119*, 101–113. [CrossRef]
7. Rahimi, M.; Noruzi, E.B.; Sheykhsaran, E.; Ebadi, B.; Kariminezhad, Z.; Molaparast, M.; Mehrabani, M.G.; Mehramouz, B.; Yousefi, M.; Ahmadi, R. Carbohydrate polymer-based silver nanocomposites: Recent progress in the antimicrobial wound dressings. *Carbohydr. Polym.* **2020**, *231*, 115696. [CrossRef]
8. Shahvalizadeh, R.; Ahmadi, R.; Davandeh, I.; Pezeshki, A.; Moslemi, S.A.S.; Karimi, S.; Rahimi, M.; Hamishehkar, H.; Mohammadi, M. Antimicrobial bio-nanocomposite films based on gelatin, tragacanth, and zinc oxide nanoparticles–Microstructural, mechanical, thermo-physical, and barrier properties. *Food Chem.* **2021**, *354*, 129492. [CrossRef]
9. Eivazzadeh-Keihan, R.; Noruzi, E.B.; Mehrban, S.F.; Aliabadi, H.A.M.; Karimi, M.; Mohammadi, A.; Maleki, A.; Mahdavi, M.; Larijani, B.; Shalan, A.E. The latest advances in biomedical applications of chitosan hydrogel as a powerful natural structure with eye-catching biological properties. *J. Mater. Sci.* **2022**, *57*, 3855–3891. [CrossRef]

10. Eivazzadeh-Keihan, R.; Choopani, L.; Aliabadi, H.A.M.; Ganjali, F.; Kashtiaray, A.; Maleki, A.; Cohan, R.A.; Bani, M.S.; Komijani, S.; Ahadian, M.M. Magnetic carboxymethyl cellulose/silk fibroin hydrogel embedded with halloysite nanotubes as a biocompatible nanobiocomposite with hyperthermia application. *Mater. Chem. Phys.* **2022**, *287*, 126347. [CrossRef]
11. Chen, Z.; Wu, H.; Wang, H.; Zaldivar-Silva, D.; Agüero, L.; Liu, Y.; Zhang, Z.; Yin, Y.; Qiu, B.; Zhao, J. An injectable anti-microbial and adhesive hydrogel for the effective noncompressible visceral hemostasis and wound repair. *Mater. Sci. Eng.* **2021**, *129*, 112422. [CrossRef] [PubMed]
12. Zhi, C.; Song, X.; Soh, W.W.M.; Wen, Y.; Zhu, J.; Zhang, M.; Li, J. In Situ Synthesis of Magnetic Poly (DMAEAB-co-NIPAm)@ Fe_3O_4 Composite Hydrogel for Removal of Dye from Water. *Gels* **2021**, *7*, 201.
13. Zhang, Y.; Zhu, C.; Zhang, Z.; Zhao, J.; Yuan, Y.; Wang, S. Oxidation triggered formation of polydopamine-modified carboxymethyl cellulose hydrogel for anti-recurrence of tumor. *Colloids Surf. B Biointerfaces* **2021**, *207*, 112025. [CrossRef] [PubMed]
14. Huang, C.; Xu, X.; Fu, J.; Yu, D.-G.; Liu, Y. Recent progress in electrospun polyacrylonitrile nanofiber-based wound dressing. *Polymers* **2022**, *14*, 3266. [CrossRef] [PubMed]
15. Xu, H.; Zhang, F.; Wang, M.; Lv, H.; Yu, D.-G.; Liu, X.; Shen, H. Electrospun hierarchical structural films for effective wound healing. *Biomater. Adv.* **2022**, *136*, 212795. [CrossRef]
16. Lei, W.; Suzuki, N.; Terashima, C.; Fujishima, A. Hydrogel photocatalysts for efficient energy conversion and environmental treatment. *Front. Energy* **2021**, *15*, 577–595. [CrossRef]
17. Farris, S.; Schaich, K.M.; Liu, L.; Cooke, P.H.; Piergiovanni, L.; Yam, K.L. Gelatin–pectin composite films from polyion-complex hydrogels. *Food Hydrocoll.* **2011**, *25*, 61–70. [CrossRef]
18. Chen, Q.; Liang, S.; Thouas, G.A. Elastomeric biomaterials for tissue engineering. *Prog. Polym. Sci.* **2013**, *38*, 584–671. [CrossRef]
19. Santos, E.; Pedraz, J.L.; Hernández, R.M.; Orive, G. Therapeutic cell encapsulation: Ten steps towards clinical translation. *J. Control. Release* **2013**, *170*, 1–14. [CrossRef]
20. Jayani, R.S.; Saxena, S.; Gupta, R. Microbial pectinolytic enzymes: A review. *Process Biochem.* **2005**, *40*, 2931–2944. [CrossRef]
21. May, C.D. Industrial pectins: Sources, production and applications. *Carbohydr. Polym.* **1990**, *12*, 79–99. [CrossRef]
22. Lampugnani, E.R.; Khan, G.A.; Somssich, M.; Persson, S. Building a plant cell wall at a glance. *J. Cell Sci.* **2018**, *131*, 207373. [CrossRef] [PubMed]
23. Eivazzadeh-Keihan, R.; Noruzi, E.B.; Aliabadi, H.A.M.; Sheikhaleslami, S.; Akbarzadeh, A.R.; Hashemi, S.M.; Gorab, M.G.; Maleki, A.; Cohan, R.A.; Mahdavi, M. Recent advances on biomedical applications of pectin-containing biomaterials. *Int. J. Biol. Macromol.* **2022**, *217*, 1–18. [CrossRef] [PubMed]
24. Li, D.-Q.; Li, J.; Dong, H.-L.; Li, X.; Zhang, J.-Q.; Ramaswamy, S.; Xu, F. Pectin in biomedical and drug delivery applications: A review. *Int. J. Biol. Macromol.* **2021**, *185*, 49–65. [CrossRef] [PubMed]
25. Pacheco, M.T.; Villamiel, M.; Moreno, R.; Moreno, F.J. Structural and rheological properties of pectins extracted from industrial sugar beet by-products. *Molecules* **2019**, *24*, 392. [CrossRef]
26. Picot-Allain, M.C.N.; Ramasawmy, B.; Emmambux, M.N. Extraction, characterisation, and application of pectin from tropical and sub-tropical fruits: A review. *Food Rev. Int.* **2022**, *38*, 282–312. [CrossRef]
27. Cho, E.-H.; Jung, H.-T.; Lee, B.-H.; Kim, H.-S.; Rhee, J.-K.; Yoo, S.-H. Green process development for apple-peel pectin production by organic acid extraction. *Carbohydr. Polym.* **2019**, *204*, 97–103. [CrossRef]
28. Pena-Pereira, F.; Tobiszewski, M. *The Application of Green Solvents in Separation Processes*, 1st ed.; Elsevier: Amsterdam, Netherlands, 2017; pp. 533–545.
29. Ilghami, A.; Ghanbarzadeh, S.; Hamishehkar, H. Optimization of the ultrasonic-assisted extraction of phenolic compounds, ferric reducing activity and antioxidant activity of the Beta vulgaris using response surface methodology. *Pharm. Sci.* **2015**, *21*, 46–50. [CrossRef]
30. Hosseini, S.S.; Khodaiyan, F.; Yarmand, M.S. Optimization of microwave assisted extraction of pectin from sour orange peel and its physicochemical properties. *Carbohydr. Polym.* **2016**, *140*, 59–65. [CrossRef]
31. Maran, J.P.; Sivakumar, V.; Thirugnanasambandham, K.; Sridhar, R. Optimization of microwave assisted extraction of pectin from orange peel. *Carbohydr. Polym.* **2013**, *97*, 703–709. [CrossRef]
32. Yang, Y.; Wang, Z.; Hu, D.; Xiao, K.; Wu, J.-Y. Efficient extraction of pectin from sisal waste by combined enzymatic and ultrasonic process. *Food Hydrocoll.* **2018**, *79*, 189–196. [CrossRef]
33. Marić, M.; Grassino, A.N.; Zhu, Z.; Barba, F.J.; Brnčić, M.; Brnčić, S.R. An overview of the traditional and innovative approaches for pectin extraction from plant food wastes and by-products: Ultrasound-, microwaves-, and enzyme-assisted extraction. *Trends Food Sci. Technol.* **2018**, *76*, 28–37. [CrossRef]
34. Guo, Z.; Zhao, B.; Li, H.; Miao, S.; Zheng, B. Optimization of ultrasound-microwave synergistic extraction of prebiotic oligosaccharides from sweet potatoes (Ipomoea batatas L.). *Innov. Food Sci. Emerg. Technol.* **2019**, *54*, 51–63. [CrossRef]
35. Yang, J.-S.; Mu, T.-H.; Ma, M.-M. Optimization of ultrasound-microwave assisted acid extraction of pectin from potato pulp by response surface methodology and its characterization. *Food Chem.* **2019**, *289*, 351–359. [CrossRef]
36. Xu, S.-Y.; Liu, J.-P.; Huang, X.; Du, L.-P.; Shi, F.-L.; Dong, R.; Huang, X.-T.; Zheng, K.; Liu, Y.; Cheong, K.-L. Ultrasonic-microwave assisted extraction, characterization and biological activity of pectin from jackfruit peel. *Lwt* **2018**, *90*, 577–582. [CrossRef]
37. Yapo, B.M. Pineapple and banana pectins comprise fewer homogalacturonan building blocks with a smaller degree of polymerization as compared with yellow passion fruit and lemon pectins: Implication for gelling properties. *Biomacromolecules* **2009**, *10*, 717–721. [CrossRef]

38. Ahlawat, S.; Battan, B.; Dhiman, S.S.; Sharma, J.; Mandhan, R.P. Production of thermostable pectinase and xylanase for their potential application in bleaching of kraft pulp. *J. Ind. Microbiol. Biotechnol.* **2007**, *34*, 763–770. [CrossRef]
39. Mohnen, D. Pectin structure and biosynthesis. *Curr. Opin. Plant Biol.* **2008**, *11*, 266–277. [CrossRef]
40. Laurent, M.; Boulenguer, P. Stabilization mechanism of acid dairy drinks (ADD) induced by pectin. *Food Hydrocoll.* **2003**, *17*, 445–454. [CrossRef]
41. Sriamornsak, P. Chemistry of pectin and its pharmaceutical uses: A review. *Int. J.* **2003**, *3*, 206–228.
42. Sriamornsak, P. Analysis of Selected Physico-Chemical Properties of Pectin and Alginate Gels Intended for Drug Delivery. Ph.D. Thesis, Charles Sturt University, Wagga Wagga, Australia, 2002.
43. Bourne, M.C. Applications of chemical kinetic theory to the rate of thermal softening of vegetable tissue. *ACS Symp. Ser.* **1989**, *405*, 98–110.
44. Plaschina, I.G.; Braudo, E.E.; Tolstoguzov, V.B. Circular-dichroism studies of pectin solutions. *Carbohydr. Res.* **1978**, *60*, 1–8. [CrossRef]
45. BeMiller, J. Acid-catalyzed hydrolysis of glycosides. *Adv. Carbohydr. Chem.* **1967**, *22*, 25–108.
46. Thakur, S.; Chaudhary, J.; Kumar, V.; Thakur, V.K. Progress in pectin based hydrogels for water purification: Trends and challenges. *J. Environ. Manag.* **2019**, *238*, 210–223. [CrossRef]
47. Rial-Hermida, M.I.; Rey-Rico, A.; Blanco-Fernandez, B.; Carballo-Pedrares, N.; Byrne, E.M.; Mano, J.F. Recent progress on polysaccharide-based hydrogels for controlled delivery of therapeutic biomolecules. *ACS Biomater. Sci. Eng.* **2021**, *7*, 4102–4127. [CrossRef]
48. Manzoor, A.; Dar, A.H.; Pandey, V.K.; Shams, R.; Khan, S.; Panesar, P.S.; Kennedy, J.F.; Fayaz, U.; Khan, S.A. Recent insights into polysaccharide-based hydrogels and their potential applications in food sector: A review. *Int. J. Biol. Macromol.* **2022**, *213*, 987–1006. [CrossRef]
49. Zhou, M.; Hu, Q.; Wang, T.; Xue, J.; Luo, Y. Alginate hydrogel beads as a carrier of low density lipoprotein/pectin nanogels for potential oral delivery applications. *Int. J. Biol. Macromol.* **2018**, *120*, 859–864. [CrossRef]
50. Babaladimath, G.; Badalamoole, V. Magnetic nanoparticles embedded in pectin-based hydrogel for the sustained release of diclofenac sodium. *Polym. Int.* **2018**, *67*, 983–992. [CrossRef]
51. Ahadi, F.; Khorshidi, S.; Karkhaneh, A. A hydrogel/fiber scaffold based on silk fibroin/oxidized pectin with sustainable release of vancomycin hydrochloride. *Eur. Polym. J.* **2019**, *118*, 265–274. [CrossRef]
52. Moghaddam, R.H.; Dadfarnia, S.; Shabani, A.M.H.; Moghaddam, Z.H.; Tavakol, M. Electron beam irradiation synthesis of porous and non-porous pectin based hydrogels for a tetracycline drug delivery system. *Mater. Sci. Eng. C* **2019**, *102*, 391–404. [CrossRef]
53. Abbasi, M.; Sohail, M.; Minhas, M.U.; Khan, S.; Hussain, Z.; Mahmood, A.; Shah, S.A.; Kousar, M. Novel biodegradable pH-sensitive hydrogels: An efficient controlled release system to manage ulcerative colitis. *Int. J. Biol. Macromol.* **2019**, *136*, 83–96. [CrossRef]
54. Gautam, M.; Santhiya, D. In-situ mineralization of calcium carbonate in pectin based edible hydrogel for the delivery of protein at colon. *J. Drug Deliv. Sci. Technol.* **2019**, *53*, 101137. [CrossRef]
55. Ullah, K.; Sohail, M.; Buabeid, M.A.; Murtaza, G.; Ullah, A.; Rashid, H.; Khan, M.A.; Khan, S.A. Pectin-based (LA-co-MAA) semi-IPNS as a potential biomaterial for colonic delivery of oxaliplatin. *Int. J. Pharm.* **2019**, *569*, 118557. [CrossRef]
56. Bera, H.; Abbasi, Y.F.; Yoke, F.F.; Seng, P.M.; Kakoti, B.B.; Ahmmed, S.K.M.; Bhatnagar, P. Ziprasidone-loaded arabic gum modified montmorillonite-tailor-made pectin based gastroretentive composites. *Int. J. Biol. Macromol.* **2019**, *129*, 552–563. [CrossRef] [PubMed]
57. Güner, O.Z.; Kocaaga, B.; Batirel, S.; Kurkcuoglu, O.; Güner, F.S. 2-Thiobarbituric acid addition improves structural integrity and controlled drug delivery of biocompatible pectin hydrogels. *Int. J. Polym. Mater. Polym. Biomater.* **2021**, *70*, 703–711. [CrossRef]
58. Li, D.-Q.; Wang, S.-Y.; Meng, Y.-J.; Li, J.-F.; Li, J. An injectable, self-healing hydrogel system from oxidized pectin/chitosan/γ-Fe2O3. *Int. J. Biol. Macromol.* **2020**, *164*, 164–4566. [CrossRef] [PubMed]
59. Bashir, S.; Zafar, N.; Lebaz, N.; Mahmood, A.; Elaissari, A. Hydroxypropyl methylcellulose-based hydrogel copolymeric for controlled delivery of galantamine hydrobromide in Dementia. *Processes* **2020**, *8*, 1350. [CrossRef]
60. Gazzi, R.P.; Frank, L.A.; Onzi, G.; Pohlmann, A.R.; Guterres, S.S. New pectin-based hydrogel containing imiquimod-loaded polymeric nanocapsules for melanoma treatment. *Drug Deliv. Transl. Res.* **2020**, *10*, 1829–1840. [CrossRef]
61. Kaushik, P.; Priyadarshini, E.; Rawat, K.; Rajamani, P.; Bohidar, H.B. pH responsive doxorubucin loaded zein nanoparticle crosslinked pectin hydrogel as effective site-specific anticancer substrates. *Int. J. Biol. Macromol.* **2020**, *152*, 1027–1037. [CrossRef]
62. Pandey, M.; Choudhury, H.; D/O Segar Singh, S.K.; Chetty Annan, N.; Bhattamisra, S.K.; Gorain, B.; Mohd Amin, M.C.I. Budesonide-loaded pectin/polyacrylamide hydrogel for sustained delivery: Fabrication, characterization and in vitro release kinetics. *Molecules* **2021**, *26*, 2704. [CrossRef]
63. Balahura, L.R.; Dinescu, S.; Balaș, M.; Cernencu, A.; Lungu, A.; Vlăsceanu, G.M.; Iovu, H.; Costache, M. Cellulose nanofiber-based hydrogels embedding 5-FU promote pyroptosis activation in breast cancer cells and support human adipose-derived stem cell proliferation, opening new perspectives for breast tissue engineering. *Pharmaceutics* **2021**, *13*, 1189. [CrossRef] [PubMed]
64. O-chongpian, P.; Na Takuathung, M.; Chittasupho, C.; Ruksiriwanich, W.; Chaiwarit, T.; Baipaywad, P.; Jantrawut, P. Composite Nanocellulose Fibers-Based Hydrogels Loading Clindamycin HCl with Ca2+ and Citric Acid as Crosslinking Agents for Pharmaceutical Applications. *Polymers* **2021**, *13*, 4423. [CrossRef] [PubMed]

65. Li, D.-Q.; Wang, S.-Y.; Meng, Y.-J.; Guo, Z.-W.; Cheng, M.-M.; Li, J. Fabrication of self-healing pectin/chitosan hybrid hydrogel via Diels-Alder reactions for drug delivery with high swelling property, pH-responsiveness, and cytocompatibility. *Carbohydr. Polym.* **2021**, *268*, 118244. [CrossRef] [PubMed]
66. Lemos, T.S.; de Souza, J.F.; Fajardo, A.R. Magnetic microspheres based on pectin coated by chitosan towards smart drug release. *Carbohydr. Polym.* **2021**, *265*, 118013. [CrossRef] [PubMed]
67. An, H.; Yang, Y.; Zhou, Z.; Bo, Y.; Wang, Y.; He, Y.; Wang, D.; Qin, J. Pectin-based injectable and biodegradable self-healing hydrogels for enhanced synergistic anticancer therapy. *Acta Biomater.* **2021**, *131*, 149–161. [CrossRef]
68. Bostanudin, M.F. Amphiphilic Alkylated Pectin Hydrogels for Enhanced Topical Delivery of Fusidic Acid: Formulation and In Vitro Investigation. *Sci. Pharm.* **2022**, *90*, 13. [CrossRef]
69. Cai, R.; Pan, S.; Li, R.; Xu, X.; Pan, S.; Liu, F. Curcumin loading and colon release of pectin gel beads: Effect of different de-esterification method. *Food Chem.* **2022**, *389*, 133130. [CrossRef]
70. Günter, E.A.; Popeyko, O.V. Delivery system for grape seed extract based on biodegradable pectin-Zn-alginate gel particles. *Int. J. Biol. Macromol.* **2022**, *219*, 1021–1033. [CrossRef]
71. Özkahraman, B.; Özbaş, Z.; Yaşayan, G.; Akgüner, Z.P.; Yarımcan, F.; Alarçin, E.; Bal-Öztürk, A. Development of mucoadhesive modified kappa-carrageenan/pectin patches for controlled delivery of drug in the buccal cavity. *J. Biomed. Mater. Res. Part B Appl. Biomater.* **2022**, *110*, 787–798. [CrossRef]
72. Zheng, C.; Huang, W.; Zou, Y.; Huang, W.; Peng Fei, P.; Guoguang Zhang, G. Fabrication of phenylalanine amidated pectin using ultra-low temperature enzymatic method and its hydrogel properties in drug sustained release application. *Int. J. Biol. Macromol.* **2022**, *216*, 263–271. [CrossRef]
73. Kocaaga, B.; Guner, F.S.; Kurkcuoglu, O. Molecular dynamics simulations can predict the optimum drug loading amount in pectin hydrogels for controlled release. *Mater. Today Commun.* **2022**, *31*, 103268. [CrossRef]
74. Nalini, T.; Khaleel Basha, S.; Mohamed Sadiq, A.; Sugantha Kumari, V. Pectin/chitosan nanoparticle beads as potential carriers for quercetin release. *Mater. Today Commun.* **2022**, *33*, 104172. [CrossRef]
75. Zhou, Z.; Wang, Z.; Liu, X.; Zhao, Z.; Heng An, H.; Wang, Y.; He, Y.; Qin, J. Pectin-based self-healing hydrogel through acylhydrazone connection for controlled drug release and enhanced tumor therapy. *J. Drug Deliv. Sci. Technol.* **2022**, *70*, 103210. [CrossRef]
76. Nabipour, H.; Batool, S.; Hu, Y. Pectin-coated baclofen-layered zinc hydroxide nanohybrid as a bio-based nanocomposite carrier for the oral delivery. *IEEE Trans. NanoBioscience* **2022**, 1. [CrossRef]
77. Chang, L.; Chang, R.; Shen, J.; Wang, Y.; Song, H.; Kang, X.; Zhao, Y.; Guo, S.; Qin, J. Self-healing pectin/cellulose hydrogel loaded with limonin as TMEM16A inhibitor for lung adenocarcinoma treatment. *Int. J. Biol. Macromol.* **2022**, *219*, 754–766. [CrossRef]
78. Almutairi, T.M.; Al-Rasheed, H.H.; Monier, M.; Alatawi, F.S.; Elsayed, N.H. Synthesis and characterization of photo-crosslinkable cinnamate-functionalized pectin. *Int. J. Biol. Macromol.* **2022**, *210*, 208–217. [CrossRef]

gels

MDPI

Article

Development of Efficient Sodium Alginate/Polysuccinimide-Based Hydrogels as Biodegradable Acetaminophen Delivery Systems

Long Toan Trinh, Saebin Lim, Hyun Jong Lee * and Il Tae Kim *

Department of Chemical and Biological Engineering, Gachon University, Seongnam-si 13120, Republic of Korea; longtt.hust@gmail.com (L.T.T.); binivini1999@gmail.com (S.L.)
* Correspondence: hjlee2@gachon.ac.kr (H.J.L.); itkim@gachon.ac.kr (I.T.K.)

Abstract: Efficient drug delivery systems are essential for improving patient outcomes. Acetaminophen (AP), which is a kind of oral administration, is a commonly used pain reliever and fever reducer. However, oral administration carries various health risks, especially overdose and frequent use; for instance, AP is administered approximately 4 times per day. Therefore, the aim of this study is to develop an efficient delivery system for once-daily administration by combining sodium alginate and polysuccinimide (PSI) hydrogels to delay the release of analgesic AP. PSI is a biodegradable polymer that can be used safely and effectively in drug delivery systems because it is eliminated by hydrolysis in the intestine. The use of PSI also improves the mechanical properties of hydrogels and prolongs drug release. In this study, hydrogel characterizations such as mechanical properties, drug dissolution ability, and biodegradability were measured to evaluate the hydrolysis of PSI in the intestine. Based on the results, hydrogels could be designed to improve the structural mechanical properties and to allow the drug to be completely dissolved, and eliminated from the body through PSI hydrolysis in the intestines. In addition, the release profiles of AP in the hydrogels were evaluated, and the hydrogels provided continuous release of AP for 24 h. Our research suggests that sodium alginate/PSI hydrogels can potentially serve as biodegradable delivery systems for AP. These findings may have significant implications for developing efficient drug delivery systems for other classes of drugs.

Keywords: acetaminophen; bioavailability; drug delivery; polysuccinimide; prolonged; sodium alginate

Citation: Trinh, L.T.; Lim, S.; Lee, H.J.; Kim, I.T. Development of Efficient Sodium Alginate/Polysuccinimide-Based Hydrogels as Biodegradable Acetaminophen Delivery Systems. *Gels* **2023**, *9*, 980. https://doi.org/10.3390/gels9120980

Academic Editors: Ying Huang, Zhengwei Huang and Xuanjuan Zhang

Received: 29 November 2023
Revised: 13 December 2023
Accepted: 13 December 2023
Published: 14 December 2023

1. Introduction

Acetaminophen (AP), a commonly used analgesic and antipyretic, is a component of several prescribed and over-the-counter medicines [1]. However, use of an inappropriate dosage or continuous use of the drug may lead to patient overdose. The recommended dosage of this drug is 500–1000 mg every 4–6 h, with a maximum of 4000 mg every 24 h for both adults and children (≥12 years) [2]. Excess AP can cause acute liver damage. Moreover, AP is one of the most commonly used medicines causing acute liver injury and is the leading cause of acute liver failure as its elimination half-life is approximately 2 h for therapeutic doses [3–6]. Regular use of AP is essential for analgesic and antipyretic support in patients with fever and discomfort. Although AP is considered a relatively safe drug in the human gastroduodenum, current epidemiological studies have shown that large doses of this drug may increase the risk of bleeding and gastric ulcers [7]. AP is also known to be well absorbed in a wide range of intestinal locations [8]. Therefore, an efficient method for AP delivery to the intestinal tract that retains its analgesic and antipyretic effects for a long period is necessary to reduce its continuous intake and prevent the harmful effects of AP on the body. Nowadays, hydrogels are an interesting class of targeted delivery systems and have been used in numerous biomaterials, including drug delivery systems [9].

Hydrogels are three-dimensional polymeric networks that are hydrophilic, insoluble, and capable of retaining significant volumes of water and biological fluids. They can be used for various biomedical applications, such as the sustained release of therapeutic agents, owing to their biocompatibility and biodegradability [10,11]. Hydrogels have been studied for various biomedical, industrial, and agricultural applications. In the medical field, hydrogels have also been used to enhance recovery and lessen scarring. Hydrogels are excellent materials for injury healing since they have been proven to improve the mechanical and functional properties of animal tissues [12]. They can absorb and store various drugs and facilitate their release at an appropriate rate over time. They exhibit low toxicity and high biocompatibility and biodegradability. Hydrogels can also be completely broken down into monomers or small molecules [13], making them suitable as drug delivery systems to promote sustained and regulated local drug release, improve the therapeutic efficacy of drugs, and decrease the adverse side effects of drugs [14]. Multiple delivery systems based on polymers, such as alginate, are currently being investigated for intestine-targeted delivery of drugs. Digestive system enzymes degrade and release drugs [15–17].

Sodium alginate (NaAlg), a natural polymer derived from brown seaweed, is non-toxic and biodegradable. NaAlg is composed of various quantities of d-mannuronic acid and L-guluronic acid. NaAlg has been used for controlled drug release as it rapidly creates three-dimensional networks in the presence of multivalent metal cations [18]. The sodium salt of alginic acid, composed of d-mannuronic acid and L-guluronic acid, chemically creates a hydrogel with the addition of crosslinkers, such as calcium chloride, via interactions with a divalent cation [19]. Under acidic conditions, NaAlg transforms into porous alginic acid, which becomes a soluble layer in the colon [20]. The prolonged release of AP from NaAlg-based controlled-release tablets, as investigated by Rubio and Ghaly, has been the subject of numerous studies on the use of alginate gels for sustained drug release [21]. Sharma et al. tested an in situ quick-gelling formulation for an oral sustained alginate gel for patients with dysphagia [22]. Furthermore, incorporating organic and inorganic elements into biopolymer matrices for reinforcement often improves the mechanical and physical properties of the matrix, which can increase drug encapsulation [23]. Recently, other bioinorganic components, including montmorillonite and $CaCO_3$ [24–26], have been applied to alginate matrices to generate various alginate–inorganic composite particles that support long-term drug release. Therefore, NaAlg is a suitable drug delivery system.

Polysuccinimide (PSI) is a non-toxic aspartic acid (ASP) polycondensation product [27]. PSI is biodegradable and pH responsive [28]. It is stable under acidic pH and is rapidly soluble at neutral pH due to the hydrolysis of PSI in the succinimidyl primary sequence. This is comparable to the gastrointestinal environment, which has an acidic pH in the stomach and neutral pH in the intestine [29]. A study on the use of PSI in conjunction with AgNPs to boost antibacterial activity and extend the paracetamol release period yielded positive results [30]. PSI nanoparticles (NPs) were shown to hydrolyze slowly under physiological pH 7.4 conditions while being stable under acidic conditions. The ability of NPs to be dissolved under neutral conditions while maintaining stability in acidic environments is a highly desirable property for oral delivery technologies [31]. Therefore, it is considered that PSI exhibits potential for the development of medication delivery systems. The control of drug carrier material degradation in the acidic environment of the stomach using the combination of PSI and NaAlg may aid in prolonging the drug release time in the gastrointestinal environment. PSI is mainly hydrolyzed by the intestinal environment. Therefore, its release in the intestine is important for efficient drug delivery [32]. Furthermore, it is well known that substances that are biologically persistent or that degrade too slowly can cause undesirable biological reactions in the long term [33]. As a result, it is also important to design a drug delivery system that can break it down and release it in the body in a timely manner. The aforementioned qualities and properties of PSI make it a candidate for ensuring the ability of a drug delivery system to hydrolyze substances to extend the drug release time and prevent drug carrier substances from remaining in the body for a long period of time. In this study, we developed an efficient hydrogel system using NaAlg substrate combined with AP-containing PSI to prolong AP release while minimizing drug release in the stomach and maximizing drug release in the intestine. The NaAlg substrate was combined with PSI and crosslinked with Ca^{2+} to form a hydrogel. The biocompatibility and prolonged AP release efficiency of this hydrogel were further studied and compared with those of other hydrogels prepared using only a NaAlg substrate.

2. Results and Discussion

2.1. Synthesis and Characterization of Alg-PSI

The hydrogel synthesis process comprises two steps, as shown in Figure 1. In the first step, PSI was synthesized by the bulk polycondensation of ASP with an o-phosphoric acid catalyst. ^{1}H NMR spectroscopy was used to confirm PSI production (Figure S1). The methine proton signal (a) at 5.3 ppm and the methylene proton signal (b) at 2.7 and 3.2 ppm are visible [34]. The molecular weight of PSI was estimated using gel permeation chromatography (Figure S2). The number average molecular weight (M_n) and weight average molecular weight (M_w) were 20,867 and 34,724 g mol^{-1}, respectively, with a PDI of 1.66.

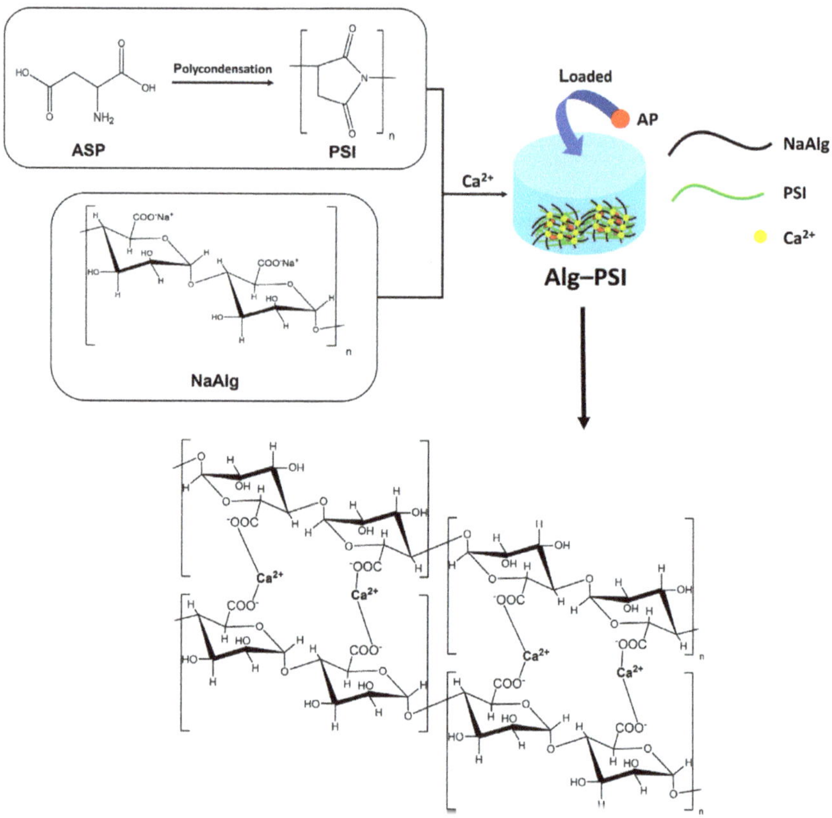

Figure 1. Scheme of hydrogel synthesis using PSI and Alg.

The second step involved the synthesis of the Alg-PSI hydrogel. The color of NaAlg and PSI precursor solutions changed markedly when dissolved together (Figure 2A–D). The higher the PSI concentration, the more opaque the NaAlg-PSI mixture. Hydrogel formation was detected using the tube inversion method (Figure 2A–D). The vials were inverted and the flowability of the solution was determined. Gel formation was observed immediately after adding $CaCl_2$ solution to the mixture (Figure 2A–D).

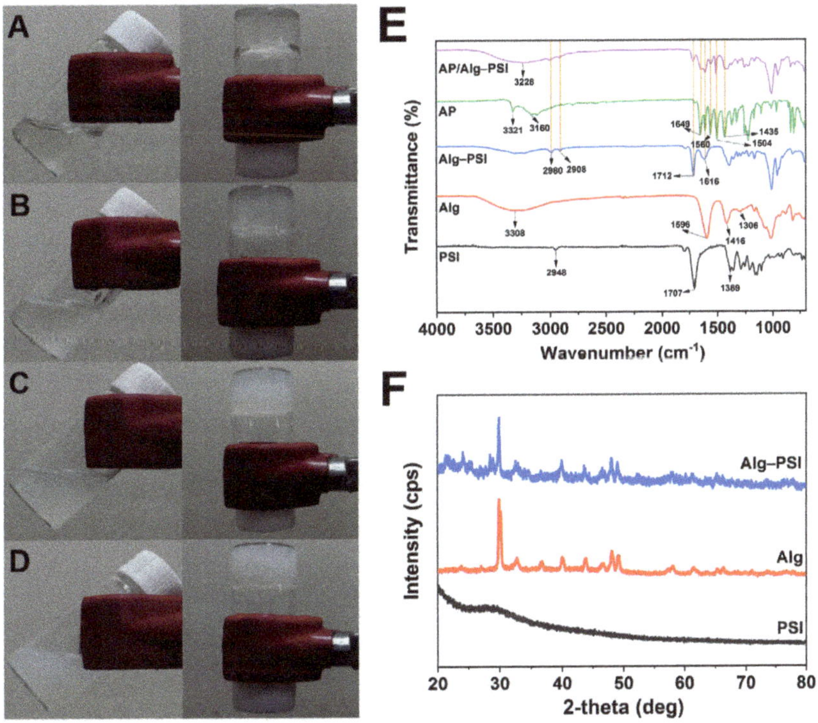

Figure 2. (**A–D**) Formation of Alg-PSI hydrogels with different concentrations of PSI: (**A**) 0%, (**B**) 0.5%, (**C**) 1.5%, and (**D**) 3% PSI (w/v). (**E**) Fourier-transform infrared (FT-IR) spectra of PSI, Alg, Alg-PSI, AP, and AP/Alg-PSI. (**F**) X-ray diffraction (XRD) patterns of PSI, Alg, and Alg-PSI.

FT-IR spectroscopy was used to examine the functional groups of PSI, Alg, and Alg-PSI samples (Figure 2E). For the PSI sample, characteristic bands of the C=O and C-N groups of the imide ring were observed at 1707 and 1389 cm^{-1}, respectively [35]. The peak at 2948 cm^{-1} represents -CH_2 [36]. For the Alg sample, the peaks at 3308, 1596, 1416, and 1306 cm^{-1} represent O-H, -COO- asymmetric, and C-O stretching, respectively [18]. For the Alg-PSI sample, the characteristic peak of Alg was shifted from 1596 to 1616 cm^{-1} and the peak of PSI from 1707 to 1712 cm^{-1}, illustrating the combination between Alg and PSI. In addition, the spectra exhibited obvious changes in the characteristic peaks. New peaks at 2980 and 2908 cm^{-1} indicate the presence of the -CH_2. As a result, the incorporation of PSI into the alginate hydrogel structure is visible. It is noted that when looking into the FT-IR spectrum of Alg-PSI, all characteristic peaks from NaAlg and PSI are still observed even though they show some shifts in peak positions. This may illustrate that there are not any additional chemical reactions between Alg and PSI, but the physical properties have been changed due to the coexistence of Alg and PSI. Meanwhile, the vibrational peaks for O-H and CH_3 stretching were detected at 3321 cm^{-1} and 3160 cm^{-1}, respectively, in the FT-IR spectrum of AP. The vibrational peak at 1649 cm^{-1} was attributed to C=O. At 1560 cm^{-1}, the N–H amide II bending was observed. The C–C stretching peak came at 1435 cm^{-1}, while the asymmetrical bending in the C–H bond appeared at 1504 cm^{-1} [37]. The AP spectrum reveals an exceptional absorption at around 3321 cm^{-1}, which is associated with O–H and not visible in the AP/Alg-PSI spectrum. Rather, a wide peak at 3228 cm^{-1} can be observed. Furthermore, the distinctive peaks for both AP and the Alg-PSI hydrogel are still visible in the AP/Alg-PSI sample, demonstrating that AP and the hydrogel do not react chemically to form byproducts or deteriorate the drug, ensuring the medicinal benefits of AP. It shows that AP was successfully loaded into the hydrogel matrix.

Differences in the structures of PSI, Alg, and Alg-PSI were determined using their X-ray diffraction (XRD) patterns, as shown in Figure 2F. No peaks for PSI are visible in the XRD pattern because PSI is an amorphous polymer [38]. XRD patterns of the Alg and Alg-PSI samples exhibited no significant differences. However, the Alg-PSI sample showed more noise patterns than Alg, possibly due to the incorporation of amorphous PSI.

Figure 3 shows SEM images of freeze-dried Alg and Alg-PSI at different PSI concentrations. The surface of the Alg sample showed a similar morphology compared to previous studies (Figure 3A) [39,40]. The Alg surface is rough and contains nanometer-sized particles due to water separation occurring during the freeze-drying process. This causes the polymer networks to break down, forming a rough surface [41]. Here, the Alg-PSI hydrogels had rougher surfaces than those of Alg as PSI was blended into Alg. Surface roughness became more complex as the applied amount of PSI increased from 0.5 to 3% PSI (*w/v*). This can be explained by the insolubility of PSI in water. When PSI is exposed to water, its structure collapses instantaneously, and polymer precipitation is evenly dispersed and blended in NaAlg before this mixture forms a hydrogel with $CaCl_2$ crosslinking, leading to a rough surface [31]. This demonstrates that PSI was incorporated into the hydrogel.

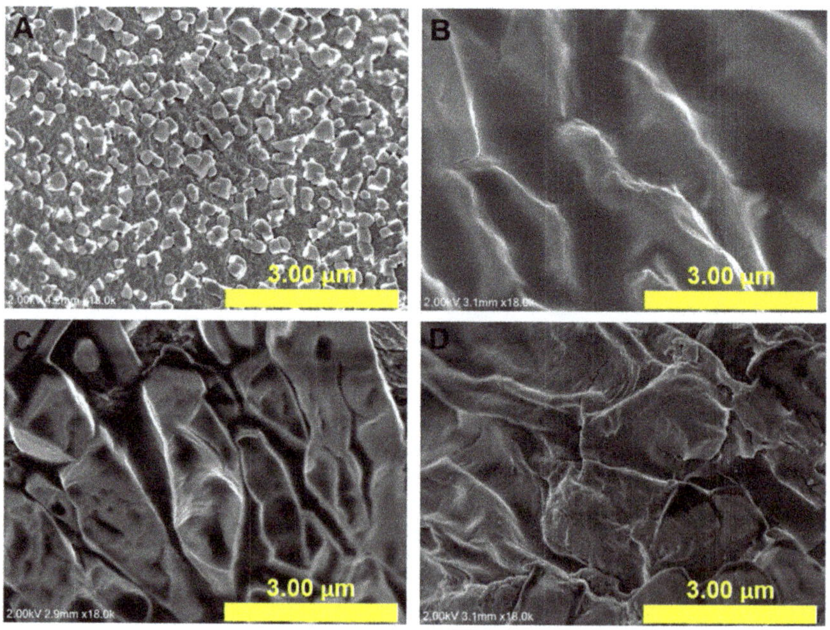

Figure 3. Scanning electron microscopy (SEM) images of hydrogels containing different concentrations of PSI: (**A**) 0%, (**B**) 0.5%, (**C**) 1.5%, and (**D**) 3% PSI (*w/v*).

2.2. Rheological Properties of Hydrogels

Next, the mechanical properties, stability, and viscosity of the hydrogels were investigated using a rheological test method. Gelation occurred when Ca^{2+} cations were added to Alg, forming an Alg-PSI complex. After 24 h of rest and stabilization, the rheological properties of the hydrogels were analyzed based on the rheological characteristics of the material and frequency variation. Figure 4A–D shows the storage modulus (G′) and loss modulus (G″) of the hydrogels as functions of frequency. G′ was greater than G″ in all samples, indicating that Ca^{2+} cations replaced Na^+ ion sites to form a stable hydrogel network. The average value of G′ of the hydrogels is calculated in Figure 4E. G′ increased with increasing PSI concentrations in the complex. Moreover, the combination of PSI with Alg made the hydrogel structure more solid and stable, increasing its application stability as a drug delivery system. Figure 4F shows the viscosity of the hydrogels. Viscosity decreased with increasing cutting speed and increased with increasing PSI concentration in the hydrogel network. PSI acted as a thickener in the hydrogel system. PSI may have contributed to the increase in viscosity by improving the interactions with Alg in the hydrogel matrix and increasing hydrogel entanglement.

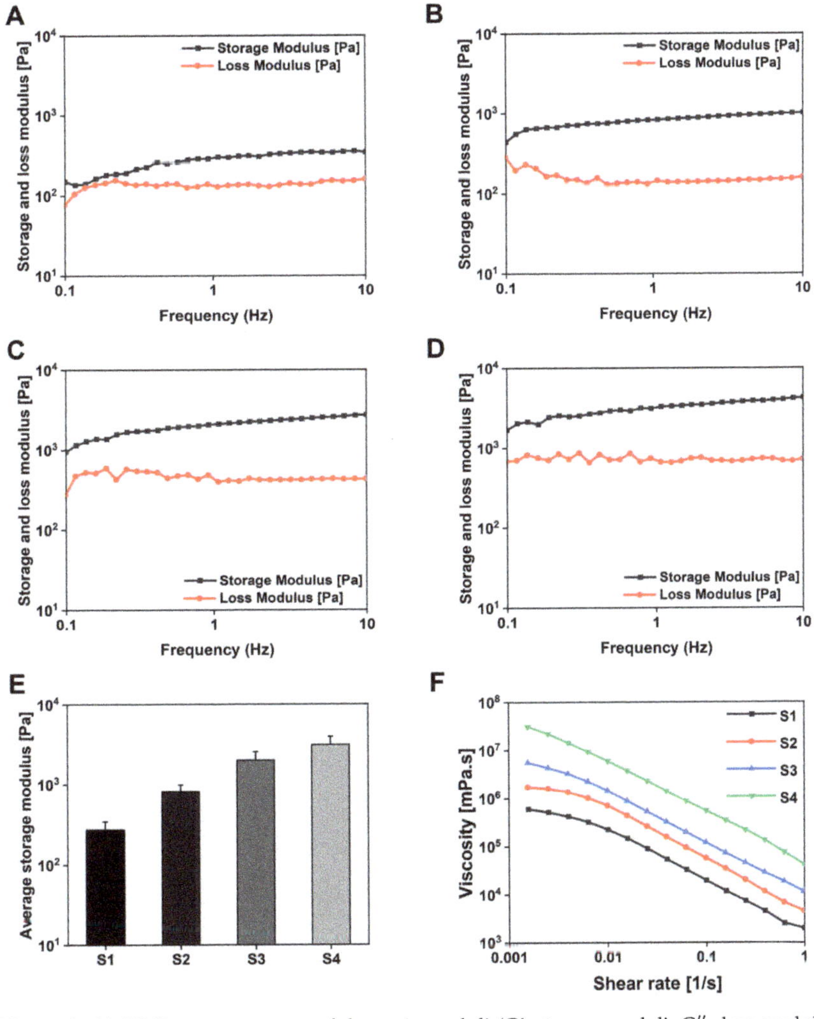

Figure 4. (**A**–**D**) Frequency sweep of dynamic moduli (G′: storage moduli; G″: loss moduli) of Alg-PSI hydrogels at 37 °C. Frequency = 0.1–10 Hz, strain = 0.1%. (**A**) 0%, (**B**) 0.5%, (**C**) 1.5%, and (**D**) 3% PSI (*w/v*). (**E**) Average G′ values of Alg-PSI hydrogels: (S1) 0%, (S2) 0.5%, (S3) 1.5%, and (S4) 3% PSI (*w/v*). Significant differences were observed between all of the groups ($p < 0.01$). (**F**) Viscosity curves of Alg-PSI hydrogels: (S1) 0%, (S2) 0.5%, (S3) 1.5%, and (S4) 3% PSI (*w/v*).

2.3. Drug Loading Efficiency and Degradation Tests to Assess Drug Release

Figure 5B shows the determined loading efficiency of AP in the hydrogel. The loading efficiency of the S1 hydrogel was ~60%, and the drug loss was mostly caused by the porous nature of the alginate network which allowed the drug molecules to diffuse rapidly during the gelation step. On the other hand, the loading efficiency of the S4 hydrogel increased by ~70% with increasing PSI content in the hydrogel structure. This characteristic is comparable to other research that used alginate hydrogels to encapsulate calcium phosphate for boosting medication loading effect [42]. The diffusion ability of the drug was hindered by the presence of PSI in the alginate matrix. As a result, the loss of the model drug during the preparation process was reduced, and ultimately, the loading efficiency was dramatically improved. The drug release rate may be also reduced by such an impact [43].

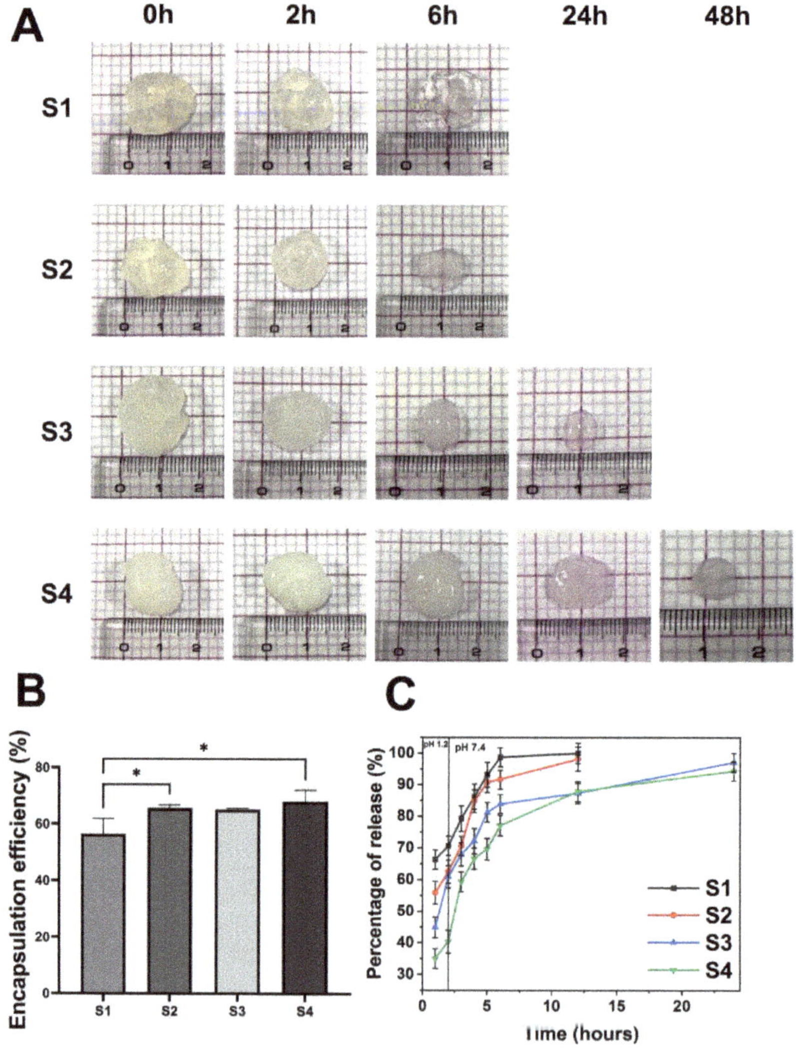

Figure 5. (**A**) Images of Alg-PSI hydrogels with different concentrations of PSI as a function of time: (S1) 0%, (S2) 0.5%, (S3) 1.5%, and (S4) 3% PSI (*w/v*). (**B**) Percentage of AP loaded into Alg-PSI hydrogels with different concentrations of PSI: (S1) 0%, (S2) 0.5%, (S3) 1.5%, and (S4) 3% PSI (*w/v*). The symbol * indicates a significant difference ($p < 0.05$). (**C**) Drug release profile of Alg-PSI hydrogels with different concentrations of PSI at pH 1.2 and 7.4: (S1) 0%, (S2) 0.5%, (S3) 1.5%, and (S4) 3% PSI (*w/v*).

The degradation of hydrogels is similar to the two basic processes that lead to the degradation of polymer-based materials: surface and bulk degradation [44]. Surface degradation is defined as occurring only at the surface, producing particles of smaller size while retaining the core of the material [45]. Bulk deterioration, on the other hand, causes continuous erosion throughout the material, resulting in fragments as byproducts. Therefore, these two mechanisms can be used together to characterize the degradation mechanism observed in the hydrogel [46] (Figure 5A). Alg is known to undergo proton-catalyzed hydrolysis [47]. Thus, the crosslinked Alg substrate, when exposed to an acidic environment, can be converted to alginic acid, which can lead to a reduced degree of crosslinking and thus faster decomposition. When the hydrogels were exposed to HCl, alginate was hydrolyzed to alginic acid, the COO^- group was converted to a bonded carboxylic group, and the electrostatic interactions between the Ca^{2+} and COO^- ions almost disappeared. Furthermore, ion exchange can occur between the H+ and free Ca^{2+} ions within the hydrogels [48]. The hydrogels swelled quicker when the solution medium was changed from HCl to phosphate buffer at pH 7.4 [48]. However, it did not attain a high water absorption value because of the loose topology of the hydrogels. Furthermore, ion exchange between H+ ions (produced by carboxylic group ionization in a buffer at pH 7.4) and Na^+ ions present in the buffer may occur, resulting in carboxylic group absorption. In addition, Alg began to dissolve in the substrate, lowering the weight of the hydrogels and finally fully dissolving. As a result, hydrogels with greater PSI concentrations have a longer hydrogel collapse time, as shown in Figure 5A, because PSI promotes the entanglement of the polymer network. Simultaneously, PSI was not hydrolyzed in the acidic medium, resulting in PSI partially preventing the hydrogel from converting Alg to alginic acid during formulation. As a result, when the hydrogel network was in PBS at pH 7.4, it remained more stable than hydrogels with only Alg substrates. Notably, the hydrogel exhibited symptoms of collapse even at pH 7.4 owing to the alkaline hydrolysis capabilities of PSI. This hydrolysis may persist for up to 24 h in hydrogels with higher PSI values (S3 and S4). The duration for which a material remains in the body is one of the most critical characteristics of drug delivery methods. Their duration should not be too long nor too short to have a therapeutic effect. The duration required to extend the active ingredient and remove the drug delivery material from the body by approximately 24 h is highly reasonable for a medication delivery system for patients with high fever and pain alleviation. These results indicate that S3 and S4 hydrogels have great potential as AP drug delivery systems.

Drug release from the hydrogel was primarily determined by the characteristics of the Alg-PSI matrix and drug encapsulation in the polymer system. The mechanism of drug release from the hydrogel system can be divided into two major pathways: drug release via alginate network breakdown, and drug diffusion via the alginate network [49]. The hydrogels were released differently in media of acidic pH 1.2 and in PBS (pH 7.4), as illustrated in Figure 5C. The AP release of the released hydrogel S1 was greatest at pH 1.2, reaching 70% at 2 h. Drug release via alginate network degradation to alginic acid and hydrogel degradation released the drug from the hydrogel structure. In contrast, in the case of hydrogels S2, S3, and S4, the drug release was somewhat reduced because the presence of PSI inhibited the structural deterioration of the networks, making the hydrogel networks more stable and minimizing the rapid release of AP into the environment. When the medium was changed from pH 1.2 to pH 7.4, the hydrogels had a loose structure due to degradation in the acidic environment, and they could be dissolved in the medium. After 6 h, the S1 and S2 hydrogels begin to dissolve completely. This dissolution took approximately 24 h longer for S3 and over 48 h for S4 due to higher concentrations of PSI in S3 and S4. It is noted that during the first 2 h in the gastric environment, S4 allows significantly less AP release than the other hydrogels and allows for higher release when transported into the intestinal-like environment, improving AP efficacy. Owing to PSI's ability to be bio-hydrolyzed, it was added to the alginate hydrogel network to extend the time during which AP was released while also assisting in the removal of extra hydrogel from the body.

2.4. Cell Viability of Hydrogels

An MTT assay was performed to confirm the biocompatibility of Alg-PSI hydrogels at different concentrations. The hydrogels were stabilized for 24 h after gelation, sterilized, and adapted to the cell culture medium. Then, the hydrogels were placed in a hanging insert for indirect interactions with the cells, avoiding direct contact between the hydrogels and cells. The relative cell proliferation (%) was expressed by dividing the absorbance of the treatment groups by the absorbance of the control group. Figure 6 shows slightly higher cell proliferation in the groups treated with Alg-PSI hydrogels than in the control group. These results suggest that the Alg-PSI hydrogels had no cytotoxic effects on cells. In particular, Ca^{2+} ions were used to crosslink the Alg-PSI hydrogels, which altered the cell culture environment but had no adverse effects on the cells. PSI is non-cytotoxic and biocompatible with NIH/3T3 cells; therefore, Alg-PSI hydrogels have excellent cell biocompatibility regardless of the addition ratio [36]. Although not investigated in this study, the number of M-blocks in the alginate structure has been reported to promote cytokine production and affect immunogenicity [50]. These effects can potentially be used to modulate the immunity and control the release behavior of AP and develop new drug delivery systems in the future.

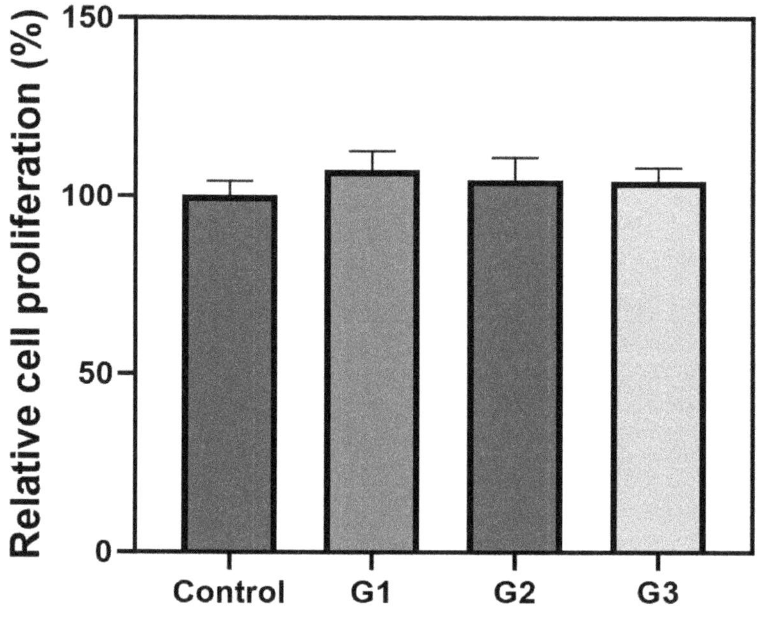

Figure 6. Effects of Alg-PSI hydrogels on the proliferation of NIH/3T3 fibroblasts after 24 h using the 3-(4,5-dimethylthiazol-2-yl)-2,5-diphenyltetrazolium bromide (MTT) assay. (G1) 0.5%, (G2) 1.5%, (G3) 3% PSI (*w*/*v*), and control group. No statistical differences were observed between groups.

3. Conclusions

In this study, we developed an AP analgesic delivery system comprising a PSI-blended alginate hydrogel network. Alginate- and PSI-based hydrogels can be used as long-lasting and biodegradable drug delivery systems. Here, MTT assay revealed that the developed hydrogels were biocompatible. By changing the concentration of PSI, we enhanced the mechanical properties and controlled the biodegradation and release of AP in the hydrogel system. AP was released slowly by the developed system and remained stable for approximately one day. However, to accurately control the encapsulation and release of the drug, the study of detailed physical interactions between PSI and Alg is essential, and further studies through molecular simulations should be performed. This system can reduce the daily intake frequency while maintaining the pain-relieving effect of drugs. Moreover, the hydrogel can be easily removed from the body via PSI bio-hydrolysis in the intestine. Therefore, the developed hydrogel can prolong the analgesic effects of drugs and rapidly eliminate the drug delivery materials from the body to prevent toxicity. Based on the studies mentioned above, additional short-term drug delivery methods can be created and implemented. In the future, it would be possible to further improve the biodegradation and drug release period of Alg-PSI hydrogels, making them suitable for long-term drug delivery systems.

4. Materials and Methods

4.1. Materials

AP (98.0–102.0%), ASP (≥99%), dimethyl sulfoxide (DMSO, ≥99.9%), hydrochloric acid (HCl, 36.5–38.0%), and methanol (CH_3OH, ≥99.9%) were purchased from Sigma-Aldrich (Seoul, Republic of Korea). Calcium chloride ($CaCl_2.2H_2O$), Dulbecco's modified Eagle's medium (DMEM), penicillin–streptomycin, and trypsin-EDTA solution were purchased from WelGene, Inc. (Daegu, Republic of Korea). NaAlg, sulfolane (99%), mesitylene (98%), phosphoric acid (o-H_3PO_4, 85%), fetal bovine serum (FBS), phosphate-buffered saline (PBS, pH 7.4), Dulbecco's phosphate-buffered saline (DPBS), and 3-(4,5-dimethylthiazol-2-yl)-2,5-diphenyltetrazolium bromide (MTT) were purchased from Thermo Fisher Scientific (Waltham, MA, USA).

4.2. Synthesis of Alginate (Alg) Hydrogels

4.2.1. Synthesis of Alg

The alginate hydrogel fabrication method was the same as described above, except for the addition of PSI. Briefly, 2 mL $CaCl_2$ solution (1.5%, *w/v*) was added to 2.5 mL NaAlg (1.5%, *w/v*). The gel was stabilized at room temperature for 24 h and washed with distilled water.

4.2.2. Synthesis of Acetaminophen-Loaded Alg (AP/Alg) Hydrogels

Briefly, 0.005 g AP was dissolved in the 2.5 mL NaAlg (1.5%, *w/v*). Then, 2 mL $CaCl_2$ solution (1.5%, *w/v*) was added, and the gel was washed with distilled water after allowing it to stabilize for 24 h at room temperature.

4.3. Synthesis of Alg-PSI Hydrogels

4.3.1. Synthesis of PSI

PSI was prepared as previously described [51]. Briefly, 17.5 g ASP and 500 μL o-phosphoric acid (85%) were added to 34 g solvent (mesitylene/sulfolane, 7:3) in a three-necked flask and refluxed under nitrogen pressure. After 4.5 h at 162 °C, the water created in the reaction mixture was removed using a Dean–Stark trap, and the obtained solid was washed with CH_3OH and water. PSI was dried at 70 °C in an oven. The percentage yield of PSI was over 90%.

4.3.2. Synthesis of Complex Alg-PSI Hydrogels

NaAlg solution (1.5%, *w/v*) was prepared in distilled water, and PSI (0.5, 1.5, and 3% *w/v*) was dissolved in DMSO. Then, 2 mL NaAlg solution was added to 0.5 mL PSI solution and stirred for 30 min at 20 °C. After homogenization (NaAlg-PSI), the forming gels were prepared by adding 2 mL calcium chloride ($CaCl_2$; 1.5%, *w/v*) to 2.5 mL NaAlg-PSI solution. After allowing the gels (Alg-PSI) to stabilize for 24 h at room temperature, they were washed with distilled water and dried at room temperature for subsequent experiments. Different Alg-PSI hydrogels were established using the same steps by immobilizing the NaAlg concentration and varying the PSI concentration (0.5, 1.5, and 3% *w/v*).

4.3.3. Synthesis of AP-Loaded Complex Alg-PSI (AP/Alg-PSI) Hydrogels

After preparing NaAlg-PSI solutions, AP (0.005 g) was dissolved in the solution to develop an AP/NaAlg-PSI mixed solution. Then, hydrogels containing the drug were obtained by adding 2 mL $CaCl_2$ solution (1.5%, *w/v*) to 2.5 mL AP/NaAlg-PSI solution. The gels (AP/Alg-PSI) were stored for 24 h for stabilization, washed with distilled water, and dried at room temperature. The hydrogel samples were sterilized by immersing them in a 70% ethanol solution for 10 min, followed by transfer to DPBS for an additional 1 h. The characterization was performed after the sterilization process.

4.4. Characterization

^1H NMR (Jeol, JNM-ECZR, 500 MHz) was used to characterize PSI. DMSO-D6 was used as the solvent to prepare the ^1H NMR samples. The number average molecular weight (M_n), weight average molecular weight (M_w), and polydispersity index (PDI) of PSI were obtained using GPC (Agilent Technologies 1100). Dimethylformamide was used as the eluent. The Fourier-transform infrared (FT-IR) spectra of Alg, PSI, and Alg-PSI were recorded separately using an FT-IR spectrometer (PerkinElmer Frontier, Shelton, USA) in the range of 4000–700 cm^{-1}. X-ray diffraction of PSI, Alg, and Alg-PSI were collected with X-ray diffractometer (Rigaku (SmartLab), Wilmington, MA, USA) with a Cu Kα and produced materials in the 2-theta range from 20° to 80°. Scanning electron microscopy (SEM; Hitachi SU8600, Tokyo, Japan) at the Smart Materials Research Center for IoT (Gachon University, Republic of Korea) was used to examine the morphologies of Alg and Alg-PSI. Prior to scanning, the SEM sample was subjected to a platinum ion coating applied using a sputter coater for 120 s under vacuum.

4.5. Rheological Tests of Hydrogels

The stability of the hydrogels was investigated by their dynamic rheological behavior using a rheometer (MCR92, Anton Paar, Graz, Austria) equipped with a parallel-plate geometry and a flat-bottom plate with a diameter of 25 mm. After preparation, the gels of each group were washed with distilled water, and the water was removed. First, the storage (G') and loss (G'') moduli were measured by a frequency sweep (0.1–10 Hz) under the linear viscoelastic region (LVR) with a fixed strain (0.1%) at 37 °C to confirm the hydrogel. In addition, the viscosity was confirmed according to the shear rate change (0.001–1 s^{-1}).

4.6. Degradation Analyses of Hydrogels

The degradation rates of the hydrogels were measured in a test medium to correlate the observed drug release with the rate of hydrolysis. The hydrogels were transferred to a water bath in a HCl medium (pH 1.2) and subjected to shaking for 2 h. The hydrogels were then placed in PBS (pH 7.4). All tests were carried out at 37 °C. The surface of the hydrogel was then dried and images of the hydrogel were recorded.

4.7. AP Encapsulation Efficiency

AP encapsulation efficiency was determined by extracting the entire drug from the hydrogels. The drug-containing hydrogels were completely dissolved in PBS solution (pH 7.4) by magnetic stirring for 72 h and then completely sonicated to obtain a homogeneous solution (to ensure that the hydrogels had completely degraded and released AP into the PBS solution). The amount of AP was estimated using a UV spectrophotometer at 243 nm and the drug encapsulation efficiency was calculated according to the following equation. The results were obtained from three independent experiments.

$$\%\ \text{Drug encapsulation} = \frac{W_1}{W_2} \times 100$$

where W_1 is the actual amount of AP in the hydrogel and W_2 is the amount of AP theoretically encapsulated.

4.8. Acetaminophen Release Tests

The drug release behavior was observed by loading the loaded hydrogels in HCl (pH 1.2) and PBS (pH 7.4) solutions in an orbital shaker bath at 37 °C set at 100 rpm to simulate the digestive environment. At regular intervals, the extracts (1 mL) were removed and immediately replaced with an equal volume of fresh solution. AP release was determined by analyzing the periodic percentage of the extract using UV spectroscopy. The UV-visible absorbance of AP was measured at 243 nm.

4.9. In Vitro Cytotoxicity

NIH/3T3 cells were obtained from the Korean Cell Line Bank (Seoul, Republic of Korea). NIH/3T3 fibroblasts were cultured in culture medium (DMEM containing 10% *v*/*v* FBS and 1% *v/v* penicillin–streptomycin) at 37 °C, 5% CO_2, and 95% air. The biocompatibility of the hydrogels according to the concentration of PSI (0, 0.5, 1.5, and 3% *w*/*v*) was evaluated by MTT assay. Fibroblasts were seeded in 24-well plates at a density of 2×10^4 cells/well and incubated with the culture medium for 24 h. The hydrogel samples were prepared in a size of 25 μL. After 24 h of seeding, the hydrogels were placed on a hanging insert to contact the cells. A sufficient amount of the medium was added to the hanging insert containing the gel to allow it to sink. The control group was not treated with the hydrogel, whereas the other groups (G1, G2, and G3) were treated with hydrogel, respectively. After 24 h, the medium was replaced with culture medium containing 10% *v*/*v* MTT solution (6 mg/mL). After 1 h, the formazan crystals were dissolved in DMSO, and the absorbance was measured using a microplate reader (Agilent Technologies, Santa Clara, CA, USA). The relative cell proliferation (%) was expressed by dividing the absorbance of the treatment groups by the absorbance of the control group.

4.10. Statistical Analysis

All data are expressed as mean ± standard deviation. Each experiment was repeated three times, unless otherwise indicated. Statistical evaluation was performed using one-way analysis of variance (ANOVA). A p value < 0.05 was considered statistically significant. Statistical analysis was performed using the statistical software GraphPad Prism 9.

Supplementary Materials: The following supporting information can be downloaded at: https://www.mdpi.com/article/10.3390/gels9120980/s1, Figure S1: ^{1}H NMR spectroscopy of PSI; Figure S2: GPC of PSI.

Author Contributions: L.T.T.: conceptualization, data curation, formal analysis, validation, and writing—original draft. S.L.: formal analysis and validation. H.J.L.: funding acquisition, and writing—review and editing. I.T.K.: funding acquisition, project administration, and writing—review and editing. The authors declare no competing financial interests. All authors have read and agreed to the published version of the manuscript.

Funding: This work was supported by the Gachon University research fund of 2023. (GCU-202307880001). This study was also supported by the National Research Foundation of Korea (NRF) grant funded by the Korean government (MSIT) (NRF-2021R1A6A1A03038996).

Institutional Review Board Statement: Not applicable.

Informed Consent Statement: Not applicable.

Data Availability Statement: The raw/processed data required to reproduce these findings cannot be shared at this time as the data also form part of an ongoing study.

Conflicts of Interest: The authors declare no conflict of interest.

References

1. Luo, G.; Huang, L.; Zhang, Z. The Molecular Mechanisms of Acetaminophen-Induced Hepatotoxicity and Its Potential Therapeutic Targets. *Exp. Biol. Med.* **2023**, *248*, 412–424. [CrossRef]
2. Adhikary, S.; Al Hoque, A.; Ray, M.; Paul, S.; Hossain, A.; Goswami, S.; Dey, R. Investigation of Paracetamol Entrapped Nanoporous Silica Nanoparticles in Transdermal Drug Delivery System. *Appl. Biochem. Biotechnol.* **2023**, *195*, 4712–4727. [CrossRef]
3. Chen, C.; Wu, H.; Li, Q.; Liu, M.; Yin, F.; Wu, M.; Wei, X.; Wang, H.; Zha, Z.; Wang, F. Manganese Prussian Blue Nanozymes with Antioxidant Capacity Prevent Acetaminophen-Induced Acute Liver Injury. *Biomater. Sci.* **2023**, *11*, 2348–2358. [CrossRef]
4. Jaeschke, H.; Adelusi, O.B.; Akakpo, J.Y.; Nguyen, N.T.; Sanchez-Guerrero, G.; Umbaugh, D.S.; Ding, W.X.; Ramachandran, A. Recommendations for the Use of the Acetaminophen Hepatotoxicity Model for Mechanistic Studies and How to Avoid Common Pitfalls. *Acta Pharm. Sin. B* **2021**, *11*, 3740–3755. [CrossRef]
5. Bernal, W.; Auzinger, G.; Dhawan, A.; Wendon, J. Acute Liver Failure. *Lancet* **2010**, *376*, 190–201. [CrossRef]
6. Kulsoom, R.; Sarfraz, M.; Afzal, A.; Farooq, M.; Adnan, S.; Ashraf, M.U.; Khan, S.A. Synthesis of Calcium Carbonate-Quince Bio-Composite for Programmed and on-Demand Drug Release of Paracetamol at Target Site: A Green Chemistry Approach. *Polym. Bull.* **2023**, *80*, 6965–6988. [CrossRef]
7. Rainsford, K.D.; Whitehouse, M.W. Paracetamol [Acetaminophen]-Induced Gastrotoxicity: Revealed by Induced Hyperacidity in Combination with Acute or Chronic Inflammation. *Inflammopharmacology* **2006**, *14*, 150–154. [CrossRef] [PubMed]
8. Shimono, N.; Takatori, T.; Ueda, M.; Mori, M.; Higashi, Y.; Nakamura, Y. Chitosan Dispersed System for Colon-Specific Drug Delivery. *Int. J. Pharm.* **2002**, *245*, 45–54. [CrossRef] [PubMed]
9. Bordbar-Khiabani, A.; Gasik, M. Smart Hydrogels for Advanced Drug Delivery Systems. *Int. J. Mol. Sci.* **2022**, *23*, 3665. [CrossRef] [PubMed]
10. Trinh, T.A.; Duy Le, T.M.; Ho, H.G.V.; To, T.C.T.; Nguyen, V.V.L.; Huynh, D.P.; Lee, D.S. A Novel Injectable pH-Temperature Sensitive Hydrogel Containing Chitosan-Insulin Electrosprayed Nanosphere Composite for an Insulin Delivery System in Type i Diabetes Treatment. *Biomater. Sci.* **2020**, *8*, 3830–3843. [CrossRef] [PubMed]
11. Yu, Y.; Xu, S.; Li, S.; Pan, H. Genipin-Cross-Linked Hydrogels Based on Biomaterials for Drug Delivery: A Review. *Biomater. Sci.* **2021**, *9*, 1583–1597. [CrossRef] [PubMed]
12. Levkovskyi, I.O.; Mochizuki, S.; Zheng, A.; Zhang, X.; Zhang, F. Lipoic Acid-Based Poly(Disulfide)s: Synthesis and Biomedical Applications. *Nano TransMed* **2023**, *2*, 100006. [CrossRef]
13. Jing, S.; Liu, Y.; Ye, Z.; Ghaleb Al-bashari, A.A.; Zhou, H.; He, Y. Ferrostatin-1 Loaded Gelatin Methacrylate Scaffold Promotes Recovery from Spinal Cord Injury via Inhibiting Apoptosis and Ferroptosis. *Nano TransMed* **2023**, *2*, 100005. [CrossRef]
14. Dattilo, M.; Patitucci, F.; Prete, S.; Parisi, O.I.; Puoci, F. Polysaccharide-Based Hydrogels and Their Application as Drug Delivery Systems in Cancer Treatment: A Review. *J. Funct. Biomater.* **2023**, *14*, 55. [CrossRef]
15. Pérez-Madrigal, M.M.; Shaw, J.E.; Arno, M.C.; Hoyland, J.A.; Richardson, S.M.; Dove, A.P. Robust Alginate/Hyaluronic Acid Thiol-Yne Click-Hydrogel Scaffolds with Superior Mechanical Performance and Stability for Load-Bearing Soft Tissue Engineering. *Biomater. Sci.* **2020**, *8*, 405–412. [CrossRef] [PubMed]
16. Hegde, V.; Uthappa, U.T.; Altalhi, T.; Jung, H.Y.; Han, S.S.; Kurkuri, M.D. Alginate Based Polymeric Systems for Drug Delivery, Antibacterial/Microbial, and Wound Dressing Applications. *Mater. Today Commun.* **2022**, *33*, 104813. [CrossRef]
17. Mahmood, A.; Mahmood, A.; Ibrahim, M.A.; Hussain, Z.; Ashraf, M.U.; Salem-Bekhit, M.M.; Elbagory, I. Development and Evaluation of Sodium Alginate/Carbopol 934P-Co-Poly (Methacrylate) Hydrogels for Localized Drug Delivery. *Polymers* **2023**, *15*, 311. [CrossRef]
18. Treenate, P.; Monvisade, P. In Vitro Drug Release Profiles of PH-Sensitive Hydroxyethylacryl Chitosan/Sodium Alginate Hydrogels Using Paracetamol as a Soluble Model Drug. *Int. J. Biol. Macromol.* **2017**, *99*, 71–78. [CrossRef]
19. Pervez, S.; Nawaz, M.A.; Aman, A.; Qayyum, S.; Nawaz, F.; Qader, S.A.U. Agarose Hydrogel Beads: An Effective Approach to Improve the Catalytic Activity, Stability and Reusability of Fungal Amyloglucosidase of GH15 Family. *Catal. Lett.* **2018**, *148*, 2643–2653. [CrossRef]
20. George, M.; Abraham, T.E. Polyionic Hydrocolloids for the Intestinal Delivery of Protein Drugs: Alginate and Chitosan—A Review. *J. Control. Release* **2006**, *114*, 1–14. [CrossRef]
21. Rubio, M.R.; Ghaly, E.S. In-Vitro Release of Acetaminophen from Sodium Alginate Controlled Release Pellets. *Drug Dev. Ind. Pharm.* **1994**, *20*, 1239–1251. [CrossRef]
22. Sharma, S.; Sarkar, G.; Srestha, B.; Chattopadhyay, D.; Bhowmik, M. In-Situ Fast Gelling Formulation for Oral Sustained Drug Delivery of Paracetamol to Dysphagic Patients. *Int. J. Biol. Macromol.* **2019**, *134*, 864–868. [CrossRef] [PubMed]
23. Hasnain, M.S.; Nayak, A.K. Alginate-Inorganic Composite Particles as Sustained Drug Delivery Matrices. In *Applications of Nanocomposite Materials in Drug Delivery*; Elsevier: Amsterdam, The Netherlands, 2018; pp. 39–74. ISBN 9780128137581.
24. Jain, S.; Datta, M. Montmorillonite-Alginate Microspheres as a Delivery Vehicle for Oral Extended Release of Venlafaxine Hydrochloride. *J. Drug Deliv. Sci. Technol.* **2016**, *33*, 149–156. [CrossRef]
25. Iliescu, R.I.; Andronescu, E.; Ghitulica, C.D.; Voicu, G.; Ficai, A.; Hoteteu, M. Montmorillonite-Alginate Nanocomposite as a Drug Delivery System—Incorporation and in Vitro Release of Irinotecan. *Int. J. Pharm.* **2014**, *463*, 184–192. [CrossRef] [PubMed]

26. Wu, J.L.; Wang, C.Q.; Zhuo, R.X.; Cheng, S.X. Multi-Drug Delivery System Based on Alginate/Calcium Carbonate Hybrid Nanoparticles for Combination Chemotherapy. *Colloids Surf. B Biointerfaces* **2014**, *123*, 498–505. [CrossRef] [PubMed]
27. Currie, S.; Kim, S.; Gu, X.; Ren, X.; Lin, F.; Liu, S.; Yang, C.; Kim, J.H.; Liu, S. Mucus-Penetrating PEGylated Polysuccinimide-Based Nanocarrier for Intravaginal Delivery of SiRNA Battling Sexually Transmitted Infections. *Colloids Surf. B Biointerfaces* **2020**, *196*, 111287. [CrossRef] [PubMed]
28. Hill, M.R.; MacKrell, E.J.; Forsthoefel, C.P.; Jensen, S.P.; Chen, M.; Moore, G.A.; He, Z.L.; Sumerlin, B.S. Biodegradable and PH-Responsive Nanoparticles Designed for Site-Specific Delivery in Agriculture. *Biomacromolecules* **2015**, *16*, 1276–1282. [CrossRef] [PubMed]
29. Wu, H.; Hu, P.; Xu, Y.; Xiao, C.; Chen, Z.; Liu, X.; Jia, J.; Xu, H. Phloem Delivery of Fludioxonil by Plant Amino Acid Transporter-Mediated Polysuccinimide Nanocarriers for Controlling Fusarium Wilt in Banana. *J. Agric. Food Chem.* **2021**, *69*, 2668–2678. [CrossRef]
30. Barczikai, D.; Kacsari, V.; Domokos, J.; Szabó, D.; Jedlovszky-Hajdu, A. Interaction of Silver Nanoparticle and Commonly Used Anti-Inflammatory Drug within a Poly(Amino Acid) Derivative Fibrous Mesh. *J. Mol. Liq.* **2021**, *322*, 114575. [CrossRef]
31. Adelnia, H.; Blakey, I.; Little, P.J.; Ta, H.T. Poly(Succinimide) Nanoparticles as Reservoirs for Spontaneous and Sustained Synthesis of Poly(Aspartic Acid) under Physiological Conditions: Potential for Vascular Calcification Therapy and Oral Drug Delivery. *J. Mater. Chem. B* **2023**, *11*, 2650–2662. [CrossRef]
32. Heading, R.C.; Nimmo, J.; Prescott, L.F.; Tothill, P. The Dependence of Paracetamol Absorption on the Rate of Gastric Emptying. *Br. J. Pharmacol.* **1973**, *47*, 415–421. [CrossRef] [PubMed]
33. Sannino, A.; Demitri, C.; Madaghiele, M. Biodegradable Cellulose-Based Hydrogels: Design and Applications. *Materials* **2009**, *2*, 353–373. [CrossRef]
34. Lim, C.W.; Kim, D. Bone Targeting Nano-Aggregates Prepared from Self-Assembled Polyaspartamide Graft Copolymers for PH Sensitive DOX Delivery. *Biomater. Sci.* **2021**, *9*, 1660–1667. [CrossRef]
35. Kim, M.; Shin, S.W.; Lim, C.W.; Kim, J.; Um, S.H.; Kim, D. Polyaspartamide-Based Graft Copolymers Encapsulating Iron Oxide Nanoparticles for Imaging and Fluorescence Labelling of Immune Cells. *Biomater. Sci.* **2017**, *5*, 305–312. [CrossRef]
36. Jalalvandi, E.; Shavandi, A. Polysuccinimide and Its Derivatives: Degradable and Water Soluble Polymers (Review). *Eur. Polym. J.* **2018**, *109*, 43–54. [CrossRef]
37. Almurisi, S.H.; Doolaanea, A.A.; Akkawi, M.E.; Chatterjee, B.; Sarker, M.Z.I. Taste Masking of Paracetamol Encapsulated in Chitosan-Coated Alginate Beads. *J. Drug Deliv. Sci. Technol.* **2020**, *56*, 101520. [CrossRef]
38. Tóth, K.; Fekete, N.; Klaudia Simon, V.; Tóth, B.; Ferenc Kovács, Á.; Pállinger, É.; Antal, I.; Kőhidai, L.; Jedlovszky-Hajdú, A.; Juriga, D.; et al. Different Implantable Electrospun Meshes for Simultaneous Application of Prednisone and Doxorubicin. *J. Mol. Liq.* **2023**, *381*, 121854. [CrossRef]
39. Hua, S.; Ma, H.; Li, X.; Yang, H.; Wang, A. PH-Sensitive Sodium Alginate/Poly(Vinyl Alcohol) Hydrogel Beads Prepared by Combined Ca2+ Crosslinking and Freeze-Thawing Cycles for Controlled Release of Diclofenac Sodium. *Int. J. Biol. Macromol.* **2010**, *46*, 517–523. [CrossRef]
40. Pasparakis, G.; Bouropoulos, N. Swelling Studies and in Vitro Release of Verapamil from Calcium Alginate and Calcium Alginate-Chitosan Beads. *Int. J. Pharm.* **2006**, *323*, 34–42. [CrossRef]
41. Weng, L.; Tessier, S.N.; Swei, A.; Stott, S.L.; Toner, M. Controlled Ice Nucleation Using Freeze-Dried Pseudomonas Syringae Encapsulated in Alginate Beads. *Cryobiology* **2017**, *75*, 1–6. [CrossRef]
42. Popescu, I.N.; Poinescu, A.A.; Ungureanu, D.N.; Picu, A. Novel Developments in Advanced Materials Fields: Porous and Non-Porous Biomaterials Used in Regenerative Medicine and Tissue Engineering. *Sci. Bull. Valahia Univ.-Mater. Mech.* **2023**, *19*, 42–52. [CrossRef]
43. Zhang, H.; Shi, Y.; Xu, X.; Zhang, M.; Ma, L. Structure Regulation of Bentonite-Alginate Nanocomposites for Controlled Release of Imidacloprid. *ACS Omega* **2020**, *5*, 10068–10076. [CrossRef] [PubMed]
44. Gijpferich, A. Mechanisms of Polymer Degradation and Erosion. In *The Biomaterials: Silver Jubilee Compendium*; Elsevier: Amsterdam, The Netherlands, 1996; Volume 17.
45. Yoshioka, T.; Kawazoe, N.; Tateishi, T.; Chen, G. In Vitro Evaluation of Biodegradation of Poly(Lactic-Co-Glycolic Acid) Sponges. *Biomaterials* **2008**, *29*, 3438–3443. [CrossRef] [PubMed]
46. Martinez, J.O.; Chiappini, C.; Ziemys, A.; Faust, A.M.; Kojic, M.; Liu, X.; Ferrari, M.; Tasciotti, E. Engineering Multi-Stage Nanovectors for Controlled Degradation Andtunable Release Kinetics. *Biomaterials* **2013**, *34*, 8469–8477. [CrossRef] [PubMed]
47. Gombotz, W.R.; Fong Wee, S. Protein Release from Alginate Matrices. In *Advanced Drug Delivery Reviews*; Elsevier: Amsterdam, The Netherlands, 1998; Volume 31.
48. Bajpai, S.K.; Sharma, S. Investigation of Swelling/Degradation Behaviour of Alginate Beads Crosslinked with Ca^{2+} and Ba^{2+} Ions. *React. Funct. Polym.* **2004**, *59*, 129–140. [CrossRef]
49. Marković, D.; Zarubica, A.; Stojković, N.; Vasić, M.; Cakić, M.; Nikolić, G. Alginates and similar exopolysaccharides in biomedical application and pharmacy: Controled delivery of drugs. *Adv. Technol.* **2016**, *5*, 39–52. [CrossRef]

50. Abasalizadeh, F.; Moghaddam, S.V.; Alizadeh, E.; Akbari, E.; Kashani, E.; Fazljou, S.M.B.; Torbati, M.; Akbarzadeh, A. Alginate-Based Hydrogels as Drug Delivery Vehicles in Cancer Treatment and Their Applications in Wound Dressing and 3D Bioprinting. *J. Biol. Eng.* **2020**, *14*, 8.
51. Tomida, M.; Nakato, T.; Matsunami, S.; Kakuchi, T. Convenient Synthesis of High Molecular Weight Poly(Succinimide) by Acid-Catalyses Polycondensation of L-Aspartic Acid. *Polymer* **1997**, *18*, 4733–4736. [CrossRef]

gels

MDPI

Article

Design and Development of a Topical Nanogel Formulation Comprising of a Unani Medicinal Agent for the Management of Pain

Amit Sah [1], Geeta Aggarwal [2], Gaurav K. Jain [1], Syed Mohammad Abbas Zaidi [3], Punnoth Poonkuzhi Naseef [4], Mohamed S. Kuruniyan [5] and Foziyah Zakir [6,*]

1. Department of Pharmaceutics, Delhi Institute of Pharmaceutical Sciences and Research, Delhi Pharmaceutical Sciences and Research University, Sector-3, M.B. Road, Pushp Vihar, New Delhi 110017, India
2. Department of Pharmaceutics, School of Pharmaceutical Sciences, Delhi Pharmaceutical Sciences and Research University, Sector-3, M.B. Road, Pushp Vihar, New Delhi 110017, India; geetaaggarwal17@gmail.com
3. Department of Moalajat (Internal Medicine), Hakim Syed Zia ul Hasan (HSZH) Govt. Unani Medical College, Bhopal 462003, India
4. Department of Pharmaceutics, Moulana College of Pharmacy, Perinthalmanna Kerala 679321, India; drnaseefpp@gmail.com
5. Department of Dental Technology, College of Applied Medical Sciences, King Khalid University, Abha 61421, Saudi Arabia; mkurunian@kku.edu.sa
6. Department of B.Pharm (Ayurveda), School of Pharmaceutical Sciences, Delhi Pharmaceutical Sciences and Research University, Sector-3, M.B. Road, Pushp Vihar, New Delhi 110017, India

* Correspondence: foziyahzakir@gmail.com

Abstract: The oil of the Unani medicinal herb Baboona (*Matricaria chamomilla*) has shown potential in the management of pain. However, predicaments such as poor skin penetration, skin sensitization, liable to degradation, and volatile nature restrict its use. Therefore, our group for the first time has developed a carrier-based delivery system to facilitate the direct application of chamomile oil to the forehead. The developed nanogel was characterized for physical parameters such as compatibility, TEM, and stability studies. Further, it was also evaluated for pH, viscosity, spread ability, and extrudability, as well as through texture analyses, in vitro studies, and skin irritation tests. The formulation was successfully developed with all the necessary attributes. The in vitro studies revealed the enhanced skin penetration of chamomile oil nanogel. The in vivo studies were also performed in chemically induced pain models, mimicking migraine. The studies show significant improvement of the pain threshold for chamomile nanogel when compared to the positive control group and the results were comparable to marketed diclofenac formulations. Finally, the encapsulation into nanogel reduced the skin irritation property. The nanogel formulation showed promising effects in the pain management of migraine.

Keywords: Baboona; chamomile; *Matricaria chamomilla*; migraine; nanogel; nanoemulsion; pain; topical

Citation: Sah, A.; Aggarwal, G.; Jain, G.K.; Zaidi, S.M.A.; Naseef, P.P.; Kuruniyan, M.S.; Zakir, F. Design and Development of a Topical Nanogel Formulation Comprising of a Unani Medicinal Agent for the Management of Pain. *Gels* **2023**, *9*, 794. https://doi.org/10.3390/gels9100794

Academic Editors: Ying Huang, Zhengwei Huang and Xuanjuan Zhang

Received: 3 April 2023
Revised: 16 May 2023
Accepted: 7 June 2023
Published: 2 October 2023

1. Introduction

Migraine is a common neural condition characterized by chronic, pulsating pain, mostly experienced on one side of the head, i.e., unilateral attack [1]. The severity of headaches ranges from moderate to severe, usually persists for 4–24 h, and is associated with symptoms such as nausea, vomiting, photophobia, or phonophobia. It is most common in children and younger individuals, especially in women [2].

According to the World Health Organization, migraine affects approximately 12% of the global population, 17% of women, and 6% of men [3]. It severely affects quality of life and puts a burden on individuals due to the high treatment costs. Although many treatments are available for pain management, they exhibit limited efficacy due to significant side effects, toxicity, and unfavorable physicochemical characteristics [4]. In the search for alternate therapies, herbal plants and oils have been explored as they contain comparatively far safer compounds. The popularity of natural medicine is continuously increasing, and it has emerged as a globally sought-after approach. Unani is an ancient medicinal system of medicine that is widely practiced in India as well as in certain other countries. The application of herbal drugs into nanotechnology is a novel approach to magnify the solubility, absorption rate, and permeation of herbal medicine, which demonstrates the high bioavailability and therapeutic potential of these medicinal plants [5].

Baboona (*Matricaria chamomilla*) is a very promising Unani medicinal herb used orally as well as locally to treat many disorders including various kinds of headaches, particularly migraine [6]. Research regarding its formulation and application in medicine is rare, and it is used for its soothing qualities as a sedative, mild analgesic, and sleep medication. The main constituents in chamomile essential oil (CO) are chamazulene, apigenin, and bisabolol which possess properties of pain management. There are many reports which claim that CO possesses properties for the treatment of migraine [7]. The apigenin and its derivatives act as a COX-2 inhibitor and possess anti-inflammatory effects due to the inhibition of endogenous prostaglandin E2 (PGE2) levels in RAW 264.7 macrophages [8]. It also acts as a neuroprotective agent because of the inhibition of nitric oxide synthase and subsequently leads to inhibited nitric oxide levels, which play an important role in inducing acute to chronic migraine attack [9].

However, pure essential oil has irritation potential and cannot be applied directly to the forehead. Additionally, CO is volatile, with poor tissue permeability and high degradation. Therefore, its beneficial therapeutic properties require a delivery approach that is able to improve its penetrative power, negate its volatility, and enhance the therapeutic effectiveness of CO [6]. This could be achieved by delivering CO in a sustained manner by using a delivery carrier such as nanoemulsion-based gel. Nanoemulsion formulations possess unique properties such as high biocompatibility, ease of surface modification, and smaller size [10]. Nanoemulsion-based formulations are effective fortreating migraine, as they enhance the therapeutic efficacy and tolerability of antimigraine drugs [11]. Moreover, a nanoemulsion-based nanogel offers high patient acceptability, as it is non-invasive and easily applicable [12]. Despite the numerous therapeutic benefits of CO in migraine, nobody has ever attempted to enhance its efficacy using a suitable carrier system.

In our work, we propose the formulation of a CO-loaded nanoemulsion (CON) and its conversion into gel (COG), which is expected to reduce skin irritation and enhance the therapeutic activity of CO. The study will be supplemented with characterization studies to confirm the suitability of the formulation, as well as in vitro release and permeation studies and in vivo pharmacodynamic studies in order to demonstrate performance attributes, followed by in vitro skin irritation and stability studies, which will address the physical stability and acceptability of the formulation.

2. Results and Discussion

2.1. Construction of Pseudo-Ternary Phase Diagrams

All the nanoemulsions were prepared usingthe aqueous titration method, as shown in Figure 1. The nanoemulsion forming zone was marked on pseudo-ternary phase diagrams. The pseudo-ternary phase diagrams with the largest nanoemulsion-forming zones were selected for the preparation of nanoemulsions. It was observed that the Smix, comprising PEG 400/Tween 20 at a ratio of 1:1, revealed the formation of a nanoemulsion. Further, the most prominent nanoemulsion zone was seen in the oil:Smix ratio of 1:5.

2.2. Thermodynamic Stability Studies of Developed Nanoemulsion

The stability studies of different chamomile-loaded nanoemulsion formulations are shown in Table 1. The CONs that were physically unstable showed signs of phase separation and were therefore rejected.

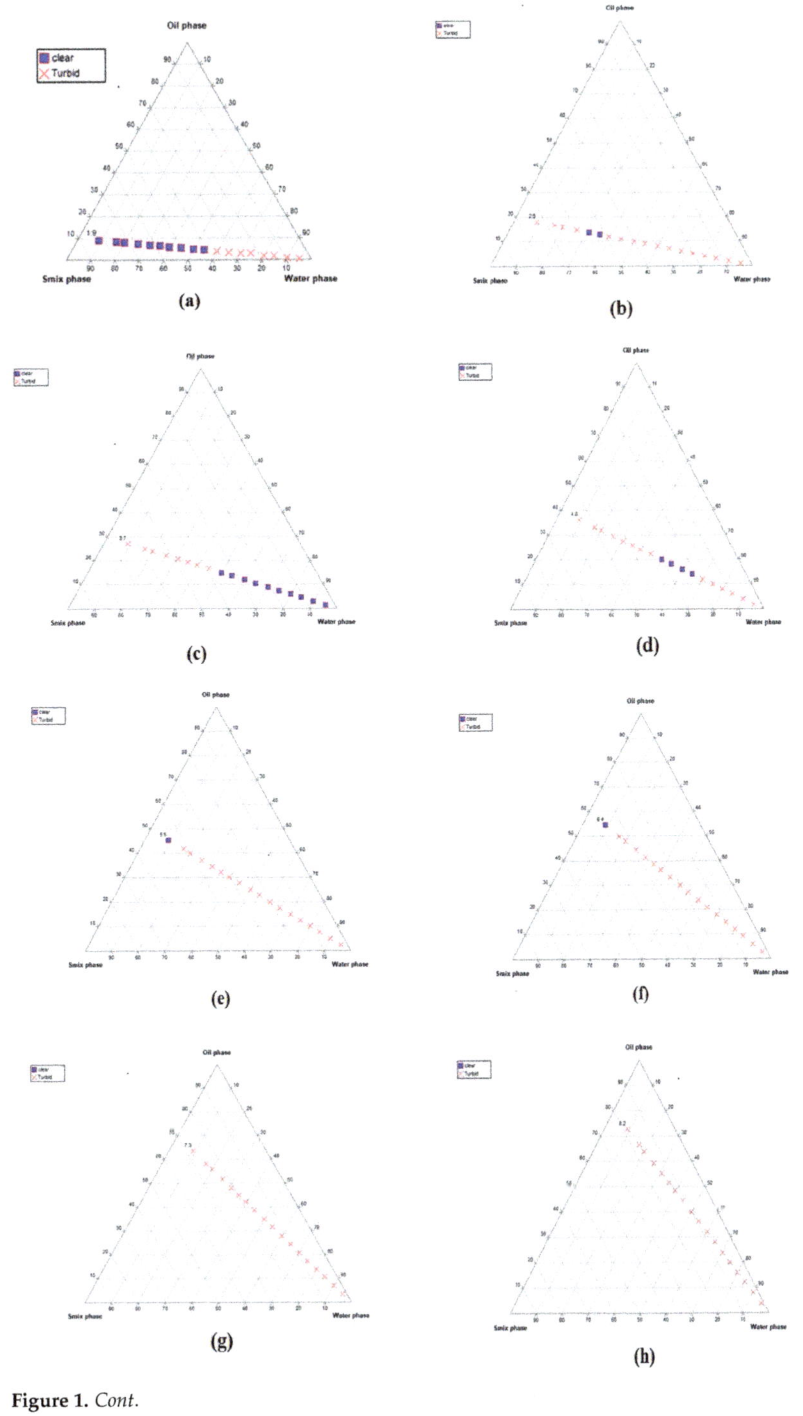

Figure 1. *Cont.*

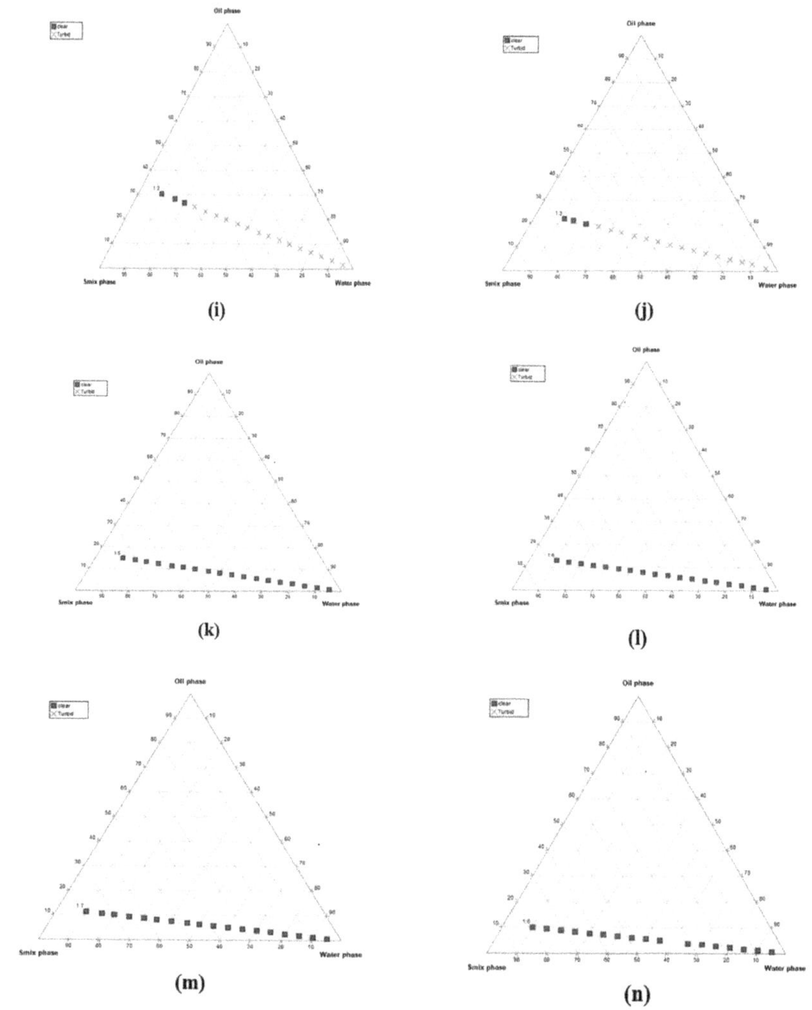

Figure 1. Pseudo-ternary phase diagrams of the developed nanoemulsion formulations using different oil:Smix ratios: (**a**) 1:9, (**b**) 2:8, (**c**) 3:7, (**d**) 4:6, (**e**) 5:5, (**f**) 6:4, (**g**) 7:3, (**h**) 8:2, (**i**) 1:2, (**j**) 1:3, (**k**) 1:5, (**l**) 1:6, (**m**) 1:7, and (**n**) 1:8. The figure shows oil, surfactant:cosurfatant, and water in each corner with 100% of each component. The blue dots indicate clear dispersion, whereas red crosses denote turbidity. It can be seen that the nanoemulsion forming zone (represented by blue dots) is larger in images (**k–n**), suggesting stable formulation.

Table 1. Thermodynamic stability studies of selected nanoemulsion formulations.

Oil:Smix	Heating–Cooling Cycles	Centrifugation	Freeze–Thaw	Inference
1:5	Clear	Clear	Clear	Stable
1:6	PS	PS	PS	Unstable
1:7	PS	PS	PS	Unstable
1:8	PS	PS	PS	Unstable

2.3. Characterization of Selected Nanoemulsion Formulations

2.3.1. Particle Size, Zeta Potential, and Polydispersity Index (PDI)

The particle size, zeta potential, and polydispersity index of the optimized CONs are 22.95 nm, −6.27 mV, and 0.335, respectively, as represented in Figure 2.

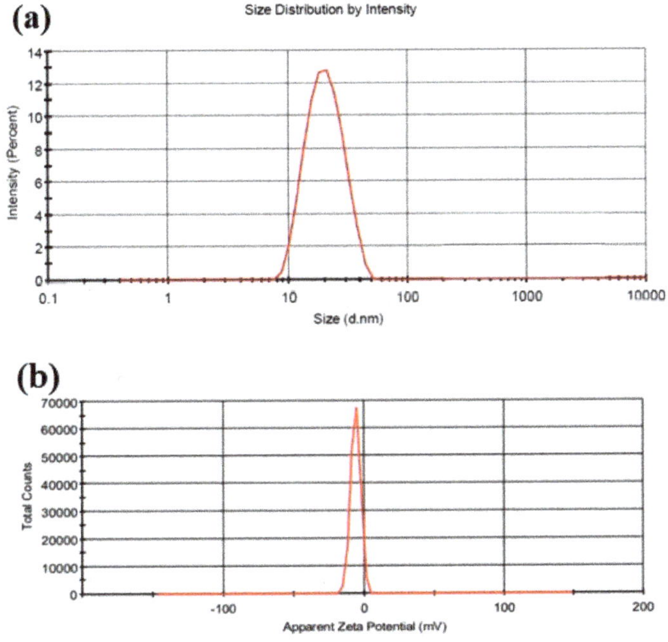

Figure 2. (**a**) Droplet size distribution by intensity; (**b**) zeta potential graph.

2.3.2. Transmission Electron Microscopy

Transmission electron microscopy (TEM) was performed in order to study the morphology and structure of the optimized CONs. The formulations were observed at magnification 45,000× and 11,000× under TEM. From the TEM images, we can conclude that globules were well-formed and spherical. No aggregation was observed. The globules observed in the TEM images were found to be less than 100 nm (Figure 3).

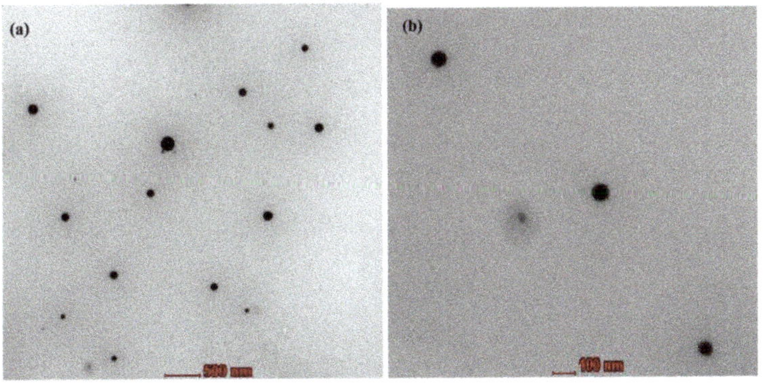

Figure 3. Transmission electron microscope images of nanoemulsion droplets at a magnification of (**a**) 11,000× (scale in the image represents 500 nm) and (**b**) 45,000× (scale in the image represents 500 nm).

2.3.3. Compatibility Studies

Structural compatibility was further observed for CON and CO with the help of the FT-IR spectra (Table 2). The FT-IR spectrum of pure CO was compared with that of CON. Since there was no major shift in the peaks of CO-loaded nanoemulsion, this indicates no chemical incompatibility or interactions in the drug-excipient combination, as shown in Figure 4a,b.

Table 2. FT-IR spectra peaks of chamomile oil and nanoemulsion. The figures represent the wave number values for different functional groups.

S. No	Groups	Actual Value (cm^{-1})	Observed Values	
			Chamomile Oil	Nanoemulsion
1.	$-CH_2-$ (aliphatic asymmetric)	2926	2929.21	2925.55
2.	C=O	1730–1750	1731.81	1732.42
3.	=C-H (Scissor)	Approx. 1465	1456.51	1457.14
4.	C-O	1000–1300	1374.96	1355.23

2.4. *Preparation and Evaluation of Nanogel*

The prepared nanogel was evaluated for various parameters to uncover whether the final optimized COG was suitable for topical application.

2.4.1. pH Measurement

The pH value of the nanogel was found to be 5.8 ± 0.03, as determined by the pH meter.

2.4.2. Viscosity

The values for viscosity shown in Figure 5 indicate a decrease in viscosity upon increasing the shear rate from 1 to 100 s^{-1}. This indicates the shear-thinning or pseudoplastic behavior of the prepared formulations.

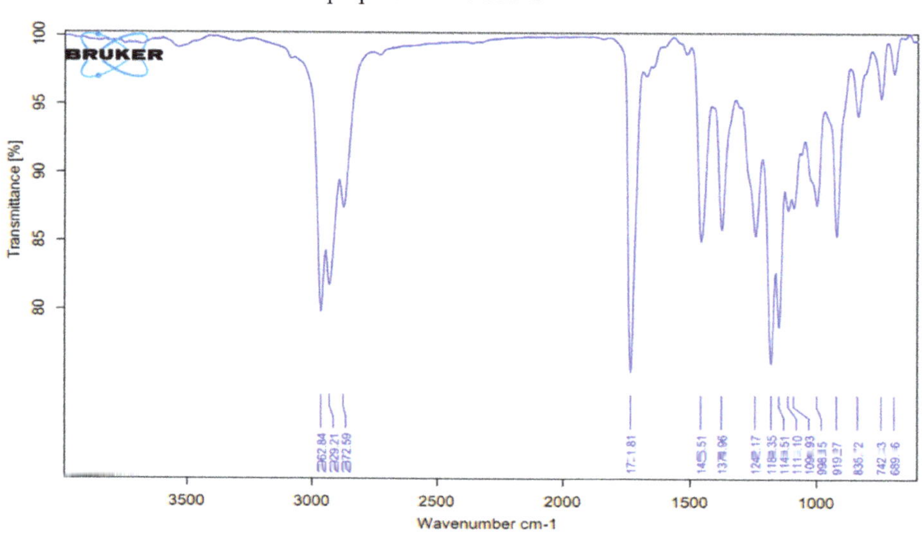

(**a**)

Figure 4. *Cont.*

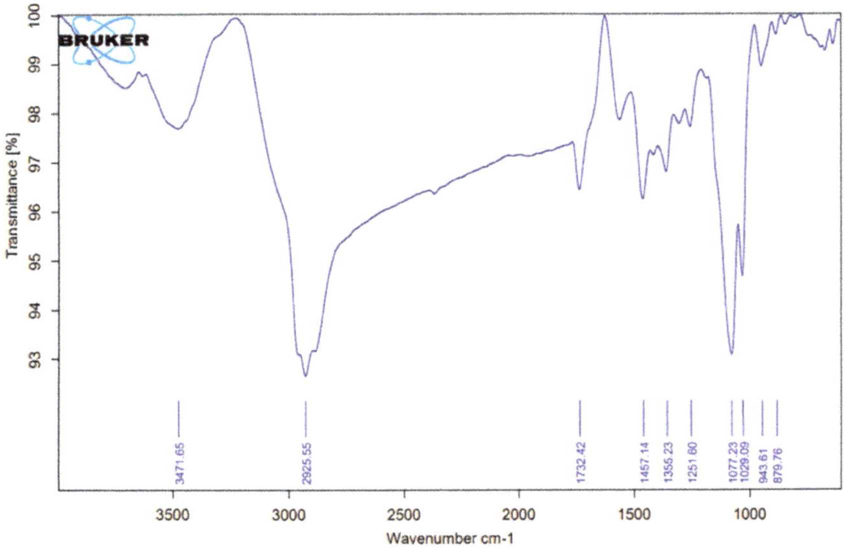

(**b**)

Figure 4. Fourier transform infrared spectra of (**a**) pure chamomile oil and (**b**) chamomile oil-loaded nanoemulsion.

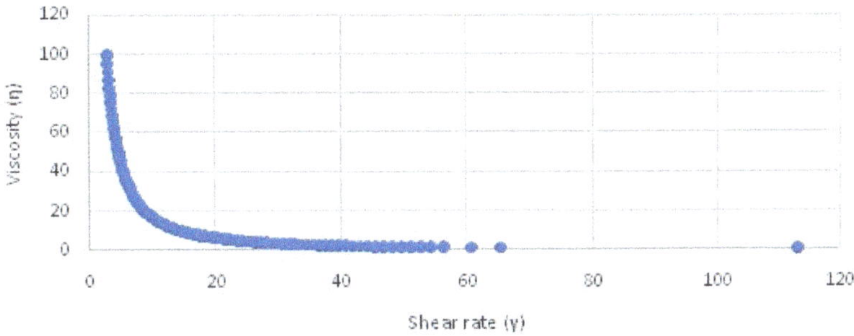

Figure 5. Viscosity vs. shear rate graph of the developed chamomile nanogel.

2.4.3. Spreadability and Extrudability Tests

It was observed that the gel formulation showed good spreadability (21.66 g·cm/s), which is required for easier topical application. Similarly, the extrudability of the formulation was found to be 89.9%. Sufficient extrudability is an important requirement for the easy removal of the gel from the tube while it is being used by the patient.

2.4.4. Texture Analysis

The firmness, consistency, cohesiveness, and cohesion function were assessed as texture parameters of the formulations, as shown in Figure 6, and were found to be 161.14 g, 618.77 g·s, −86.20 g, and −392.94 g·s, respectively.

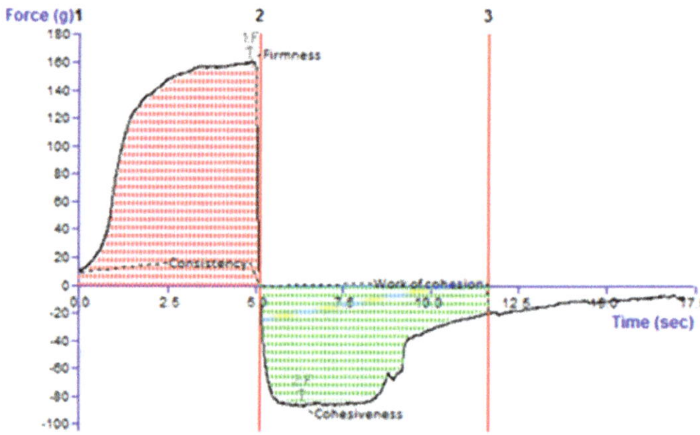

Figure 6. Texture profile analysis graph of the developed chamomile oil nanogel. Line 2 indicates the firmness value when pointing up, and when pointing down, it indicates cohesiveness. The area between 1 and 2 is consistency and between 2 and 3 is cohesiveness.

2.5. Analytical Method Development

The calibration curve was constructed for apigenin, the major chemical component present in CO. The linear regression analysis data showed a good linear relationship ($r^2 = 0.9979$) concerning the peak area in the concentration range of 10–50 ng/µL per spot. The values of slope and intercept are 0.0006 and 0.0001, respectively.

2.6. In Vitro Drug Diffusion Study

An in vitro diffusion study was carried out in a dialysis membrane bag for 24 h and showed maximum drug release (99.5 ± 9.1%) for CONs, as shown in Figure 7. When the formulation was loaded into the gel, then the release was sustained with only a 60 ± 11.3% release over 24 h.

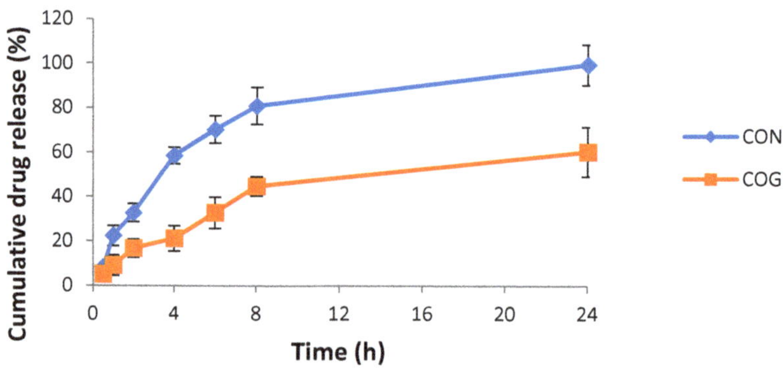

Figure 7. Cumulative percent drug release vs. time graph of the chamomile oil nanogel (COG) vs. chamomile oil nanoemulsion (CON) in a phosphate buffer of pH 6.

2.7. Drug-Release Kinetics

In vitro study data were fitted in different models, such as zero-order and first-order, as well as into a Higuchi plot, Korsmeyer–Peppas plot, and Hixson–Crowell plot. The Korsmeyer–Peppas plot out of all the above showed the highest linearity, which was interpreted via regression coefficients, as shown in Table 3. It was observed from the analysis that an R^2 value of 0.9051 was obtained in the Korsmeyer–Peppas plot with an n value of 0.323, which reveals the Fickian diffusion as shown.

Table 3. Drug-release kinetic analysis of the nanogel formulation using different models.

Zero-Order		First-Order		Higuchi		Korsmeyer–Peppas			Hixson–Crowell	
K_0 (intercept)	R^2	K_1 (intercept)	R^2	K_H (intercept)	R^2	K_{KP} (intercept)	N	R^2	K_{HC}	R^2
0.054	0.1065	0.001	0.1124	0.237	0.8763	0.290	0.417	0.9051	0.000	0.1104

2.8. Skin Penetration Study

The skin penetration ability of the CO in the formulation was also assessed. It was observed that only about 14 ± 2.1 ng of CO was penetrated when used alone(Figure 8). On the other hand, this number rose considerably when incorporated into a nanogel formulation, with about 70% of the CO penetrating the skin within 24 h. This shows the success of the formulation.

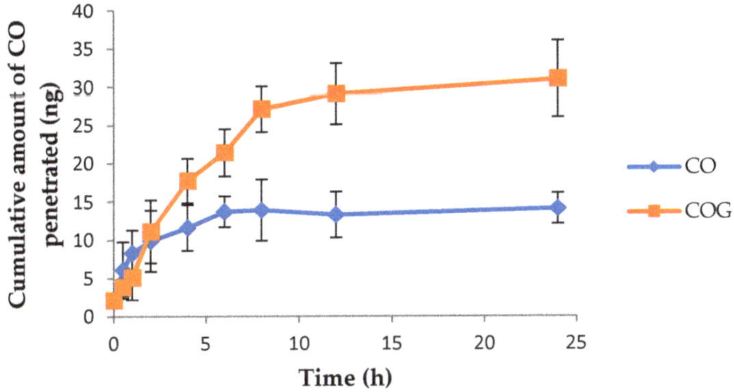

Figure 8. Graph depicting the cumulative amount of pure chamomile oil (CO) and chamomile oil from nanogel (COG) penetrating excised rat skin vs. time.

2.9. In Vivo Studies

2.9.1. Tail Flick Test

The animals in the control group were subjected to a heat source and the time for latency was noted throughout the test period. No significant difference in reaction time was noted. The second group of animals was treated with diclofenac sodium and the latency time was increased considerably (7.21 ± 0.97), which was suggestive of an increased pain threshold. Furthermore, the group treated with a topical chamomile formulation was also tested and showed a significant increase (6.88 ± 0.81, $p < 0.001$) in tail flick latency, somewhat similar to the response given by the standard group (Table 4). However, the insignificant difference between standard treatment and chamomile formulation as depicted by the percent inhibition in analgesia suggests comparable analgesic activity.

Table 4. The time for latency before treatment and post-treatment using different formulations during the tail flick test.

Groups	Dose	Time (S)		Percent Inhibition
		Pre-Treatment	Post-Treatment	
Control (saline)	1 mL/kg, IP	2.88 ± 0.75	2.81 ± 1.12	-
Standard treatment (diclofenac)	20 mg/kg, IP	2.77 ± 0.31	7.21 ± 0.97	36.3
Test treatment (COG)	100 mg/rat, topical	2.58 ± 0.93	6.88 ± 0.81	34.62

2.9.2. Acetic Acid-Induced Writhing Test

The administration of acetic acid in Wistar rats induces pain, which had been correlated with pain suffered in migraine. The irritation generated viaacetic acid in the abdominal region of rats causes it to constrict their abdomen or stretch their hind limbs, which is known as writhing. It was observed that there was a significant writhing count after the rats were exposed to acetic acid during the experiment, confirming the successful induction of migraine. However, upon treatment with standard and test formulations, the exposed rats showed a significant decrease ($p < 0.001$) in writhing count compared to the acetic acid-exposed group (control), as shown in Table 5. We can conclude that the application of chamomile oil nanogel resulted in significantly ($p < 0.001$) reduced writhing, as compared to the control group. The response shown by the test group was almost similar to the responses exhibited through application of the standard treatment, diclofenac sodium, as suggested by their protection percentages.

Table 5. The writhing count exhibited as a result of treatment with different formulations during the acetic acid-induced writhing test.

Groups	Dose	No. of Writhes	% Protection
Control (saline)	0.3%, 10 mL/kg Ip	24.88 ± 0.98	
Standard (diclofenac sodium)	20 mg/kg Ip	11.27 ± 1.1	54.7%
Test (COG)	100 mg/rat, topical	14.75 ± 0.95	40.71%

2.9.3. Light/Dark Box Model

Nitroglycerine exposure to experimental animals showed a marked rise in the time spent in the dark box (420 ± 18.27; $p < 0.01$) and a decline in time spent in the light box (114 ± 18.27; $p < 0.001$), compared to negative control rats; stating that nitroglycerine successfully induced ananxiolytic effect (Figure 9). However, the administration of the COG formulation showed a significant decline in the time spent in the dark box (228 ± 28.65; $p < 0.001$), as well as a rise in the time spent in the light box (360 ± 28.65; $p < 0.05$) upon comparison with the positive control group. The number of transitions during the experiment was the highest in the control group and showed a significant decline in thepositive control group, suggestive of depressive behavior. However, the topical COG showed a significant elevation in the transitions.

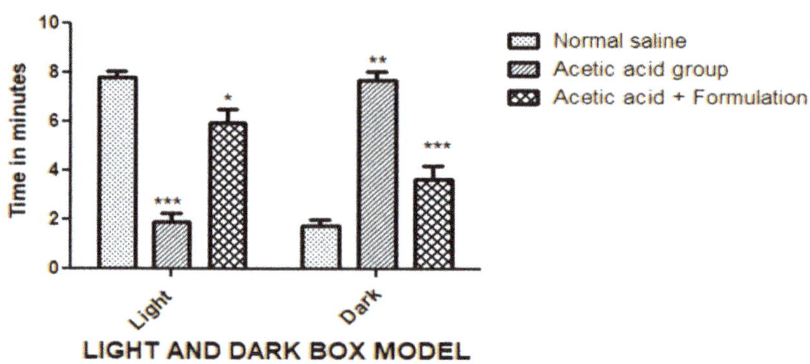

Figure 9. Anxiolytic effect via the light and dark box model. The duration of stay ofthe rats in both light and dark compartments was tested for different treatment groups. The data represented are mean ± SEM. Where *** $p < 0.001$, ** $p < 0.01$, and * $p < 0.05$. A two-way ANOVA with Tukey's post hoc test was used for statistical analysis.

2.10. Skin Irritation Test

From the graph (Figure 10), it is clear that maximum viability is observed for negative control suggestive of negligible skin irritation. On the other hand, positive control showed a viability of only 5%, which suggests toxicity to human epidermal cells. CO exhibited a reduced viability score of 36%, which depicts its irritant behavior. The loading of CO into gel significantly reduced its irritancy, which suggests that it is safe for topical application.

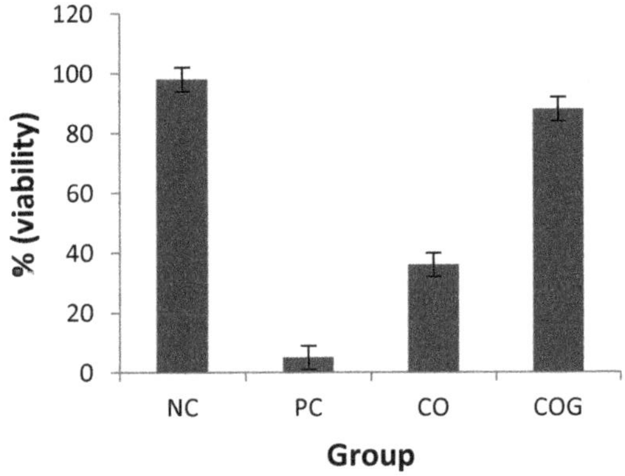

Figure 10. The in vitro skin irritation test expressed as percentage viability for different treatment groups; NC, negative control; PC, positive control; CO, pure chamomile oil; COG, chamomile oil nanogel.

2.11. Stability Study

The appearance, phase separation, pH, and percent transmittance of formulated nanogel were observed to be consistent with no signs of separation and deterioration over a period of 60 days (Table 6). The results indicated that the optimized formulation was found to be physically and chemically stable during the study period.

Table 6. Stability studies of optimized nanogel at different time intervals and temperature/humidity conditions.

Storage Conditions		Parameters			
Time	Temperature/Relative Humidity	Appearance	Phase Separation	pH	Percent Transmittance
0 day	25 ± 2 °C/60 ± 5% RH	Good	No	5.9	96.33%
	40 ± 2 °C/75 ± 5% RH	Good	No	5.9	96.41%
30 days	25 ± 2 °C/60 ± 5% RH	Good	No	5.8	95.18%
	40 ± 2 °C/75 ± 5% RH	Good	No	6.0	95.12%
60 days	25 ± 2 °C/60 ± 5% RH	Good	No	6.1	95.89%
	40 ± 2 °C/75 ± 5% RH	Good	No	6.0	95.93%

2.12. Discussion

For the development of nanoemulsion, PEG 400 and Tween 20 were selected as the surfactants and co-surfactants, respectively, as they exhibited good solubility with CO. Different Smix ratios and oil:Smix ratios were used for the development of pseudo-ternary phase diagrams. Only four oil:Smix ratios with an Smix component of 1:1 showed nanoemulsion formation, namely, 1:5, 1:6, 1:7, and 1:8. The phase diagrams were compared in order to identify a maximum nanoemulsion region to select the optimized formulation, which was found to be the 1:5 oil to Smix ratio. The selected formulations were further screened using thermodynamic stability testing to choose the most stable formulation and eliminate metastable formulations. The optimized formulation (1:5) was further subjected to the characterization of various parameters. The particle size, PdI, and zeta potential were found to be satisfactory. The nanoemulsion of droplet size 22 nm was further confirmed with TEM studies. A chamomile oil-excipient compatibility study is a crucial step to decide the long term stability of the formulation. From the FTIR spectra, it is evident that the characteristic absorption peaks of chamomile oil were also observed in nanoemulsion, such as 2925.55 cm^{-1} (-CH2- aliphatic asymmetric), 1732.42 cm^{-1} (C=O), 1457.14 cm^{-1} (=C-H scissor), and 1355.23 cm^{-1} (C-O) [13], which strongly suggest that the chemical structure of chamomile oil is not altered and thus reveals the absence of any interaction.

The next step in formulation development was the preparation of nanoemulsion-based gel, i.e., nanogel. Carbopol 940 was selected as the gelling agent for the preparation of gel because of its excellent gel-forming property at low concentrations. Additionally, it is biodegradable, biocompatible, and non-toxic to the human body [14]. Nanoemulsions as such are not suitable for topical application because of their higher flowability and poor viscosity, which may also affect the contact time of the formulation with the skin. The conversion of a nanoemulsion into gel would improve the viscosity, consistency, and applicability of the formulation at the site of application. This would also improve the contact time and the overall penetration of the nanoemulsion formulation into the skin and ultimately improve the efficacy of the formulation. A 0.5% *w/w* Carbopol gel was incorporated into the nanoemulsion at a ratio of 1:1. The viscosity analysis of the developed nanogel has shown shear thinning behavior, which is favorable. A viscous gel-like consistency on rubbing action will reduce its viscosity, increase spreadability, and thus enhanced penetration can be expected which is proven by skin penetration studies. The nanosize of the carrier system coupled with the skin penetration properties of nanoemulsion components and Carbopol gel have facilitated enhanced penetration through the skin barrier. Moreover, the sustained release of CO from the nanoemulsion gel will enable the longer residence of CO in the skin.

Furthermore, the pharmacological activity of the developed formulation was tested against migraine. The tail flick method was used as an indication of centrally acting analgesic activity [15]. Since migraine is associated with throbbing pain, a thermally induced hyperalgesia method was used to monitor the protective effects of the chamomile formulation. It can be seen that, in comparison to the control, a significant delay in the reaction time of with drawing the tail was noted for both standard and test formulations. This is suggestive of the anti-nociceptive property of chamomile.

Acetic acid is known to induce pain via the release of chemicals that trigger nociceptors, characterized by episodes of retraction of the abdomen and the stretching of hind limbs. The signal is transmitted to the central nervous system, which further causes the release of prostaglandin and contributes to the increased sensitivity to nociceptors [16]. This model is used for detecting the peripheral analgesic activity. A significantly reduced number of writhings observed after the application of the chamomile formulation is suggestive of analgesic activity. It is known that migraine patients are often hypersensitive to light; therefore, a light–dark box model was a requisite model for studying the anti-migraine properties of the developed formulation. Nitroglycerine, a known nitric oxide donor, evokes the trigeminovascular system and develops hyperalgesia and photophobia, which are characteristic of migraine [17]. During the study, it was found that pre-treatment with chamomile formulation reduced photophobia as compared to other groups, proving that chamomile suppresses nitric oxide, which is responsible for the inflammation and induction of migraine. Our results confirm previous findings that chamomile possesses antimigraine potential through the inhibition of nitric oxide release [7,18].

Further, chamomile also increased the tolerance to pain via both central and peripheral mechanisms. Since migraine is associated with both peripheral and centrally acting sensitizations [19], topical chamomile nanogel formulation can be effectively used for the treatment of migraine. However, a major drawback of chamomile oil is skin sensitization potential. Therefore, the oil was encapsulated in a nanogel formulation. Skin irritation potential was tested using a reconstituted human epidermal skin model, as per the OECD TG 439 protocol. The principle for this test is an MTT assay, where viable cells convert MTT into blue formazan salt [20]. Therefore, the higher the color intensity at 570 nm, the higher the viability of cells. The results of the test conclude that the formulation is safe for topical use.

3. Conclusions

The chamomile oil-loaded nanogel formulation was successfully developed with a uniform droplet size of less than 100 nm. The release studies show that the incorporation of chamomile oil intonanoemulsion has improved aqueous solubilization and restricted release when converted into a gel. This has provided a sustained release of chamomile oil. Increased skin penetration was observed due to the nanosize of the droplets coupled with the presence of skin permeation ingredients in the formulation. The shear thinning properties of the gel facilitated better dermal penetration due to the rubbing action, which reduced the viscosity. Finally, the in vivo studies revealed significant anti-nociceptive properties and an increased pain threshold for the developed formulation, suggesting analgesic activity. It was also observed that the formulation reduced photophobia in nitrolycerine-induced migraine models, which suggests thatprevents migraine by inhibiting the release of nitric oxide. However, further studies are still needed to confirm anti-migraine properties. Nonetheless, the studies conclude that the developed topical chamomile oil nanogel is safe for application to the forehead and can be used for analgesia in migraine.

4. Materials and Methods

4.1. Materials

Chamomile oil was procured from Vaadi Herbals, Pvt. Ltd. (Delhi, India) Apigenin, acetic acid, chloroform, toluene, and Tween 20 were procured from Sigma-Aldrich (St. Louis, MO, USA). Carbopol 940, Polyethylene glycol (PEG 400), triethanolamine, and methanol were procured from LobaChemie, Pvt. Ltd. (Mumbai, India). The dialysis membrane (size: 14,000 Da; diameter 17.5 mm) was procured from Hi-Media Laboratories, Pvt. Ltd. (Kennett Square, PA, USA).

4.2. Method of Preparation

4.2.1. Selection of the Nanoemulsion-Forming Zone viaPseudo-Ternary Phase Diagrams

For the pseudo-ternary phase diagrams, surfactant/co-surfactant mixtures (Smix) of Tween 80/span 80, Cremophor® RH40/span 80, and PEG 400/Tween 20 at different ratios (1:1, 1:2, 2:1, 1:3, and 3:1) were prepared. These pseudo-ternary phase diagrams were composed of fixed ratios of chamomile oil and Smix, namely, 1:9, 1:8, 1:7, 1:6, 1:5, 2:8 (1:4), 1:5, 1:3, 1:2, 3:7 (1:2.3), 4:6 (1:1.5), 5:5 (1:1), 6:4 (1:0.7), 7:3 (1:0.43), 8:2 (1:0.25), and 9:1 (1:0.1) with water. The mixtures of the Smix ratio were selected based on their combined HLB values. According to the method, CO and Smix were mixed together at room temperature to obtain the organic phase. Then, the water was added dropwise to the organic phase using a vortex shaker and was visualized against a light and dark background and observations were carried out for transparent formulation [21]. The results were plotted in pseudo-ternary phase diagrams using PCP-triangular software.

4.2.2. Preparation of Nanoemulsion

Pseudo-ternary phase diagrams that resulted in a maximum nanoemulsion formation zone were selected for the preparation of nanoemulsions. The composition of nanoemulsions in terms of CO:Smix:water ratios were selected from the nanoemulsion forming zones obtained from the pseudo-ternary phase diagrams. All the CON formulations were prepared via the aqueous titration method [21].

4.2.3. Thermodynamic Stability Studies of Developed Nanoemulsions

The thermodynamic stability testing of the developed CONs was carried out via aheating–cooling cycle, centrifugation, and a freeze–thaw cycle [21].

(1) Heating–cooling cycle: The samples were subjected to six storage cycles at alternate temperatures, i.e., refrigerator temperatures of 4 °C and at 45 °C, for 48 h each, and the formulated CONs were examined for stability (transparent with no phase separation).

(2) Centrifugation test: The formulated CON was centrifuged at 3500 rpm for 30 min and observed for transparency and the absence of phase separation.

(3) Freeze–thaw cycle: Three freeze–thaw cycles of the CONs between −21 °C and +25 °C for 48 h were performed and observed for transparency and the absence of phase separation.

4.2.4. Characterization of Optimized Nanoemulsion Formulations

Particle Size, Zeta Potential, and Polydispersity Index

The particle size, zeta potential, and polydispersity index (PDI) of samples were evaluated by the zetasizer apparatus (Malvern zeta sizer; Nano-ZS90). The evaluation was carried out at 25 °C at an angle of 90°. The droplet size isshownas the z-average diameter (d.nm) and the particle size distribution wasevaluated via thepolydispersity index (PDI). Samples were analyzed in triplicate [22].

Transmission Electron Microscopy (TEM)

For morphological aspects, CON formulations were further studied using transmission electron microscopy (TEM). The samples were placed on a carbon-coated grid and then stained with 1% phosphotungstic acid, following which theywereleft at room temperature for drying. Images at different magnifications under TEM were observed [23].

Compatibility Studies

The compatibility between CO and other components of the nanoemulsion was studied viaFT-IR spectroscopy (Agilent Technologies Cary 630). The spectrum of CO and CON was recorded as being inthe region of 4000 to 400 cm^{-1} [24].

4.2.5. Preparation of the Nanogel

A total of 0.5% *w*/*w* Carbopol 940 gel was prepared viasoaking in distilled water and was left overnight. Later, 1–3 drops of triethanolamine were added throughstirring, and then, the prepared nanoemulsion was incorporated ata ratio of 1:1. Methylparaben sodium (0.2% *w*/*w*) was also added via constant stirring [22].

4.2.6. Characterization and Evaluation of Nanoemulsion-Based Chamomile Nanogel

pH Measurement

The pH of COG was determined using a calibrated digital pH meter.

Viscosity

The viscosity was measured by a Rheolab QC rheometer using Rheoplus/32, v 3.61 software. The viscosity of the developed COG formulations was measured using a shear rate ranging from 1 to 100 s^{-1} for 3 min using 5 g of gel at 25 ± 2 °C. The shear strain exerted by the formulations due to the application of shear stress can be calculated using the following equation [25].

$$\text{Viscosity} = \text{Shear stress} / \text{Shear strain}$$

Spreadability

A sample of 5 g of developed COG was pressed between two slides and left for 5 min. The diameters of spread circles were measured in cm and the spreadability was calculated using the following formula:

$$S = ML/T$$

where S indicates the spreadability (g·cm/s), M indicates the mass (g), L indicates the length (cm), and T indicates the time (s) [26].

Extrudability

The developed COG formulations were placedinto clean collapsible tubes with 5 mm openings. These were then sealed with the help of heat. A constant load of 500 g was placed on the tube to release the gel from the tube. The amount of the extruded gel was collected, weighed, and the extrudability percentagewas calculated [26].

Texture Analysis

Texture profile analysis was carried out using a texture analyzer. The textural/mechanical properties of different COG formulations were measured using a TA.XT PLUS texture analyzer (Stable Microsystem, Surrey, UK) in texture profile analysis mode. This helps in the determination of textural parameters, such as firmness, consistency, and cohesiveness [27].

4.2.7. Analytical Method Development

Pre-coated HPTLC plates (silica gel 60 F254) of size 20 × 20 cm were used for chromatographic development. The twin trough glass chamber (CAMAG) was pre-saturated with a mobile phase consisting of toluene, ethyl acetate, and formic acid at a ratio of 4.5:3.5:0.2 (*v*/*v*/*v*) for 20 min to ensure the uniform distribution of solvent vapors. Samples were applied using a CAMAG Linomat V applicator viaa microliter syringe, witha capacity of 100 µL as narrow bands of 6 mm in width. The chromatogram was developed to up to 80% of plate height through the linear ascending development technique at room temperature (25 ± 2 °C) and a relative humidity of 55 ± 5%. The plate was removed, air-dried, and then derivatized with anisaldehyde sulfuric acid spraying reagent. The prepared chromatogram was dried in an oven at 60 °C for 5 min. Densitometric analysis was performed at 570 nm with a Camag TLC scanner III. The slit dimensions were 5 mm × 0.45 mm and the scanning speed of 20 mm s^{-1}. Different volumes of stock solution 2, 4, 6, 8, and 10 µL were spotted in triplicate on a TLC plate to obtain concentrations of 10, 20, 30, 40, and 50 ng per spot of apigenin. The data of peak areas and corresponding concentrations were treated via linear least-square regression analysis [28,29].

4.2.8. In Vitro Studies

The in vitro release of the developed COG and CON was performed using a dialysis membrane. A beaker placed on a magnetic stirrer was filled with phosphate buffer of pH 6 (150 mL) and kept at 37 ± 1 °C. A 1 g gel sample was placed in a dialysis membrane bag. A total of 1 mL of each sample was removedat 0.5, 1, 2, 4, 6, 8, and 24 h and at the same time replaced with an equal volume of dissolution medium. All samples were analyzed for drug content by using HPTLC at a wavelength of 570 nm [30].

4.2.9. Drug-Release Kinetics

The release kinetics of the developed COG was obtained from the in vitro drug-release data plotted in various kinetic models using the DD Solver add-in program. The drug-release kinetics were studied for various models and plots, such as zero-order, first-order, Higuchi, Korsmeyer–Peppas, and Hixson–Crowell [30].

4.2.10. Skin Penetration Study

The skin penetration study was carried out as per a previously reported procedure with slight modifications [31]. The study was performed using a Franz diffusion cell with a diffusion area of 3.3 cm^2 and a volume of 60 mL using the dorsal skin of Wistar rats. PBS of pH 7.4 was used as dissolution medium, at a temperature of 37 °C and a stirring speed of 300 rpm. The excised skin sample was sandwiched between the donor and receptor compartments. The formulation (1 g) was placed on the donor area and the samples were withdrawn at pre-determined time intervals and analyzed using HPTLC.

4.2.11. Animal Studies

Adult albino Wistar rats of either sex weighing 200–250 g were selected for the study. The rats were acclimatized for a week before the initiation of the experiment. The rats were kept under laboratory conditions, i.e., temperatures of 25 ± 2 °C, relative humidity of 45 ± 5%, and a photoperiod of 12 h. For the topical administration of nanogel, the hair of the interscapular region of rats (3 cm^2) was removed using hair removal cream. After the application of COG to the shaved region, the area was protected using a nylon mesh.

Tail Flick Test

The rats were divided into three groups, with six rats in each group, and administered the following respective treatments: the control group received only normal saline (1 mL/kg I.P.), the standard group received diclofenac gel (5%), and the test group received topical COG (100 mg/rat). The test was performed with a digital tail-flick instrument that allowed the automatic recording of the latency of the tail-flick response to radiant heat [32]. The number of seconds elapsing between the activation of the heat source and the rat flicking its tail away (latency) was recorded. To minimize tissue damage, a maximum latency of 15 s was imposed. The test was performed before and 30 min after treatment. The inhibition percentagein analgesia was calculated using the following formula:

$$\text{Percentinhibition} = \frac{\text{Posttreatmentlatency} - \text{pretreatmentlatency}}{\text{cutofftime} - \text{pretreatmentlatency}} \times 100$$

Acetic Acid-Induced Writhing Test

The analgesic activity of formulation was evaluated using the acetic acid-induced writhing method [33]. For the study, twenty four Wistar rats were randomly divided into three groups and administered the respective treatments: the control group was administered normal saline (1 mL/kg I.P.), the standard group was administered diclofenac gel (5%), and the test group was administered topical COG (100 mg/rat). After 30 min, writhing was generated via acetic acid administration 0.3% (10 mL/kg I.P.). After 5 min, the number of writhings was counted for a total period of 10 min. The protection percentage was calculated using the formula:

$$\text{Percentprotection} = \frac{\text{Numberofwrithingsincontrol} - \text{numberofwrithingsintest}}{\text{numberofwrithingsincontrol}} \times 100$$

Light/Dark Box Model

There are two compartments in a rectangular box, light and dark. The light compartment was illuminated with a bulb and the top was left uncovered and the dark compartment was painted completely black and covered on all sides. A gate between the two chambers allowed the easy movement of rats between the two compartments. The rats were divided into three groups and the respective treatments were administered: the negative control group received only normal saline I.P., the positive control group received nitroglycerine i.v. (10 mg/kg), and the test groupprecieved topical COG (100 mg/rat). After 30 min, intravenous nitroglycerine (10 mg/kg) was administered to rats to induce migraine [34]. The animals were then placed in the middle of the two chambers and observed in order to measure the length of time they remained in the lit compartment. The time spent was calculated usingthe ratio of time spent by the rat in the lit compartment to the total period for which was observed.Furthermore, the number of transitions between the two compartments during the experiment was also recorded [35].

4.2.12. Skin Irritation Test

The irritation potential of the COG topical formulation was evaluated by an in vitro skin irritation test. A previously reported method was followed [31], where artificial human epidermal cells were incubated as per the procedure. The test was performed in three groups comprising the negative control (using phosphate-buffered saline (PBS)); the positive control (using 5% SDS and CO) and the test group (using chamomile nanogel). The treatments were added to epidermal cells and incubated for 42 min at room temperature. They were then rinsed with PBS and transferred to a growth medium for incubation at 42 h, 37 °C, 5% CO_2, and $\geq$95% humidity. The tissues were then placed in MTT solution for 3 h, 37 °C, rinsed with 300µL PBS, and kept in 750 µL isopropanol and left overnight. The tissues were then transferred to 96-well plate and OD values were obtained at 570 nm. The mean percent viability was then calculated.

4.2.13. Stability Study

The modified ICH guidelines for stability studies were followed [36].The optimized formulations of COG were stored at different temperatures and humidity conditions and assessed for appearance, phase separation, pH, and transmittance percentage.

Author Contributions: Conceptualization, F.Z., G.A., G.K.J. and S.M.A.Z.; methodology, P.P.N.; formal analysis, P.P.N. and M.S.K.; investigation, A.S.; data curation, F.Z. and A.S.; writing—original draft preparation, S.M.A.Z.; writing—review and editing, F.Z.; supervision, F.Z., G.A. and G.K.J.; project administration, G.A.; funding acquisition, P.P.N. and M.S.K. All authors have read and agreed to the published version of the manuscript.

Funding: The authors extend their appreciation to the Deanship of Scientific Research at King Khalid University, Saudi Arabia, for funding this work through the Research Group Program under Grant No: RGP 2/348/44.

Institutional Review Board Statement: The animal study protocol was approved by the Institutional Ethics Committee of Delhi Pharmaceutical Sciences and Research University (protocol code IAEC/2022/I-33 and date of approval, 9 March 2022)and conducted according to the Committee for the Purpose of Control and Supervision of Experiments on Animals CPCSEA (Govt. of India), New Delhi, guidelines.

Informed Consent Statement: Not applicable.

Data Availability Statement: The research data will be supplied on request.

Conflicts of Interest: The authors declare no conflict of interest.

References

1. Rizzoli, P. *Pain Medicine: An Essential Review*; Yong, R., Nguyen, M., Nelson, E., Urman, R., Eds.; Springer: Cham, Switzerland, 2017; Chapter 138, p. 521.
2. Davenport, W.J.; Pringsheim, T.M. A 32-year-old woman with headache. *CMAJ* **2016**, *188*, E533–E534. [CrossRef] [PubMed]
3. Teleanu, R.I.; Vladacenco, O.; Teleanu, D.M.; Epure, D.A. Treatment of Pediatric Migraine: A Review. *Maedica* **2016**, *11*, 136–143.
4. González-Hernández, A.; Marichal-Cancino, B.A.; MaassenVanDenBrink, A.; Villalon, C.M. Side effects Associated with Current and Prospective Antimigraine Pharmacotherapies. *Expert Opin. Drug Metab. Toxicol.* **2018**, *14*, 25–41. [CrossRef] [PubMed]
5. Teja, P.K.; Mithiya, J.; Kate, A.S.; Bairwa, K.; Chauthe, S.K. Herbal Nanomedicines: Recent Advancements, Challenges, Opportunities and Regulatory Overview. *Phytomedicine* **2022**, *96*, 153890. [CrossRef]
6. Sah, A.; Naseef, P.P.; Kuruniyan, M.S.; Jain, G.K.; Zakir, F.; Aggarwal, G. A Comprehensive Study of Therapeutic Applications of Chamomile. *Pharmaceuticals* **2022**, *15*, 1284. [CrossRef] [PubMed]
7. Zargaran, A.; Borhani-Haghighi, A.; Faridi, P.; Daneshamouz, S.; Kordafshari, G.; Mohagheghzadeh, A. Potential Effect and Mechanism of Action of Topical Chamomile (*Matricaria chammomila* L.) oil on Migraine Headache: A Medical Hypothesis. *Med. Hypotheses* **2014**, *83*, 566–569. [CrossRef]

8. Kiraly, A.J.; Soliman, E.; Jenkins, A.; Van Dross, R.T. Apigenin inhibits COX-2, PGE2, and EP1 and also initiates terminal differentiation in the epidermis of tumor bearing mice. *Prostaglandins Leukot. Essent. Fat. Acids* **2016**, *104*, 44–53. [CrossRef] [PubMed]
9. Taheri, P.; Mohammadi, F.; Nazeri, M.; Zarei, M.R.; Chamani, G.; Esfahlani, M.A.; Taheri, F.; Shabani, M. Nitric oxide role in anxiety-like behavior, memory and cognitive impairments in animal model of chronic migraine. *Heliyon* **2020**, *6*, e05654. [CrossRef] [PubMed]
10. Wilson, R.J.; Li, Y.; Yang, G.; Zhao, C.X. Nanoemulsions for drug delivery. *Particuology* **2022**, *64*, 85–97. [CrossRef]
11. Ribeiro, L.N.M.; Da Silva, G.H.R.; Couto, V.M.; Castro, S.R.; Breitkreitz, M.C.; Martinez, C.S.; Igartúa, D.E.; Prieto, M.J.; De Paula, E. Functional Hybrid Nanoemulsions for Sumatriptan Intranasal Delivery. *Front. Chem.* **2020**, *8*, 589503. [CrossRef]
12. Algahtani, M.S.; Ahmad, M.Z.; Ahmad, J. Nanoemulsion loaded Polymeric Hydrogel for Topical Delivery of Curcumin in Psoriasis. *J. Drug Deliv. Sci. Technol.* **2020**, *59*, 101847. [CrossRef]
13. Maslowski, M.; Aleksieiev, A.; Miedzianowska, J.; Strzelec, K. Potential Application of Peppermint (*Mentha piperita* L.), German chamomile (*Matricaria chamomilla* L.) and Yarrow (*Achillea millefolium* L.) as Active Fillers in Natural Rubber Biocomposites. *Int. J. Mol. Sci.* **2021**, *22*, 7530. [CrossRef] [PubMed]
14. Zheng, Y.; Ouyang, W.-Q.; Wei, Y.-P.; Syed, S.F.; Hao, C.-S.; Wang, B.-Z.; Shang, Y.-H. Effects of Carbopol® 934 proportion on Nanoemulsion gel for Topical and Transdermal Drug Delivery: A Skin Permeation Study. *Int. J. Nanomed.* **2016**, *11*, 5971–5987. [CrossRef]
15. Hole, K.; Tjolsen, A. Tail flick test. In *Enclyopedia of Pain*; Schmidt, R., Willis, W., Eds.; Springer: Berlin, Germany, 2007; pp. 2392–2395.
16. Gupta, A.K.; Parasar, D.; Sagar, A.; Choudhary, V.; Chopra, B.S.; Garg, R.; Ashish; Khatri, N. Analgesic and Anti-inflammatory Properties of Gelsolin in Acetic acid induced Writhing, Tail immersion and Carrageenan induced paw edema in mice. *PLoS ONE* **2015**, *10*, e0135558. [CrossRef]
17. Liao, C.C.; Li, J.M.; Chen, C.H.; Lin, C.L.; Hsieh, C.L. Effect of Paeonia lactiflora, a Traditional Chinese herb, on Migraines based on Clinical Application and Animal Behavior Analyses. *Biomed. Pharmacother.* **2019**, *118*, 109276. [CrossRef]
18. Bhaskaran, N.; Shukla, S.; Srivastava, J.K.; Gupta, S. Chamomile: An Anti-inflammatory agent Inhibits inducible Nitric Oxide Synthase Expression by Blocking RelA/p65 Activity. *Int. J. Mol. Med.* **2010**, *26*, 935–940. [PubMed]
19. Tajti, J.; Vecsei, L. The Mechanism of Peripheral and Central Sensitization in Migraine. A Literature Review. *Neuropsychopharmacol. Hung.* **2009**, *11*, 15–21.
20. Riss, T.L.; Moravec, R.A.; Niles, A.L.; Duellman, S.; Benink, H.A.; Worzella, T.J.; Minor, L. Cell viability assay. In *Assay Guidance Manual*; Riss, T.L., Minor, L., Eds.; NCBI: Bethesda, MD, USA, 2016; pp. 1–41.
21. Zakir, F.; Ahmad, A.; Farooq, U.; Mirza, M.A.; Tripathi, A.; Singh, D.; Shakeel, F.; Mohapatra, S.; Ahmad, F.J.; Kohli, K. Design and Development of a Commercially viable in situ Nanoemulgel for the Treatment of Postmenopausal Osteoporosis. *Nanomedicine* **2020**, *15*, 1167–1187. [CrossRef] [PubMed]
22. Nawaz, A.; Latif, M.S.; Alnuwaiser, M.A.; Ullah, S.; Iqbal, M.; Alfatama, M.; Lim, V. Synthesis and Characterization of Chitosan-Decorated Nanoemulsion gel of 5-Fluorouracil for Topical Delivery. *Gels* **2022**, *8*, 412. [CrossRef]
23. Sakeena, M.H.; Muthanna, F.A.; Ghassan, Z.A.; Kanakal, M.M.; Elrashid, S.M.; Munavvar, A.S. Formulation and In vitro Evaluation of Ketoprofen in Palm oil Esters Nanoemulsion for Topical Delivery. *J. Oleo Sci.* **2010**, *59*, 223–228. [CrossRef]
24. Harjot, K.; John, N.A.M. Nanoemulsion for Migraine Prophylaxis Nasal Drug Delivery: Preparation, Characterization and In vitro evaluation. *Pharm. Nanotechnol.* **2016**, *4*, 229–241. [CrossRef] [PubMed]
25. Narawi, M.M.; Chiu, H.I.; Yong, Y.K.; Zain, N.N.M.; Ramachandran, M.R.; Tham, C.L.; Samsurrijal, S.F.; Lim, V. Biocompatible Nutmeg oil-Loaded Nanoemulsion as Phyto-repellent. *Front. Pharmacol.* **2020**, *11*, 214. [CrossRef] [PubMed]
26. Latif, M.S.; Nawaz, A.; Asmari, M.; Uddin, J.; Ullah, H.; Ahmad, S. Formulation Development and In vitro/in vivo Characterization of Methotrexate-Loaded Nanoemulsion Gel Formulations for Enhanced Topical Delivery. *Gels* **2022**, *9*, 3. [CrossRef]
27. Ahmed, S.; Gull, A.; Aqil, M.; Ansari, M.D.; Sultana, Y. Poloxamer-407 Thickened Lipid Colloidal System of Agomelatine for Brain Targeting: Characterization, Brain Pharmacokinetic Study and Behavioral Study on Wistar rats. *Colloids Surf. B Biointerfaces* **2019**, *181*, 426–436. [CrossRef]
28. Guzelmeric, E.; Vovk, I.; Yesilada, E. Development and Validation of an HPTLC method for Apigenin 7-O-glucoside in Chamomile Flowers and its Application for Fingerprint Discrimination of Chamomile-like Materials. *J. Pharm. Biomed. Anal.* **2015**, *107*, 108–118. [CrossRef]
29. Al Bratty, M.; Govindaram, L.K.; Thangavel, N.; Alhazmi, H.A.; Ibrahim, A.M.M.; Maruthamuthu, V.; Kandasamy, R. High Performance Thin-Layer Chromatography and In vitro Cytotoxic Studies on Ethanol Extract of *Matricaria chamomilla* L. (Asteraceae) Flowers. *Trop. J. Pharm. Res.* **2019**, *18*, 1969–1976. [CrossRef]
30. Rani, K.R.V.; Rajan, S.; Bhupathyraaj, M.; Priya, R.K.; Halligudi, N.; Al-Ghazali, M.A.; Sridhar, S.B.; Shareef, J.; Thomas, S.; Desai, S.M.; et al. The Effect of Polymers on Drug Release Kinetics in Nanoemulsion in situ Gel Formulation. *Polymers* **2022**, *14*, 427. [CrossRef] [PubMed]
31. Zakir, F.; Ahmad, A.; Mirza, M.A.; Kohli, K.; Ahmad, F.J. Exploration of a Transdermal Nanoemulgel as an Alternative Therapy for Postmenopausal Osteoporosis. *J. Drug Deliv. Sci. Technol.* **2021**, *65*, 102745. [CrossRef]
32. Meymandi, M.S.; Sepehri, G.; Izadi, G.; Zamiri, Z. Evidence for Antinociceptive Effects of Combined Administration of Vitamin E and Celecoxib in Tail-flick and Formalin test in Male rats. *Pharmacol. Rep.* **2019**, *71*, 457–464. [CrossRef]

33. Feng, L.; Cui, M.; Willis, W.D. Gabapentin Markedly Reduces Acetic Acid-induced Visceral Nociception. *Anesthesiology* **2003**, *98*, 729–733. [CrossRef]
34. Casili, G.; Lanza, M.; Filippone, A.; Campolo, M.; Paterniti, I.; Cuzzocrea, S.; Esposito, E. Dimethyl Fumarate Alleviates the Nitroglycerin (NTG)-induced Migraine in mice. *J. Neuroinflammation* **2020**, *17*, 59. [CrossRef] [PubMed]
35. Vuralli, D.; Wattiez, A.S.; Russo, A.F.; Bolay, H. Behavioral and Cognitive Animal Models in Headache Research. *J. Headache Pain* **2019**, *20*, 11. [CrossRef] [PubMed]
36. Indrati, O.; Martien, R.; Rohman, A.; Nugroho, A.K. Development of Nanoemulsion-Based Hydrogel containing Andrographolide: Physical Properties and Stability Evaluation. *J. Pharm. Bioallied Sci.* **2020**, *12* (Suppl. 2), S816–S820. [PubMed]

gels

MDPI

Article

Piperine-Loaded In Situ Gel: Formulation, In Vitro Characterization, and Clinical Evaluation against Periodontitis

Poornima K. Gopalakrishna [1], Rajamma Abburu Jayaramu [2], Sateesha Shivally Boregowda [1,*], Shruthi Eshwar [3], Nikhil V. Suresh [3], Amr Selim Abu Lila [4,5], Afrasim Moin [5], Hadil Faris Alotaibi [6], Ahmad J. Obaidullah [7] and El-Sayed Khafagy [8,9]

1 Department of Pharmaceutics, Acharya & BM Reddy College of Pharmacy, Bengaluru 560107, India; poornimakg98@gmail.com
2 Department of Pharmacognosy, KLE College of Pharmacy, Bengaluru 560010, India; abburjayaramu6@gmail.com
3 KLE Society's Institute of Dental Sciences, Bengaluru 560022, India; shruthy2015@gmail.com (S.E.); sureshnikhil596@gmail.com (N.V.S.)
4 Department of Pharmaceutics and Industrial Pharmacy, Faculty of Pharmacy, Zagazig University, Zagazig 44519, Egypt; a.abulila@uoh.edu.sa
5 Department of Pharmaceutics, College of Pharmacy, University of Hail, Hail 81442, Saudi Arabia; a.moinuddin@uoh.edu.sa
6 Department of Pharmaceutical Sciences, College of Pharmacy, Princess Nourah Bint AbdulRahman University, Riyadh 11671, Saudi Arabia; hfalotaibi@pnu.edu.sa
7 Department of Pharmaceutical Chemistry, College of Pharmacy, King Saud University, Riyadh 11451, Saudi Arabia; aobaidullah@ksu.edu.sa
8 Department of Pharmaceutics, College of Pharmacy, Prince Sattam Bin Abdulaziz University, Al-kharj 11942, Saudi Arabia; e.khafagy@psau.edu.sa
9 Department of Pharmaceutics and Industrial Pharmacy, Faculty of Pharmacy, Suez Canal University, Ismailia 41522, Egypt
* Correspondence: sbsateesh@gmail.com

Citation: Gopalakrishna, P.K.; Jayaramu, R.A.; Boregowda, S.S.; Eshwar, S.; Suresh, N.V.; Abu Lila, A.S.; Moin, A.; Alotaibi, H.F.; Obaidullah, A.J.; Khafagy, E.-S. Piperine-Loaded In Situ Gel: Formulation, In Vitro Characterization, and Clinical Evaluation against Periodontitis. *Gels* **2023**, *9*, 577. https://doi.org/10.3390/gels9070577

Academic Editors: Ying Huang, Zhengwei Huang and Xuanjuan Zhang

Received: 21 June 2023
Revised: 12 July 2023
Accepted: 12 July 2023
Published: 14 July 2023

Abstract: Periodontitis is an inflammatory disorder associated with dysbiosis and characterized by microbiologically related, host-mediated inflammation that leads to the damage of periodontal tissues including gingiva, connective tissues, and alveolar bone. The aim of this study was to develop an in situ gel consisting of piperine. Eight in situ gel formulations were designed by varying the concentration of deacylated gellan gum cross-linked with sodium tripolyphosphate, and poloxamer-407. The prepared gels were evaluated for gelation temperature, gelation time, viscosity, piperine-loading efficiency, and piperine release. Finally, the optimized formula was evaluated for anti-inflammatory effectiveness among human patients during a 14-day follow-up. The optimized in situ gel formulation exhibited a gelation temperature of 35 ± 1 °C, gelling of 36 ± 1 s, excellent syringeability, and piperine loading of 95.3 ± 2.3%. This formulation efficiently sustained in vitro drug release for up to 72 h. In vivo studies revealed an efficient sol-to-gel transformation of optimized in situ gel formulation at physiological conditions, permitting an efficient residence time of the formulation within a periodontitis pocket. Most importantly, a clinical study revealed that treatment with the optimized formulation elicited a significant reduction in the mean plaque score (p = 0.001), gingival index (p = 0.003), and pocket depth (p = 0.002), and exerted a potent anti-inflammatory potential, compared to the control group. Collectively, piperine-loaded in situ gel might represent a viable therapeutic approach for the management of gingival and periodontal diseases.

Keywords: anti-inflammatory; anti-plaque; in situ gel; periodontitis; piperine

1. Introduction

Periodontitis, also known as gum disease, is a serious gum infection that damages soft tissue surrounding teeth. If left untreated, periodontitis can ruin the bone that supports teeth and may cause other systemic disorders. According to the database reported by Global Burden of Disease (GBD), about 1.1 billion cases of severe periodontitis were reported globally [1]. Periodontitis ranks 77th among the most relevant human conditions which results in disability. Nevertheless, gum problems are resolvable if timely intervention is provided [2].

Generally, periodontitis is associated with dysbiosis [3] and characterized by microbially associated, host-mediated inflammation that leads to the damage of periodontal tissues including gingiva, connective tissues, and alveolar bone [4]. Ultimately, this activates host-derived proteinases, facilitating the apical migration of the junctional epithelium, the loss of marginal periodontal ligament fibers, and the anterior spread of the bacterial biofilm along the root surface [5]. Some techniques, such as guided bone regeneration (GBR), have been adopted for periodontal regeneration via creating a favorable environment for bone regeneration by preventing the migration of soft tissue cells into the defect and allowing space for bone-forming cells to populate the area [6,7]. For instance, periosteum-inspired composite membranes based on sodium alginate-hydroxyapatite nanoparticles [8] and chitosan/bioactive glass nanoparticle composite membranes [9] have been investigated for their potential use in periodontal regeneration. However, GBR represents a complex proliferation process in which different types of cells including osteoblasts, osteocytes, osteoclasts, and bone-lining cells should orchestrate to regulate bone formation.

Commonly, conventional periodontal therapy involves oral hygiene instruction, scaling with root planning, and surgical intervention [10]. Nevertheless, periodontal patients who do not respond to conventional mechanical therapy, or are suffering from acute periodontal infections with systemic manifestations are usually advised to be treated with systemic antibiotics. Various antibiotics, such as tetracyclines [11], moxifloxacin [12], doxycycline [13], etc., have been prescribed for the systemic treatment of periodontitis. Nevertheless, the usage of these drugs was associated with certain drawbacks such as the inability of systemic drugs to attain adequate gingival crevicular fluid concentration [14], the development of drug resistance [15], an increased risk of drug-related adverse effects [16], and questionable patient compliance [17].

Apart from the systemic administration of drugs, localized drug delivery in the form of fibres, strips, films, and micro-particulate systems has been widely investigated for the management of periodontitis [18–20]. Nevertheless, they show limited success in the treatment of periodontitis because of poor patient compliance, poor adhesion properties, and limited retention time [21]. Most importantly, the difficulty in removing such systems after the completion of treatment without causing inflammation and gingival redness adversely limited their widespread usage for the treatment of periodontitis.

Because of their unique biomimetic properties, hydrogels are employed as biomaterials in tissue engineering and regenerative medicine to enhance cell adhesion and promote tissue regeneration. Recently, the use of natural-origin materials, such as collagen, gelatin, polysccahrides, etc., for the formulation of hydrogels has attracted great attention due to their ability to mimic the native tissues' extracellular matrix and their biocompatibility [22]. In situ gelling systems (ISGs) are polymeric viscous hydrogels that, when applied to the human body, undergo a sol-to-gel transition in response to different stimuli such as ionic strength, temperature, or pH [23]. Recently, ISGs have emerged as a promising local drug delivery system for the management of periodontitis because of their ability to attain high drug levels in the gingival crevicular fluid for long periods of time, promoting the achievement of the desired therapeutic effect [24]. In addition, in situ gels are easy to administer and undergo rapid elimination from the cavity through normal catabolic pathways after the complete delivery of the drug [25]. In situ gel can be categorized into three classes according to its phase change performance: pH-sensitive [26], temperature-sensitive [26], and ionic-strength-sensitive [27]. Among them, ion-sensitive materials such as sodium alginate is the most commonly used. However, in recent years, deacetylase gellan gum has gained popularity [27]. Deacetylase gellan gum is an extracellular polysaccharide, produced from *Pseudomonas elodea* bacteria. It exhibits very high gel strength, adjustable gel elasticity, high gel transparency, and good compatibility. Because of its distinct characteristics, it has been employed for oral [28], ocular [29], and nasal administration systems [30].

Piperine is the primary alkaloid isolated from long pepper (*Piper longum*) and black pepper (*Piper nigrum*) [31]. Piperine has versatile biological activities including analgesic, anti-inflammatory, antitumor, anticonvulsant, and neuroprotective activities [32–34]. In addition, piperine has been screened for its antimicrobial activity against various microbes. It has been verified to be effective against *Escherichia coli, Staphylococcus aureus, Pseudomonas aeruginosa,* and *Bacillus subtilis* [35]. Furthermore, piperine has been reported to exert significant protective effects on inflammation, bone and collagen fiber degeneration, and alveolar bone loss in a rat periodontitis model [36].

The aim of this study, therefore, was to develop and optimize piperine-loaded in situ gel and clinically evaluate its potential in the treatment of periodontitis. In situ gel formulations were fabricated using the biodegradable biocompatible polymer, deacylated gellan gum, cross-linked with sodium tripolyphosphate, and poloxamer-407. The formulated gels were physico-chemically characterized, and the optimized formula was eventually evaluated for anti-inflammatory effectiveness among human patients suffering from periodontitis.

2. Results and Discussion

2.1. Pre-Formulation Studies

FTIR spectral analysis was carried out to study piperine compatibility with other formulation excipients. The FTIR spectra of pure piperine, gellan gum, STPP, poloxamer 407, and the physical mixture of the drug with excipients are depicted in Figure 1. The characteristic spectral peaks of N-H stretching at 3214.75 cm^{-1}, C=C aromatic stretching at 1489 cm^{-1}, and aromatic -C-H stretching at 2937.3 cm^{-1} were observed in the spectrum of pure piperine [37]. In the spectrum of the physical mixture of piperine with various excipients, no remarkable changes were observed in the absorption peaks at 3213.7 cm^{-1}, 1491 cm^{-1}, and 2940 cm^{-1}, corresponding to N-H stretching, C=C aromatic stretching, and aromatic -C-H stretching, compared to those of pure piperine. From the FTIR interpretation, it was evident that there was no interaction between the drug and various formulation excipients. Therefore, this study claimed that piperine was compatible with various formulation excipients.

2.2. In Situ Gel Formulation

Gellan gum is an ion-sensitive natural polysaccharide, which consists of double helical segments [38]. These segments can cross-link in the presence of monovalent or divalent cations to form biocompatible hydrogels [39]. However, the ion content of gingival crevicular fluid (GCF) in the periodontal pocket was found to be insufficient to initiate the gelation of the gellan gum. Accordingly, for the formulation of gellan-gum-based in situ gel, sodium tripolyphosphate (STPP) was adopted as a cross-linking agent that helps the formation of a stiff gel at the body temperature. The cross-linking process involves coacervation complexation through electrostatic attraction between negatively charged gellan gum with the positively charged STPP [40,41]. Therefore, various trials were conducted by varying the concentration of gellan gum (from 0.5 to 1% *w/v*) and STPP concentration (from 0.2 to 1% *w/v*), and the consistency of the prepared in situ gel formulations was visually inspected (Table 1). As summarized in Table 1, both gellan gum and STPP concentrations exerted a significant impact on the sol–gel transition of the formed in situ gel formulations. Gellan gum at concentrations of 0.5 to 0.75% *w/v* and STPP at concentrations of 0.2 to 0.4% *w/v* succeeded to produce formulations with a gel-like consistency at 35 ± 2 °C. On the other hand, regardless of STPP concentrations, formulations with gellan gum, at a concentration of 1% *w/v*, were found to be gel even at room temperature (non-physiological condition); accordingly, they were excluded from further investigations.

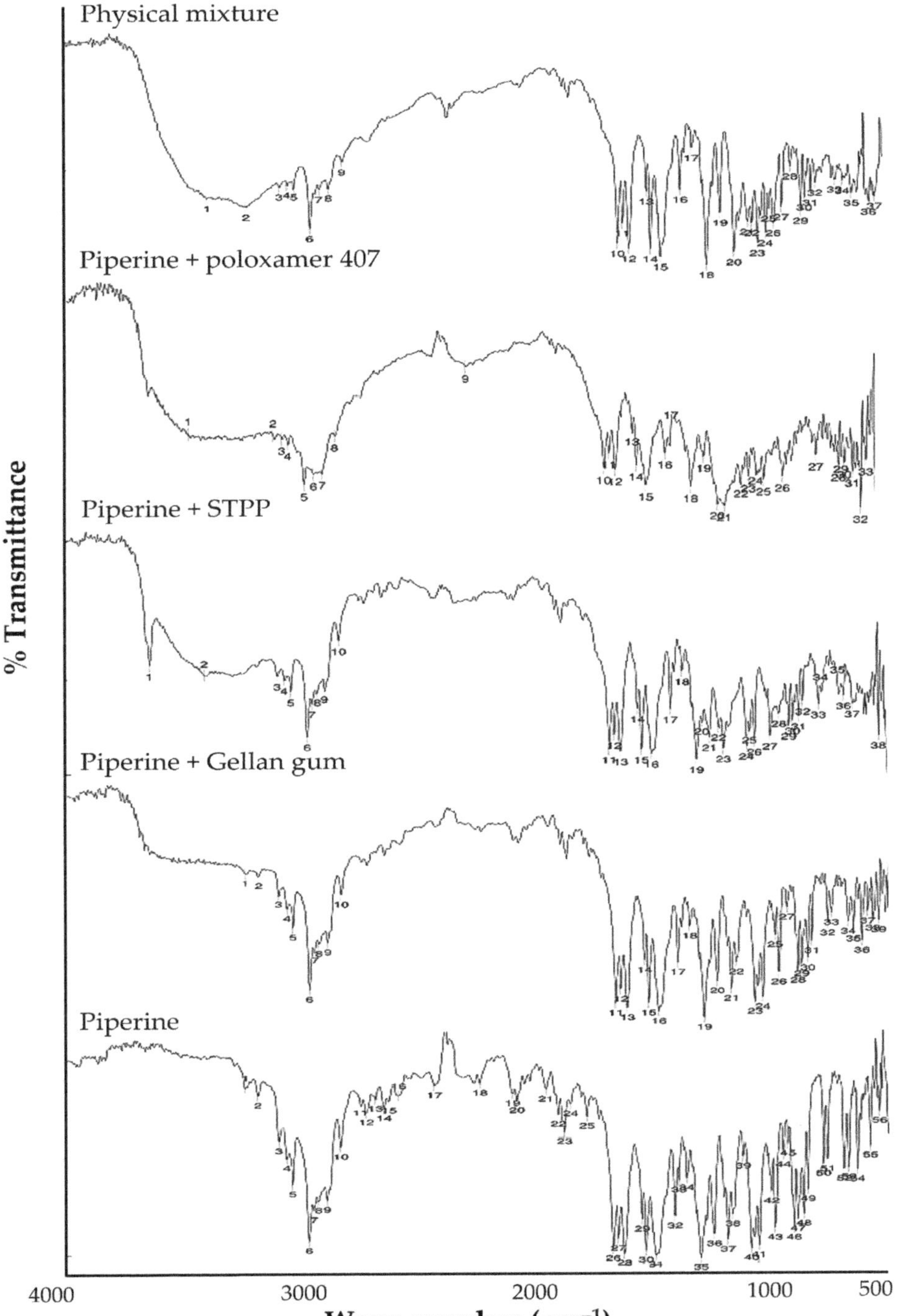

Figure 1. FTIR spectra of pure piperine and different formulation excipients.

Table 1. Polymeric combination of gellan gum and STPP and their consistency.

Trials	Gellan Gum (% *w/v*)	STPP (% *w/v*)	Consistency at 25 ± 2 °C	Consistency at 35 ± 2 °C
1	0.5	0.2	Liquid	Gel
2	0.5	0.4	Liquid	Gel
3	0.5	0.6	Liquid	Liquid
4	0.5	0.8	Liquid	Liquid
5	0.5	1.0	Liquid	Liquid
6	0.75	0.2	Liquid	Gel
7	0.75	0.4	Liquid	Gel
8	0.75	0.6	Liquid	Viscous solution
9	0.75	0.8	Liquid	Viscous solution
10	0.75	1.0	Liquid	Liquid
11	1.0	0.2	Gel	Gel
12	1.0	0.4	Gel	Gel
13	1.0	0.6	Gel	Gel
14	1.0	0.8	Gel	Gel
15	1.0	1.0	Gel	Gel

It is worth noting that physically cross-linked gellan gum hydrogels might lose their stability under physiological conditions, limiting their in vivo use. Consequently, in order to enhance the mechanical strength of the gels, the thermo-responsive polymer, poloxamer 407, was incorporated as a co-polymer during the formulation of in situ gels. Poloxamer 407 was chosen because it has unique reversible thermo-gelling properties [42]. In addition, it has an excellent safety profile and good mucoadhesive properties [43]. Furthermore, gellan gum along with poloxamer 407 can transform into stiff gel by the combined effect of body temperature and the ions available in the periodontal fluid. Accordingly, a total of eight formulations were prepared (Table 2) by incorporating poloxamer 407, at concentrations of 10 and 12% *w/v*, into STPP-cross-linked gellan gum, and, eventually, physico-chemically evaluated for the selection of an optimized formula.

Table 2. Composition and physico-chemical characteristics of piperine-loaded in situ gel formulations.

Formula	Gellan Gum	STPP	Poloxamer 407	Gelation Temperature (°C)	Gelation Time (s)	pH	Syringeability	Drug Content (%)	Viscosity at 25 °C (cps)	Viscosity at 37 °C (cps)
F1	0.5	0.2	10	31.6 ± 0.6	113.0 ± 2.1	7.2 ± 0.1	Pass	62.5 ± 1.7	58.7 ± 1.9	287.9 ± 21.3
F2	0.5	0.2	12	35.3 ± 0.6	50.0 ± 1.5	7.0 ± 0.3	Pass	87.6 ± 2.9	58.8 ± 1.4	243.0 ± 19.8
F3	0.5	0.4	10	34.3 ± 1.2	37.0 ± 1.5	7.6 ± 0.4	Pass	92.2 ± 2.1	59.4 ± 2.1	290.2 ± 22.6
F4	0.5	0.4	12	34.0 ± 1.0	35.0 ± 1.0	7.8 ± 0.1	Pass	82.5 ± 3.6	59.3 ± 1.2	287.0 ± 17.8
F5	0.75	0.2	10	30.6 ± 0.6	63.0 ± 1.2	7.0 ± 0.1	Pass	57.5 ± 1.9	59.2 ± 1.6	257.0 ± 13.4
F6	0.75	0.2	12	35.0 ± 1.0	36.0 ± 1.0	7.4 ± 0.3	Pass	95.3 ± 2.3	58.3 ± 1.7	264.0 ± 16.8
F7	0.75	0.4	10	31.7 ± 0.6	69.0 ± 2.5	7.5 ± 0.2	Pass	95.1 ± 1.8	59.6 ± 1.2	280.0 ± 19.4
F8	0.75	0.4	12	36.0 ± 1.0	43.0 ± 0.6	7.6 ± 0.2	Pass	85.1 ± 3.1	59.0 ± 0.9	263.3 ± 14.8

All data represent mean ± SD (n = 3).

2.3. *Evaluation of In Situ Gel*

2.3.1. Gelation Temperature and Gelation Time

The gelation temperature of in situ gelling systems is one of the critical characteristics that dictate their in vivo applicability. It is ideal for in situ gelling systems to exist in a sol state at room temperature to permit easy drug administration, and then be rapidly converted into a gel state once in the periodontal cavity. Furthermore, these gels should not dissolve but rather remain in a gel state for an extended period of time [44]. Generally, the gelation temperatures of in situ gel formulations have been deemed appropriate if they are within the range of body temperature [45]. The gelation temperatures of different gellan-gum-based in situ gel formulations are represented in Figure 2A and summarized in Table 2. As depicted in Table 2, all formulations (F1–F8) showed an acceptable gelation temperature, which is higher than room temperature and lower than body temperature, ranging from 30.6 ± 0.6 to 37.0 ± 1.0 °C. In addition, it was evident that the gelation temperature of formulations increased with an increase in the concentration of the thermosensitive copolymer poloxamer 407. The gelation temperature of F1 (31.6 ± 0.6 °C), prepared with a poloxamer 407 concentration of 10% *w/v*, was remarkably lower than that prepared with a poloxamer 407 concentration of 12% *w/v* (F2; 35.3 ± 0.6 °C) ($p < 0.05$). Similarly, formulations F6 and F8, containing poloxamer 407 (12% *w/v*), showed excellent gelation compared to F5 and F7, containing poloxamer 407 (10% *w/v*), respectively, due to increasing the concentration of poloxamer 407 for the same gellan gum concentration. However, the gelation temperature of F3 and F4 formulations was not influenced by poloxamer 407 concentration. Similar results were stated by Rencber et al. [46] who verified the positive effect of increasing the concentrations of the thermosensitive polymer, poloxamer 407, on the gelation temperature of gellan-gum-poloxamer-based dexamethasone mucoadhesive in situ gel.

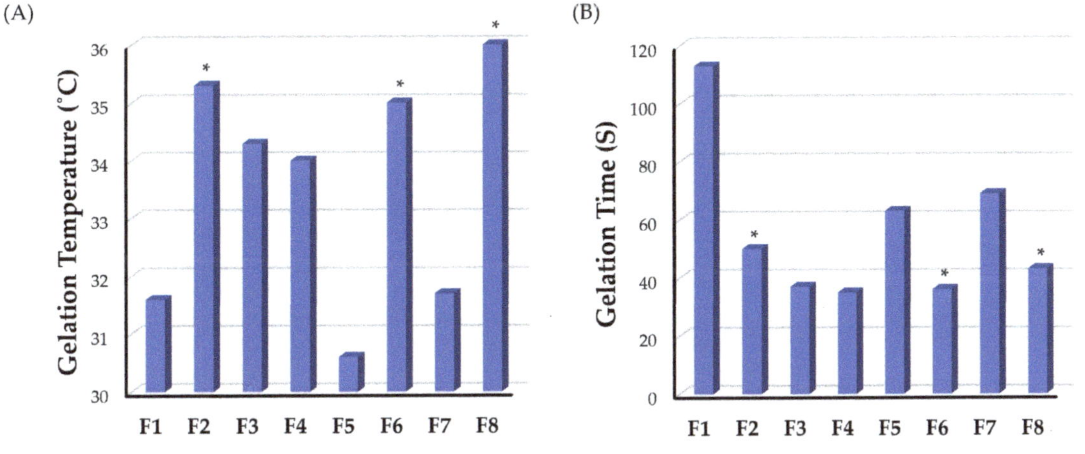

Figure 2. Impact of formulation components on (**A**) gelation temperature and (**B**) gelation time of various piperine-loaded in situ gel formulations. Comparisons were conducted among different formulations having the same gellan gum and STPP concentration. * $p < 0.05$.

Gelation time is another crucial characteristic of in situ gel systems. Generally, a shorter gelation time would be advantageous in minimizing the time necessary for the implanted dosage to turn into a viscous gel. Therefore, this would reduce the chance of the implanted formulation to be rapidly diluted by the gingival crevicular fluid and being lost with salivary secretions. The gelation time of the prepared in situ gel formulations was represented in Figure 2B and Table 2. As illustrated in Table 2, gelation times of various formulations fluctuate between 35.0 ± 1.0 s (F4) to 113.0 ± 2.1 s (F1). In addition, it was inferred that gelation times were dependent on the poloxamer 407 concentration. At the same gellan gum concentrations, increasing the poloxamer 407 concentration from 10 to 12% *w/v* resulted in an obvious reduction of gelation time of different formulations. The gelation time of F2, prepared with 12% poloxamer 407, was 50.0 ± 1.5 s, which was significantly lower than that of F1 (113.0 ± 2.1 s) prepared with 10% poloxamer 407 ($p < 0.05$). Similarly, formulations F4, F6, and F8, containing poloxamer 407 (12% *w/v*), showed remarkably lower gelation time compared to F3, F5, and F7, respectively, containing poloxamer 407 (10% *w/v*), due to increasing the concentration of poloxamer 407 for the same gellan gum concentration. Similar findings were reported for levofloxacin-poloxamer 407 gels, where gelation time were dependent on poloxamer concentrations [47].

2.3.2. Viscosity Measurement

The quantity of the gel that can be introduced into the periodontal pockets is extremely low due to small crevices. Consequently, the viscosity of the in situ gel must be low at the time of application to the periodontal pockets to ease formulation application, but high thereafter in order for the drug to remain for a sufficient amount of time in the disease site [26]. The viscosities of different formulations at both room temperature and at elevated temperatures were measured and compared. As anticipated, due to its thermo-responsiveness, the viscosities of all formulations at room temperature were significantly lower than those at elevated temperatures (37 ± 0.5 °C) (Table 2). The viscosities of test formulations at room temperature (in the sol form) were within the range of 58.3 ± 1.7 cp (F6) to 59.6 ± 1.2 cp (F7), while, at 37 ± 0.5 °C, the viscosities of test formulations fluctuated between 243.0 ± 19.8 cp (F2) and 290.2 ± 22.6 cp (F3). Such significant increase in the formulations' viscosities might be ascribed to the transition of in situ gel systems from the sol state at room temperature to the gel state at 37 ± 0.5 °C, which is higher than the gelation temperature of all test formulations. These results confirm that the formulation can be easily administered into the periodontal pockets where it will be rapidly converted into the gel state at the physiological temperature.

2.3.3. pH Measurement

In the oral cavity, the pH is maintained near neutrality (6.7–7.3) by saliva. Acidic or alkaline formulations may cause irritation to the buccal mucosa; consequently, the pH is considered as a critical parameter in the formulation of buccal dosage forms [4,26]. The pH of all in situ gel formulations containing piperine was found to fluctuate between 7.0 (F5) and 7.8 (F4) (Table 2). The reason behind the relatively basic pH of the gel is due to the cross-linking of the gellan gum with STPP. STPP is a basic component, which imparts an alkaline pH to the formulation. As depicted in Table 2, formulations cross-linked with 0.4% *w/v* STPP showed relatively higher pH values compared to those cross-linked with 0.2% *w/v* STPP. Nevertheless, since the pH of periodontal fluid is around 7.2 to 7.8, all formulations are deemed to be biologically compatible and would avoid irritation upon application in periodontal cavities.

2.3.4. Syringeability Study

A syringeability test was performed to ensure that the produced formulations had a good solution flow nature and could be utilized to deliver the formulation to the periodontal pocket. Syringeablity mainly depends on the concentration of the polymer and viscosity. Herein, all the developed formulations showed good syringeability as manifested by an easy and continuous flow through a 24-gauge needle at room temperature (Table 2). These results verify the easy application of the prepared in situ gel formulation to the disease site in vivo.

2.3.5. Drug Content Percentage

The drug content is one of the important parameters, which directly influences the drug release property. The drug content of the formulations was found to be varied from $57.5 \pm 1.9\%$ (F5) to $95.3 \pm 2.3\%$ (F6) (Table 2). Such variation in drug content percentage might be ascribed to the degree of cross-linking of gellan gum with STPP. At higher STPP concentrations, extensive cross-linked gellan gum would occur, resulting in a significant reduction in the pore size of the gel, which, in turn, would favor piperine retention within the gel structure.

2.4. In Vitro Drug Release

An in vitro release study is a pre-assessment study to understand the release pattern of the drug at body conditions. The in vitro release profiles of piperine from different in situ gel formulations are represented in Figure 3. As depicted in Figure 3, the release of piperine was in the range of $37.3 \pm 2.9\%$ (F8) to $84.3 \pm 4.6\%$ (F1) over 72 h. Variations in the percentage cumulative drug release at 72 h was directly related to formulation composition. Increasing gellan gum concentration from 0.5% *w/v* to 0.75% *w/v* significantly retarded drug release from the gel matrix. The percentage cumulative drug release from F1, prepared with 0.5% *w/v* gellan gum, was 84.3 ± 4.6%, which was significantly higher than that from F5 (62.9 ± 3.7%), prepared with 0.75% *w/v* gellan gum. In the same context, increasing poloxamer 407 concentration from 10 to 12% *w/v* remarkably slowed piperine release from the gel matrix. The percentage of the drug released from F2, prepared with 12% *w/v* poloxamer 407, was 66.2 ± 4.1%, which was significantly lower than that from F1 (84.3 ± 4.6%), prepared with 10% *w/v* poloxamer 407. Poloxamer 407 was reported to act as a release barrier within the gel matrix via reducing the number and dimension of water channels and increasing the number and size of micelles within the gel structure [24]. This clearly explains the retarding effect of poloxamer 407 on piperine release from various in situ gel formulations.

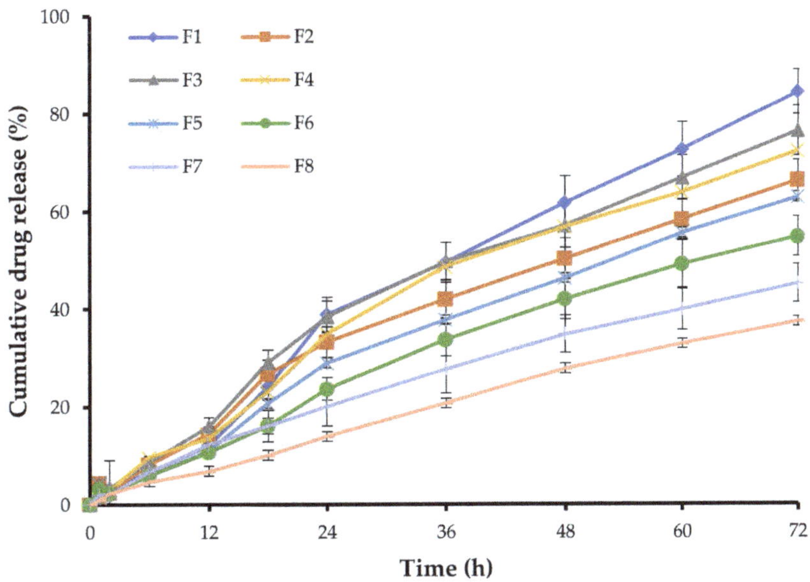

Figure 3. In vitro release profiles of piperine from various in situ gel formulations.

The in vitro release data were kinetically analyzed according to the zero-order, first-order, and Higuchi model. The relatively high correlation coefficient (R^2) values obtained with the zero-order model suggested the release of fixed drug amounts at fixed time intervals. Accordingly, it can be anticipated that, among various formulations, F6 would sustain piperine release under in vivo conditions for up to 7 days, while other formulations, such as F1, F3, and F4, would be exhausted early and may not sustain drug release for 7 days as desired.

Finally, based on various in situ gel characteristics, formulation F6, which shows reasonable physico-chemical characteristics such as good gelation temperature (35.0 ± 1.0 °C), short gelling time (36.0 ± 1.0 s), appropriate pH (7.4 ± 0.3), high drug content (95.3 ± 2.3%) along with efficient sustained drug release, was selected as an optimized formula to be used in further investigations.

2.5. DSC Studies

DSC is an important analytical tool used to study the thermal effect, physical transitions, solid–solid transitions, and compatibility of the drug with formulation ingredients. The DSC thermogram of the pure drug showed an endothermic peak at 134.9 °C, corresponding to its melting point (Figure 4) [48]. The endothermic peak of the piperine in the optimized in situ gel formulation (F6) was slightly shifted to 135.6 °C (Figure 4). This result infers that piperine is compatible with the excipients of the formulation, and no significant interaction or modification in drug properties occurred upon its incorporation in the formulation.

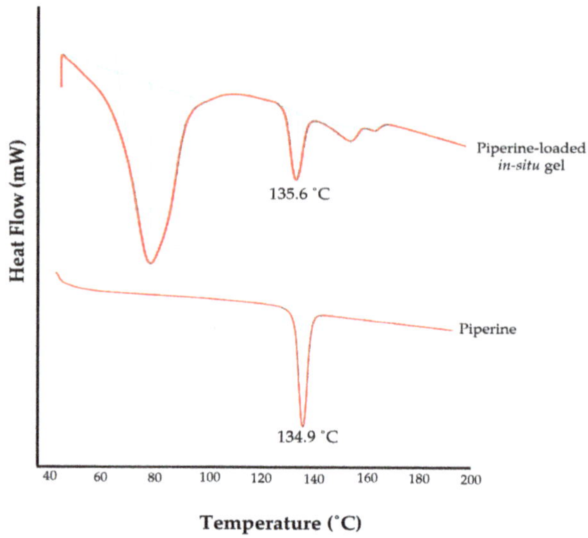

Figure 4. DSC thermograms of pure piperine and optimized piperine-loaded in situ gel formulation.

2.6. Clinical Evaluation of Piperine Gel in Human Patients

Dental deposits are the major concern for soft tissue inflammation. Recently, herbal products used in traditional medicine have gained popularity in the field of dental disease prevention. They can provide safe and long-term solutions for maintaining good dental health. Hence, in this study, the anti-plaque and anti-gingival effectiveness of piperine-loaded in situ gel was assessed in vivo. A non-randomized controlled clinical trial was conducted among 30 subjects, divided into two groups: Group 1 receiving the optimized piperine-loaded in situ gel formulation (F6) following oral prophylaxis, and Group 2 treated with oral prophylaxis alone. The demographic characters of the study participants were presented in Table 3.

Table 3. Descriptive tables of demographic variables.

Group	Gender	Frequency	Percentage	Age (Mean ± SD)
Group 1	Male Female	9 6	60% 40%	46.07 ± 10.6
Group 2	Male Female	8 7	53.3% 46.7%	44.87 ± 7.1

All the subjects were evaluated for plaque score, gingival index, and pocket depth at baseline and 14-day follow-up. At the baseline, there was no significant difference between the two groups in the mean plaque score (Group 1: 1.67 ± 0.49 vs. Group 2: 1.89 ± 0.31; p = 0.140), gingival index (Group 1: 1.70 ± 0.45 vs. Group 2: 1.99 ± 0.32; p = 0.053), and pocket depth (Group 1: 5.47 ± 0.52 vs. Group 2: 5.53 ± 0.52; p = 0.483) (Table 4). On the other hand, at the 14-day follow-up, both groups exhibited a reduction in mean plaque score, gingival index, and pocket depth, compared to baseline values (Figure 5).

Interestingly, there was a significant reduction in the clinical parameters on the 14-day follow-up in Group 1 compared to Group 2. As depicted in Table 4, at the 14-day follow-up, the plaque score showed a significant reduction in Group 1 compared to Group 2 ($p = 0.0001$). Similarly, the gingival index was significantly reduced in Group 1 ($p = 0.0003$). Most importantly, the pocket depth in Group 1 was reduced to 3.27 ± 0.59 mm at the follow-up, from 5.47 ± 0.52 mm at baseline, which was statistically lower than that in Group 2 ($p = 0.0002$). These results might be correlated with the in vitro release results, which advocate the sustained release of piperine from in situ gel formulation.

Table 4. Inter-group comparisons between baseline and 14-day follow-up scores.

Parameter	Group	Baseline		Follow-Up		*p* Value ‡
		Mean	SD	Mean	SD	
Plaque score	Control	1.89	0.31	1.55	0.28	0.0001
	Test	1.67	0.49	1.15	0.23	
	p value †	0.140				
Gingival index	Control	1.99	0.32	1.64	0.29	0.0003
	Test	1.70	0.45	1.25	0.23	
	p value	0.053				
Pocket depth	Control	5.53	0.52	4.40	0.63	0.0002
	Test	5.47	0.52	3.27	0.59	
	p value	0.483				

† *p*-value for inter-group comparisons at baseline; ‡ *p*-value for inter-group comparisons at 14-day follow-up.

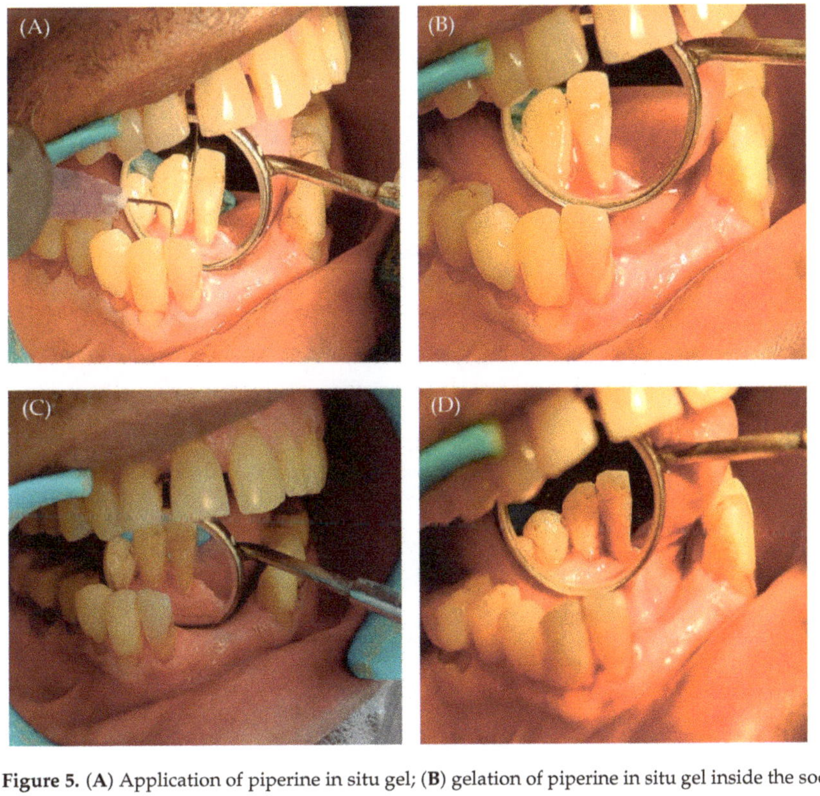

Figure 5. (**A**) Application of piperine in situ gel; (**B**) gelation of piperine in situ gel inside the socket; (**C**) intra-oral image post oral prophylaxis; and (**D**) follow-up after 14 days.

Piperine has a well-documented anti-inflammatory effect. However, few researchers have studied its potential use in periodontitis disease [49]. In this study, we confirmed the efficacy of piperine-loaded in situ gel in fighting plaque and reducing gingival inflammation and periodontal pocket depth. Such effects could be attributed to the anti-inflammatory and anti-bacterial activities of piperine. Piperine has been reported to inhibit nitric oxide and tumor necrosis factor-alpha production, both of which are known to play roles in the pathophysiology of inflammation in periodontal disease [50]. In addition, piperine has been proven to have a substantial anti-biofilm impact on *Streptococcus mutans* [51]. Nevertheless, it is worth noting that there are few limitations in our clinical study. The short study duration, along with the limited number of study participants might hinder the generalization of the study results to the whole population. However, within the limitations of the study, piperine-loaded in situ gels proved efficient anti-plaque, anti-gingival, and anti-inflammatory activities in patients with periodontitis (Figure 5 and Table 4). Further studies are required to assess the long-term benefits of our in situ gel formulation in treating periodontitis.

3. Conclusions

In this study, in situ gel formulations containing piperine was developed with a combination of gellan gum and poloxamer 407 and evaluated as a potential drug delivery system for the treatment of periodontitis. The optimized formulation exhibited an optimal phase transition temperature, short gelling time, and appropriate viscosity at room temperature, which enable the easy application of the formulation. In addition, under physiological conditions, the optimized formulation was efficiently transformed into a hard gel that could sustain drug release for up to 72 h. Most importantly, a non-randomized controlled clinical trial revealed a significant decline in mean plaque score, gingival index, and pocket depth among the test group from baseline to 14 days, whereas this was not significant in the control group. Collectively, a combination of polymers having different mechanisms of in situ gelation might represent a viable delivery vehicle for the application of drugs used for the management of periodontitis.

4. Materials and Methods

4.1. Materials

Piperine and sodium tripolyphosphate were procured from Karnataka Fine Chemicals (Bangalore, India). Poloxamer-407 was procured from Sigma Aldrich (St. Louis, CA, USA). Gellan gum (molecular weight 500 kDa, 95% deacylation degree) was a gift from Sisco Research Laboratories (Mumbai, India). All other reagents were of analytical grade.

4.2. Drug-Excipient Compatibility by Fourier-Transform Infrared (FTIR) Spectroscopy

In order to check the physico-chemical compatibility between piperine and different formulation excipients, FTIR analysis was conducted. Briefly, piperine was mixed with formulation excipients at a ratio of 1:1 and examined by BRUKER FTIR spectrometer (TENSOR; Markham, ON, Canada). The spectra of the samples were recorded and analyzed at the wavenumber range of 4000 cm^{-1} to 400 cm^{-1}.

4.3. In Situ Gel Formulation

Gellan gum (0.5–1% w/v) was dispersed in deionized water and then heated to 60–80 °C with agitation at 150 rpm for 2 h until total dissolution [52]. Sodium tripolyphosphate (STPP, 0.2–1% w/v) was added as a cross-linking agent [53] to the gellan gum solution and the dispersion was maintained at 65 °C with agitation at 150 rpm for 1 h. The obtained solution was then brought to 8 $\pm$ 2 °C on ice bath. Polaxomer-407 (10–12% w/v) was added as co-polymer [54] to the above solution and stirred at 100 rpm for 1 h until solution becomes clear. Hydro-alcoholic solution of piperine was finally added to the above dispersion with stirring for 30 min to obtain the overall concentration of piperine of 0.4% w/v. The piperine-dispersed in situ gel was then stored in refrigerator between 2 to 8 °C for further studies (Table 1).

4.4. In Vitro Evaluation of Piperine-Loaded In Situ Gel

4.4.1. Gelation Temperature

Gelation temperature is the temperature at which the preparation becomes thick, stiff, and resistant to flow. The gelation temperature of the formulation was determined by test tube inversion method [55]. Briefly, 5 mL in situ gel was placed in a test tube, containing 0.5 mL of 0.1% *w/v* NaCl, and incubated in a temperature-controlled water bath maintained at 15 ± 2 °C. The temperature of the water bath was then gradually raised to 40 °C, with an increment of 2 degrees every 5 min. At each set point, the test tube was inverted at 90°, and the temperature at which no flow was observed upon inversion was recorded as a gelation temperature.

4.4.2. Gelling Time

Gelling time of different formulations was measured by test tube inversion method [43]. Briefly, 5 mL of each formulation was added into a glass test tube containing 0.5 mL of 0.1% *w/v* NaCl. The test tube was incubated in a temperature-controlled water bath set at 35 ± 2 °C. The test tubes were periodically inverted at 90° and the time taken for the liquid to transform to gel (no flow) was recorded as gelling time in sec [56].

4.4.3. Gel pH

The pH of developed in situ gel formulation was estimated using calibrated digital pH meter (Symbiont Life Sciences, Chennai, India). The pH meter was calibrated before its use with standard buffer solutions. The pH was measured by dipping the electrode sufficiently into the formulation at room temperature. The evaluation was carried out in triplicate and results are recorded as average of three trials.

4.4.4. Gel Viscosity

Viscosity of in situ gel formulations was measured using Brookfiled DVE viscometers (Brookfield Engineering, Middleboro, MA, USA). Various trials were conducted to choose the appropriate spindle for analyzing the samples. Based on the % Torque value, spindle 61 was selected to measure the viscosity of the liquid and spindle No. 62 to measure gel viscosity. The viscosity of the liquid was measured at 25 ± 2 °C, while the gel was evaluated at 37 ± 0.5 °C.

4.4.5. Syringeability

A constant volume of 1 mL of test formulation, maintained at 25 ± 2 °C, was filled into 1 mL syringe attached with 24-gauge needle. The solution which easily passed from syringe was termed as "pass" and those which had difficulty passing through were termed as "fail" [57].

4.4.6. Drug Content of the Gel

The piperine content in the in situ gel formulation was determined by UV spectroscopy (Agilent Technologies, Santa Clara, CA, USA). Briefly, 1 mL of the gel formulation was diluted to 10 mL with ethanol and sonicated using bath sonicator until a clear solution is obtained. Solution was filtered and piperine content in the in situ gel was determined at 343 nm using ethanol as a blank [58].

4.5. Differential Scanning Calorimetric (DSC) Studies

Thermoanalytical curves of piperine and the in situ gel of piperine were obtained using a METTLER TO-143 LEDO differential scanning calorimeter (Columbus, OH, USA). Samples (4–6 mg) were placed in hermetically sealed aluminum cells. The samples were heated and scanned in the range of 20 to 200 °C at a heating rate of 10 °C min^{-1} under a nitrogen atmosphere with a flow rate of 80 mL min^{-1}. Heat flow versus temperature plots were developed to obtain the spectra [59].

4.6. Drug Release Studies

In vitro drug release profiles of in situ gel were determined using 250 mL capacity Franz diffusion cell. The dialysis membrane with an aperture size of 0.45 μ was interfaced between the donor and receptor cells. Then, 1 mL of in situ gel was placed in the donor cell. The receptor compartment was filled with 250 mL of phosphate buffer solution (PBS, pH 7.4) kept at 37 ± 0.5 °C and constantly stirred at 25 rpm. At scheduled time points (1, 2, 6, 12, 18, 24, 36, 48, 60, and 72 h), 3 mL samples were withdrawn and replenished with the same volume of pre-warmed PBS to maintain the sink condition. Drug concentration in collected samples was quantified spectrophotometrically at λ_{max} 343 nm.

4.7. Randomized Clinical Trial

4.7.1. Ethical Clearance

The study proposal was submitted for approval and clearance was obtained from the Institutional Human Ethical Committee (No. KIDS/IEC/May2023/27), K.L.E Society's Institute of Dental Sciences, Bangalore, India. Written informed consent was obtained from each of the participating subject.

4.7.2. Study Design and Study Setting

A non-randomized controlled clinical trial was conducted among 30 subjects in the age range of 20–50 years. The study was carried out on subjects aged 20 to 50 years with moderate to severe chronic periodontitis (AAP classification, 1999) and at least three nonadjacent interproximal sites with a probing pocket depth of 6 mm, and no contraindication to periodontal treatment were included in the study. Patients had to have not undergone any periodontal or antibiotic therapy in the previous 6 months and be co-operative and dedicated to maintaining good dental hygiene. Patients who smoked, used tobacco, or drank alcohol, had a systemic ailment, were pregnant or breastfeeding, had used an antimicrobial mouthwash in the previous two months, or needed periodontal surgery were excluded from the trial.

4.7.3. Study Procedure

A total of 30 patients, divided into two groups, were subjected to this study. Prior to starting the investigation, demographic characters (Table 3) were estimated through interview method. The clinical parameters including plaque index, gingival index, and periodontal pocket depth were evaluated in every individual [60]. Group 1 (test group) included 15 participants who received SRP (scaling and root planning) with piperine gel intervention. Group 2 (control group) included 15 participants who only received SRP. In the intervention group (Group 1), after finishing scaling and root planning, the in situ gel was applied into the pocket depth. For in situ gel application, a syringe with a blunt needle was used. The gel was gently released while the needle was pushed coronally from the bottom of the pocket. In the control group, SRP alone was performed, and oral hygiene instructions were given to subjects belonging to both the groups. Patients were recalled on 14th day. All clinical parameters were recorded during the recall visit.

4.8. Statistical Analysis

Data analysis was achieved using the SPSS 26.0 (SPSS Inc., Chicago, IL, USA). Statistical analysis was carried out by Student's *t*-test, paired *t*-test, and Wilcoxin test. A p value < 0.05 is considered significant.

Author Contributions: Conceptualization: S.S.B., S.E. and N.V.S.; methodology: P.K.G., R.A.J., S.E. and N.V.S.; data curation: P.K.G., R.A.J., S.E., N.V.S. and A.M.; formal analysis: A.M., H.F.A., E.-S.K. and A.J.O.; software: H.F.A. and A.J.O.; resources: H.F.A., A.J.O. and E.-S.K.; validation: S.S.B., S.E. and N.V.S.; writing the original draft: P.K.G., A.M. and S.S.B.; writing, review, and editing: A.S.A.L., H.F.A., A.J.O. and E.-S.K. All authors have read and agreed to the published version of the manuscript.

Funding: This study is supported via funding from Prince Sattam Bin Abdulaziz University project number (PSAU/2023/R/1444). This work was supported by Princess Nourah bint Abdulrahman University Researchers Supporting Project number (PNURSP2023R205), Princess Nourah bint Abdulrahman University, Riyadh, Saudi Arabia. This work was also supported by the Researchers Supporting Project number (RSPD2023R620), King Saud University, Riyadh, Saudi Arabia.

Institutional Review Board Statement: The study proposal was submitted for approval and clearance was obtained from the Institutional Human Ethical Committee (No. KIDS/IEC/May2023/27), K.L.E Society's Institute of Dental Sciences, Bangalore, India.

Informed Consent Statement: Informed consent was obtained from all subjects involved in the study.

Data Availability Statement: Not applicable.

Acknowledgments: This study is supported via funding from Prince Sattam Bin Abdulaziz University project number (PSAU/2023/R/1444). The authors extend their appreciation to the Researchers Supporting Project number (RSPD2023R620), King Saud University, Riyadh, Saudi Arabia. Additionally, the authors extend their appreciation to Princess Nourah bint Abdulrahman University, Riyadh, Saudi Arabia for funding this work under Researchers Supporting Project number (PNURSP2023R205). Authors also acknowledge the research support extended by Acharya & BM Reddy College of Pharmacy, India.

Conflicts of Interest: The authors declare no conflict of interest.

References

1. Eke, P.I.; Dye, B.A.; Wei, L.; Thornton-Evans, G.O.; Genco, R.J. Prevalence of periodontitis in adults in the United States: 2009 and 2010. *J. Dent. Res.* **2012**, *91*, 914–920. [CrossRef]
2. Tonetti, M.S.; Greenwell, H.; Kornman, K.S. Staging and grading of periodontitis: Framework and proposal of a new classification and case definition. *J. Periodontol.* **2018**, *89* (Suppl. S1), S159–S172. [CrossRef]
3. Van Dyke, T.E.; Bartold, P.M.; Reynolds, E.C. The Nexus Between Periodontal Inflammation and Dysbiosis. *Front. Immunol.* **2020**, *11*, 511. [CrossRef]
4. Hajishengallis, G. Periodontitis: From microbial immune subversion to systemic inflammation. *Nat. Rev. Immunol.* **2015**, *15*, 30–44. [CrossRef]
5. Cekici, A.; Kantarci, A.; Hasturk, H.; Van Dyke, T.E. Inflammatory and immune pathways in the pathogenesis of periodontal disease. *Periodontology 2000* **2014**, *64*, 57–80. [CrossRef]
6. Kim, Y.K.; Ku, J.K. Guided bone regeneration. *J. Korean Assoc. Oral Maxillofac. Surg.* **2020**, *46*, 361–366. [CrossRef]
7. Elgali, I.; Omar, O.; Dahlin, C.; Thomsen, P. Guided bone regeneration: Materials and biological mechanisms revisited. *Eur. J. Oral Sci.* **2017**, *125*, 315–337. [CrossRef]
8. D'Elía, N.L.; Silva, R.R.; Sartuqui, J.; Ercoli, D.; Ruso, J.; Messina, P.; Mestres, G. Development and characterization of bilayered periosteum-inspired composite membranes based on sodium alginate-hydroxyapatite nanoparticles. *J. Colloid Interface Sci.* **2020**, *572*, 408–420. [CrossRef]
9. Mota, J.; Yu, N.; Caridade, S.G.; Luz, G.M.; Gomes, M.E.; Reis, R.L.; Jansen, J.A.; Walboomers, X.F.; Mano, J.F. Chitosan/bioactive glass nanoparticle composite membranes for periodontal regeneration. *Acta Biomater.* **2012**, *8*, 4173–4180. [CrossRef]
10. Sinha, S.; Sonoo, P.R.; Siddhartha, R.; Singh, S.K.; Singh, A. Effect of Conventional Periodontal Treatment (Scaling and Root Planing) on Type-2 Diabetic Patient with Moderate Generalized Chronic Periodontitis: A Clinical Study. *J. Pharm. Bioallied Sci.* **2021**, *13*, S706–S710. [CrossRef]
11. Nadig, P.S.; Shah, M.A. Tetracycline as local drug delivery in treatment of chronic periodontitis: A systematic review and meta-analysis. *J. Indian Soc. Periodontol.* **2016**, *20*, 576–583. [CrossRef]
12. Swain, G.P.; Patel, S.; Gandhi, J.; Shah, P. Development of Moxifloxacin Hydrochloride loaded in-situ gel for the treatment of periodontitis: In-vitro drug release study and antibacterial activity. *J. Oral Biol. Craniofacial Res.* **2019**, *9*, 190–200. [CrossRef]
13. Preshaw, P.M.; Hefti, A.F.; Jepsen, S.; Etienne, D.; Walker, C.; Bradshaw, M.H. Subantimicrobial dose doxycycline as adjunctive treatment for periodontitis. A review. *J. Clin. Periodontol.* **2004**, *31*, 697–707. [CrossRef]
14. Haque, M.M.; Yerex, K.; Kelekis-Cholakis, A.; Duan, K. Advances in novel therapeutic approaches for periodontal diseases. *BMC Oral Health* **2022**, *22*, 492. [CrossRef] [PubMed]
15. Walker, C.B. The acquisition of antibiotic resistance in the periodontal microflora. *Periodontology 2000* **1996**, *10*, 79–88. [CrossRef]

16. Barca, E.; Cifcibasi, E.; Cintan, S. Adjunctive use of antibiotics in periodontal therapy. *J. Istanb. Univ. Fac. Dent.* **2015**, *49*, 55–62. [CrossRef]
17. Loesche, W.J.; Grossman, N.; Giordano, J. Metronidazole in periodontitis (IV). The effect of patient compliance on treatment parameters. *J. Clin. Periodontol.* **1993**, *20*, 96–104. [CrossRef]
18. Sholapurkar, A.; Sharma, D.; Glass, B.; Miller, C.; Nimmo, A.; Jennings, E. Professionally Delivered Local Antimicrobials in the Treatment of Patients with Periodontitis-A Narrative Review. *Dent. J.* **2021**, *9*, 2. [CrossRef]
19. Lacević, A.; Vranić, E.; Zulić, I. Endodontic-periodontal locally delivered antibiotics. *Bosn. J. Basic Med. Sci.* **2004**, *4*, 73–78. [CrossRef]
20. Rajeshwari, H.R.; Dhamecha, D.; Jagwani, S.; Rao, M.; Jadhav, K.; Shaikh, S.; Puzhankara, L.; Jalalpure, S. Local drug delivery systems in the management of periodontitis: A scientific review. *J. Control. Release* **2019**, *307*, 393–409. [CrossRef]
21. Vyas, S.P.; Sihorkar, V.; Dubey, P.K. Preparation, characterization and in vitro antimicrobial activity of metronidazole bearing lectinized liposomes for intra-periodontal pocket delivery. *Pharmazie* **2001**, *56*, 554–560. [PubMed]
22. Gomez-Florit, M.; Pardo, A.; Domingues, R.M.A.; Graça, A.L.; Babo, P.S.; Reis, R.L.; Gomes, M.E. Natural-Based Hydrogels for Tissue Engineering Applications. *Molecules* **2020**, *25*, 5858. [CrossRef] [PubMed]
23. Kouchak, M. In situ gelling systems for drug delivery. *Jundishapur J. Nat. Pharm. Prod.* **2014**, *9*, e20126. [CrossRef] [PubMed]
24. Yadav, R.; Kanwar, I.L.; Haider, T.; Pandey, V.; Gour, V.; Soni, V. In situ gel drug delivery system for periodontitis: An insight review. *Future J. Pharm. Sci.* **2020**, *6*, 33. [CrossRef]
25. Boregowda, S.S.; Maggidi, S.R.; Jayaramu, R.A.; Puttegowda, N.; Parbin, N. Development of an In situ Gel Polymer Composite for Local and Sustained Delivery of Drugs in Vaginal Cavity. *Drug Deliv. Lett.* **2019**, *9*, 211–221. [CrossRef]
26. Garala, K.; Joshi, P.; Shah, M.; Ramkishan, A.; Patel, J. Formulation and evaluation of periodontal in situ gel. *Int. J. Pharm. Investig.* **2013**, *3*, 29–41. [CrossRef]
27. Zhu, L.; Ao, J.; Li, P. A novel in situ gel base of deacetylase gellan gum for sustained ophthalmic drug delivery of ketotifen: In vitro and in vivo evaluation. *Drug Des. Dev. Ther.* **2015**, *9*, 3943–3949. [CrossRef]
28. Mahdi, M.H.; Conway, B.R.; Smith, A.M. Evaluation of gellan gum fluid gels as modified release oral liquids. *Int. J. Pharm.* **2014**, *475*, 335–343. [CrossRef]
29. Chen, Q.; Zheng, Y.; Li, Y.; Zeng, Y.; Kuang, J.; Hou, S.; Li, X. The effect of deacetylated gellan gum on aesculin distribution in the posterior segment of the eye after topical administration. *Drug Deliv.* **2012**, *19*, 194–201. [CrossRef]
30. Galgatte, U.C.; Kumbhar, A.B.; Chaudhari, P.D. Development of in situ gel for nasal delivery: Design, optimization, in vitro and in vivo evaluation. *Drug Deliv.* **2014**, *21*, 62–73. [CrossRef]
31. Gorgani, L.; Mohammadi, M.; Najafpour, G.D.; Nikzad, M. Piperine-The Bioactive Compound of Black Pepper: From Isolation to Medicinal Formulations. *Compr. Rev. Food Sci. Food Saf.* **2017**, *16*, 124–140. [CrossRef]
32. Derosa, G.; Maffioli, P.; Sahebkar, A. Piperine and Its Role in Chronic Diseases. *Adv. Exp. Med. Biol.* **2016**, *928*, 173–184. [CrossRef]
33. Jaisin, Y.; Ratanachamnong, P.; Wongsawatkul, O.; Watthammawut, A.; Malaniyom, K.; Natewong, S. Antioxidant and anti-inflammatory effects of piperine on UV-B-irradiated human HaCaT keratinocyte cells. *Life Sci.* **2020**, *263*, 118607. [CrossRef]
34. Smilkov, K.; Ackova, D.G.; Cvetkovski, A.; Ruskovska, T.; Vidovic, B.; Atalay, M. Piperine: Old Spice and New Nutraceutical? *Curr. Pharm. Des.* **2019**, *25*, 1729–1739. [CrossRef]
35. Zou, L.; Hu, Y.Y.; Chen, W.X. Antibacterial mechanism and activities of black pepper chloroform extract. *J. Food Sci. Technol.* **2015**, *52*, 8196–8203. [CrossRef]
36. Dong, Y.; Huihui, Z.; Li, C. Piperine inhibit inflammation, alveolar bone loss and collagen fibers breakdown in a rat periodontitis model. *J. Periodontal Res.* **2015**, *50*, 758–765. [CrossRef]
37. Aziz, D.M.; Hama, J.R.; Alam, S.M. Synthesising a novel derivatives of piperine from black pepper (*Piper nigrum* L.). *J. Food Meas. Charact.* **2015**, *9*, 324–331. [CrossRef]
38. Matricardi, P.; Cencetti, C.; Ria, R.; Alhaique, F.; Coviello, T. Preparation and characterization of novel gellan gum hydrogels suitable for modified drug release. *Molecules* **2009**, *14*, 3376–3391. [CrossRef]
39. Liu, L.; Wang, B.; Gao, Y.; Bai, T.-c. Chitosan fibers enhanced gellan gum hydrogels with superior mechanical properties and water-holding capacity. *Carbohydr. Polym.* **2013**, *97*, 152–158. [CrossRef] [PubMed]
40. Kirchmajer, D.M.; Steinhoff, B.; Warren, H.; Clark, R.; in het Panhuis, M. Enhanced gelation properties of purified gellan gum. *Carbohydr. Res.* **2014**, *388*, 125–129. [CrossRef] [PubMed]
41. Giuliano, E.; Paolino, D.; Fresta, M.; Cosco, D. Mucosal Applications of Poloxamer 407-Based Hydrogels: An Overview. *Pharmaceutics* **2018**, *10*, 159. [CrossRef]
42. Dumortier, G.; Grossiord, J.L.; Agnely, F.; Chaumeil, J.C. A Review of Poloxamer 407 Pharmaceutical and Pharmacological Characteristics. *Pharm. Res.* **2006**, *23*, 2709–2728. [CrossRef]
43. Gugleva, V.; Michailova, V.; Mihaylova, R.; Momekov, G.; Zaharieva, M.M.; Najdenski, H.; Petrov, P.; Rangelov, S.; Forys, A.; Trzebicka, B.; et al. Formulation and Evaluation of Hybrid Niosomal In Situ Gel for Intravesical Co-Delivery of Curcumin and Gentamicin Sulfate. *Pharmaceutics* **2022**, *14*, 747. [CrossRef]
44. Elmowafy, E.; Cespi, M.; Bonacucina, G.; Soliman, M.E. In situ composite ion-triggered gellan gum gel incorporating amino methacrylate copolymer microparticles: A therapeutic modality for buccal applicability. *Pharm. Dev. Technol.* **2019**, *24*, 1258–1271. [CrossRef]

45. Baloglu, E.; Karavana, S.Y.; Senyigit, Z.A.; Guneri, T. Rheological and mechanical properties of poloxamer mixtures as a mucoadhesive gel base. *Pharm. Dev. Technol.* **2011**, *16*, 627–636. [CrossRef]
46. Rençber, S.; Karavana, S.Y. Formulation and optimization of gellan gum-poloxamer based dexamethasone mucoadhesive in situ gel. *J. Res. Pharm.* **2020**, *201*, 529–538. [CrossRef]
47. Sapra, P.; Patel, D.; Soniwala, M. Development and optimization of in situ periodontal gel containing Levofloxacin for the treatment of periodontal diseases. *J. Sci. Innov. Res.* **2013**, *2*, 607–626.
48. Tiwari, A.; Mahadik, K.R.; Gabhe, S.Y. Piperine: A comprehensive review of methods of isolation, purification, and biological properties. *Med. Drug Discov.* **2020**, *7*, 100027. [CrossRef]
49. Pentapati, K.C.; Kukkamalla, M.A.; Siddiq, H.; Sabnis, N. Effectiveness of novel herbal dentifrice in control of plaque, gingivitis, and halitosis-Randomized controlled trial. *J. Tradit. Complement. Med.* **2020**, *10*, 565–569. [CrossRef] [PubMed]
50. Pradeep, C.R.; Kuttan, G. Effect of piperine on the inhibition of nitric oxide (NO) and TNF-alpha production. *Immunopharmacol. Immunotoxicol.* **2003**, *25*, 337–346. [CrossRef]
51. Dwivedi, D.; Singh, V. Effects of the natural compounds embelin and piperine on the biofilm-producing property of Streptococcus mutans. *J. Tradit. Complement. Med.* **2016**, *6*, 57–61. [CrossRef] [PubMed]
52. Jelkmann, M.; Leichner, C.; Zaichik, S.; Laffleur, F.; Bernkop-Schnürch, A. A gellan gum derivative as in-situ gelling cationic polymer for nasal drug delivery. *Int. J. Biol. Macromol.* **2020**, *158*, 1037–1046. [CrossRef]
53. Grasdalen, H.; Smidsrød, O. Gelation of gellan gum. *Carbohydr. Polym.* **1987**, *7*, 371–393. [CrossRef]
54. Barse, R.; Kokare, C.; Tagalpallewar, A. Influence of hydroxypropylmethylcellulose and poloxamer composite on developed ophthalmic in situ gel: Ex vivo and in vivo characterization. *J. Drug Deliv. Sci. Technol.* **2016**, *33*, 66–74. [CrossRef]
55. Jommanee, N.; Chanthad, C.; Manokruang, K. Preparation of injectable hydrogels from temperature and pH responsive grafted chitosan with tuned gelation temperature suitable for tumor acidic environment. *Carbohydr. Polym.* **2018**, *198*, 486–494. [CrossRef] [PubMed]
56. Sheshala, R.; Quah, S.Y.; Tan, G.C.; Meka, V.S.; Jnanendrappa, N.; Sahu, P.S. Investigation on solution-to-gel characteristic of thermosensitive and mucoadhesive biopolymers for the development of moxifloxacin-loaded sustained release periodontal in situ gels. *Drug Deliv. Transl. Res.* **2019**, *9*, 434–443. [CrossRef] [PubMed]
57. Shekhawat, M.; Surti, Z.; Surti, N.I. Biodegradable in situ Gel for Subcutaneous Administration of Simvastatin for Osteoporosis. *Indian J. Pharm. Sci.* **2018**, *80*, 395–399. [CrossRef]
58. Lee, J.G.; Kim, A.Y.; Kim, D.W.; Kim, Y.J. Determination and risk characterisation of bio-active piperine in black pepper and selected food containing black pepper consumed in Korea. *Food Sci. Biotechnol.* **2021**, *30*, 209–215. [CrossRef]
59. Stasiłowicz, A.; Rosiak, N.; Tykarska, E.; Kozak, M.; Jenczyk, J.; Szulc, P.; Kobus-Cisowska, J.; Lewandowska, K.; Płazińska, A.; Płaziński, W.; et al. Combinations of Piperine with Hydroxypropyl-β-Cyclodextrin as a Multifunctional System. *Int. J. Mol. Sci.* **2021**, *22*, 4195. [CrossRef]
60. Prakasam, A.; Elavarasu, S.S.; Natarajan, R.K. Antibiotics in the management of aggressive periodontitis. *J. Pharm. Bioallied Sci.* **2012**, *4*, S252–S255. [CrossRef]

gels

MDPI

Article

Enhanced Osteogenesis Potential of MG-63 Cells through Sustained Delivery of VEGF via Liposomal Hydrogel

Milton Hongli Tsai [1], Rohaya Megat Abdul Wahab [1], Shahrul Hisham Zainal Ariffin [2], Fazren Azmi [3] and Farinawati Yazid [4,*]

[1] Discipline of Orthodontics, Department of Family Oral Health, Faculty of Dentistry, Universiti Kebangsaan Malaysia, Kuala Lumpur 50300, Malaysia; p106712@siswa.ukm.edu.my (M.H.T.); rohaya_megat@ukm.edu.my (R.M.A.W.)

[2] Department of Biological Sciences and Biotechnology, Faculty of Science and Technology, Universiti Kebangsaan Malaysia, Bangi 43600, Malaysia; hisham@ukm.edu.my

[3] Faculty of Pharmacy, Universiti Kebangsaan Malaysia, Kuala Lumpur 50300, Malaysia; fazren.azmi@ukm.edu.my

[4] Discipline of Pediatric Dentistry, Department of Family Oral Health, Faculty of Dentistry, Universiti Kebangsaan Malaysia, Kuala Lumpur 50300, Malaysia

* Correspondence: drfarinawati@ukm.edu.my; Tel.: +603-928-970-76

Abstract: The challenges of using VEGF to promote osteoblastic differentiation include a short half-life and a narrow therapeutic window. A carrier system combining hydrogel and liposomes may improve the therapeutic efficacy of VEGF for bone regeneration. This study aimed to investigate the effects of delivery of VEGF via liposomal hydrogel on the osteogenesis of MG-63 cells. Liposomal hydrogel scaffold was fabricated and then characterized in terms of the morphological and chemical properties using FESEM and FTIR. In 2.5D analysis, the MG-63 cells were cultured on liposomal hydrogel + VEGF as the test group. The osteogenic effects of VEGF were compared with the control groups, i.e., hydrogel without liposomes + VEGF, osteogenic medium (OM) supplemented with a bolus of VEGF, and OM without VEGF. Cell morphology, viability, and differentiation and mineralization potential were investigated using FESEM, MTT assay, ALP activity, and Alizarin red staining. The characterization of scaffold showed no significant differences in the morphological and chemical properties between hydrogel with and without liposomes ($p > 0.05$). The final 2.5D culture demonstrated that cell proliferation, differentiation, and mineralization were significantly enhanced in the liposomal hydrogel + VEGF group compared with the control groups ($p < 0.05$). In conclusion, liposomal hydrogel can be used to deliver VEGF in a sustained manner in order to enhance the osteogenesis of MG-63 cells.

Keywords: hydrogel; liposomes; MG-63 osteoblast-like cells; osteogenesis; vascular endothelial growth factor

Citation: Tsai, M.H.; Megat Abdul Wahab, R.; Zainal Ariffin, S.H.; Azmi, F.; Yazid, F. Enhanced Osteogenesis Potential of MG-63 Cells through Sustained Delivery of VEGF via Liposomal Hydrogel. *Gels* **2023**, *9*, 562. https://doi.org/10.3390/gels9070562

Academic Editors: Ying Huang, Zhengwei Huang and Xuanjuan Zhang

Received: 27 June 2023
Revised: 8 July 2023
Accepted: 9 July 2023
Published: 10 July 2023

1. Introduction

The treatment of critical-sized bone defects (CSD) due to traumatic injuries, tumor resection, or congenital diseases has always been a challenge in the field of medicine and dentistry because spontaneous healing cannot occur in the absence of bone grafts [1]. One common example of a congenital bony defect in the orofacial region is the cleft alveolar and palate, whereby the autograft harvested from the patient's anterior iliac crest is the most reliable graft material used for closing the bone defect at the alveolar cleft, with success rates reported to be as high as 96.2% [2]. Till today, autogenous bone grafts are still considered the "gold standard", because they fulfil all the ideal requirements for bone regeneration by having osteogenic, osteoconductive, and osteoinductive properties [3,4]. However, such autograft is often associated with significant donor-site morbidity and psychological complications, which are quite a lot to endure for school-aged cleft patients [5].

Non-autogenous bone grafts such as allografts and xenografts are osteoconductive and mildly osteoinductive, but they lack the osteogenic capacity of autografts and they may carry the risk of infectivity and immune rejection [6]. To address these issues, bone tissue engineering (BTE) has developed synthetic bone substitutes as promising strategies in regenerative therapies. This type of bone substitute generally consists of natural, synthetic, or composite scaffolds, modified with osteoprogenitor cells and/or incorporated with growth factors [7,8]. Hence, the "triad" of osteogenic cells, osteoconductive scaffold materials, and osteoinductive signals (growth factors) successfully recapitulates the properties of autograft, omitting the need for invasive donor-site surgeries [9].

Nevertheless, BTE is such a complex subject of research that even though countless studies have been carried out in the field of BTE, there are still many critical variables yet to be optimized and standardized through in vitro, and subsequently in vivo, animal studies before BTE can be used safely and effectively in clinical applications [10]. One such parameter that requires further fine-tuning is the efficacy of osteogenic differentiation. Thus, it is essential to investigate the choice of signaling factors that may induce a more efficient cellular expansion and osteogenic differentiation by manipulating the culture microenvironment [11].

Among the different types of mediators in osteogenesis, vascular endothelial growth factor (VEGF) is an interesting growth factor that is capable of coupling effects of angiogenesis and osteogenesis during both bone formation and regeneration. In this coupling mechanism, VEGF performs a twofold function: inducing angiogenesis by acting on endothelial cells, which is crucial for the recruitment of osteoprogenitor cells, and promoting osteogenesis by direct action on non-endothelial cells (osteoblasts and osteoclasts) that express VEGF receptors [12,13]. Nevertheless, the use of VEGF is yet to be implemented in bone regenerative therapies in a clinical setting. Some of the challenges are the relatively short half-life and narrow therapeutic window that limit the therapeutic efficacy of VEGF for bone regeneration [14]. In other words, it is very difficult to sustain the bioavailability of the growth factor without exceeding the toxicity threshold.

The solution to this problem lies in the selection of a carrier system to improve the release kinetics of the growth factor (VEGF). Hydrogels, a unique type of extracellular matrix-like scaffold with excellent biocompatibility, can allow complete encapsulation of cells and growth factors and facilitate cell expansion and differentiation. Some of the promising hydrogel systems are chitosan, alginate, gellan gum, and hyaluronan [15–17]. Among them, thermosensitive hydrogel is produced via physical gelation method, utilizing chitosan, which is a natural polymer derived from the shells of shellfish, an abundant waste product of the seafood industry (renewable resource). Thus, the fabrication of liposomal hydrogel is economical and environment friendly without the need for complex chemical stimuli and expensive equipment [18]. Chitosan has a good biocompatibility property and forms non-toxic oligosaccharide on degradation [19]. Its thermos-responsive nature allows the hydrogel to reach the target site with minimum invasiveness to deliver preloaded cells or biomolecules and then undergo gelation in situ [20]. In addition, nanoliposomes can function as secondary carriers to further minimize the burst release effect when hydrogel is used as the sole carrier. Therefore, a potential solution is to combine both liposomes and chitosan hydrogel, making use of the advantages of both materials. This type of hybrid hydrogel has already been used for other purposes in the medical field, particularly for drug delivery [21]. Hence, the aim of the current study was to investigate the application of liposomal hydrogel as a promising strategy yet to be extended to the field of BTE.

2. Results

2.1. Characterization of Scaffold

2.1.1. Characterization of VEGF-Loaded Liposomes

The particle sizes, PDI, and zeta-potentials of the liposomes measured by the dynamic light scattering (DLS) method are summarized in Table 1. Liposomes loaded with 100 ng/mL VEGF were fabricated with an average size of 171 nm, an average polydispersity index (PDI) of 0.174, and zeta-potential of +38.5 mV. There were no significant differences in size, PDI, or zeta-potential between liposomes with and without VEGF ($p > 0.05$).

Table 1. Characterization of liposomes with and without VEGF. Data are expressed as mean ± standard deviation. ($n = 8$).

Formulation	Particle Sizes (d·nm)	PDI	Zeta-Potential (mV)	p-Value
Liposomes without VEGF	114 ± 28	0.221 ± 0.01	30.7 ± 12.8	$p > 0.05$
Liposomes with VEGF	171 ± 31	0.174 ± 0.03	38.5 ± 12.9	

The morphology of the liposomes with and without VEGF under transmission electron microscope (TEM) is shown in Figure 1. The particle sizes correlated well with measurements obtained from the nanosizer. The liposomes demonstrated spherical unilamellar morphology and dispersed nature.

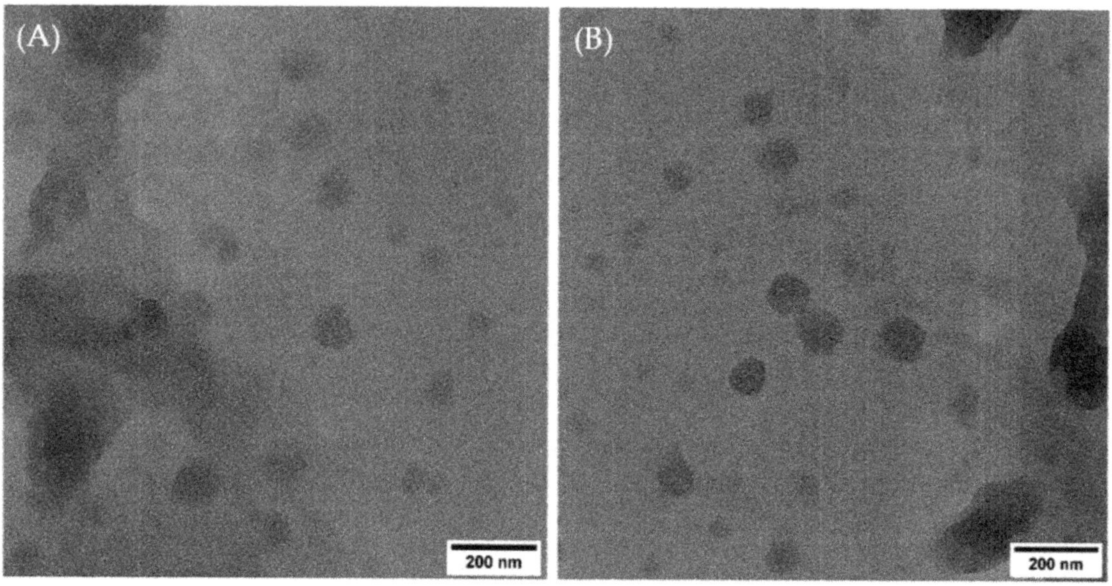

Figure 1. TEM images of (**A**) liposomes without VEGF and (**B**) liposomes with VEGF. (Scale bar: 200 nm).

2.1.2. Microstructure and Porosity of Hydrogels

The gelation time for hydrogel without liposomes was 10.5 ± 0.52 min but only 7.21 ± 0.43 min for liposomal hydrogel according to the test tube inverting method. The difference was statistically significant ($p < 0.05$). The transparent hydrogel solutions at 4 °C were transformed into non-transparent semisolid hydrogels (Figure 2a–d).

After gelation, the hydrogel without liposomes and liposomal hydrogel were freeze-dried and observed under FESEM. The hydrogels demonstrated a porous and sponge-like microstructure (Figure 3a–f). Successful incorporation of the liposomes in the hydrogel could be verified through Figure 3f, the size of which agreed with the findings from the nanosizer.

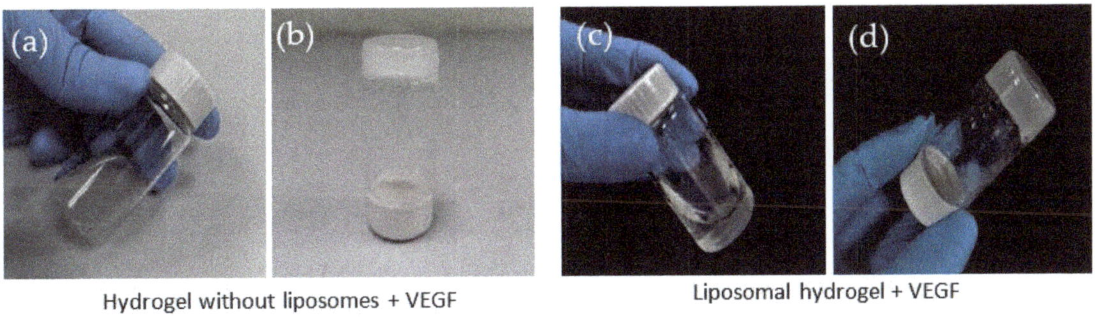

Figure 2. Photographs of the process of gelation for hydrogel without liposomes (**a**,**b**) and liposomal hydrogel (**c**,**d**). The tube inversion method was used to determine the gelation time (**b**,**d**).

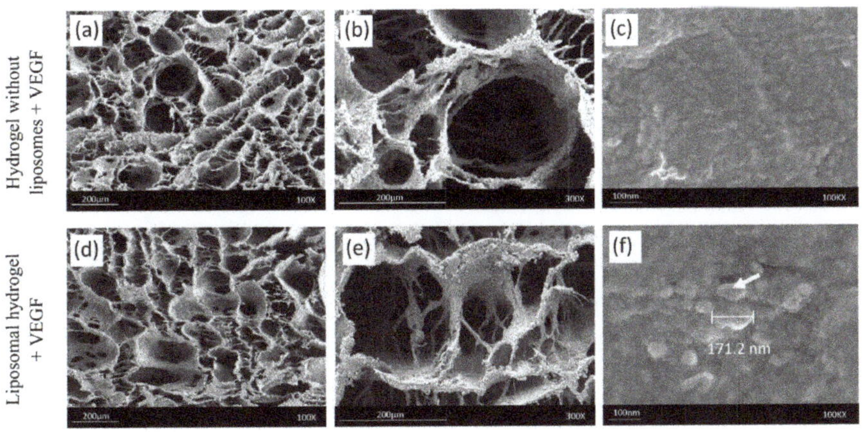

Figure 3. Field emission scanning electron microscope images of hydrogel without liposomes (**a**–**c**) and liposomal hydrogel (**d**–**f**) at 100×, 300×, and 100k× magnifications. The white arrow (**f**) indicates the presence of liposomes in the liposomal hydrogel.

The average pore size and porosity are presented in Table 2. The average pore diameters of hydrogel without liposomes and liposomal hydrogel were 201.4 ± 24.5 µm and 205.8 ± 25.5 µm, respectively. The porosity of the hydrogel without liposomes was 81.1 ± 3.3%, while the porosity of the liposomal hydrogel was 84.0 ± 2.9%. There were no significant differences in the pore size and porosity of hydrogels with or without liposomes.

Table 2. Average pore size and porosity of hydrogels with and without liposomes. Data represent mean ± standard deviation (n = 20).

Hydrogel Formulation	Pore Size (µm)	Porosity (%)	p-Value
Hydrogel without liposomes + VEGF	201.4 ± 24.5	81.1 ± 3.3	$p > 0.05$
Liposomal hydrogel + VEGF	205.8 ± 25.5	84.0 ± 2.9	

2.1.3. Chemical Properties of Hydrogels Using FTIR Spectroscopy

The FTIR spectra are displayed in Figure 4A. In general, the spectra of the hydrogel without liposomes and the liposomal hydrogel showed similar trends, except in the region between 2800 cm^{-1} and 3000 cm^{-1}, in which a peak at 2868 cm^{-1} was noted in the liposomal hydrogel spectrum due to the presence cholesterol. The two distinctive absorbance peaks at 1740 and 1591 cm^{-1} in the spectrum of chitosan (CS) represented the C=O bonding of –NHCO– (amide I) and the N–H bonding of the –NH2 (amide II) functional group, respectively. The FTIR spectrum of the lyophilized hydrogel without liposomes showed reduced absorption frequencies of amide I (1657 cm^{-1}) and amide II (1556 cm^{-1}). A similar trend was observed with regard to liposomal hydrogel, in which the intensities of the amide bands were 1568 cm^{-1} and 1411 cm^{-1}, respectively. No new peaks could be observed in either hydrogel spectra.

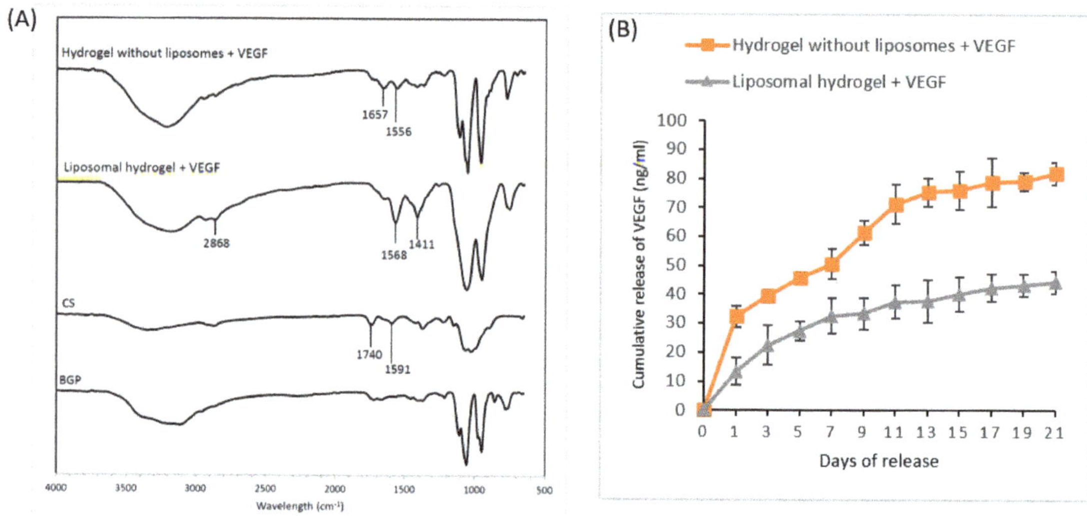

Figure 4. (**A**) The FTIR spectra of hydrogel without liposomes and liposomal hydrogel: (**B**) The cumulative release profile of VEGF from hydrogel with and without liposomes over 21 days. The cumulative release of VEGF in ng/mL was determined as a function of time by VEGF ELISA kit at 450 nm. Each value is expressed as mean ± SD (n = 3 per group). Chitosan, CS; β-glycerophosphate, BGP.

2.1.4. Release Profile of VEGF from the Hydrogels

The cumulative release graph (Figure 4B) shows an initial burst release of 32.3 ± 3.5% of VEGF in the first 24 h in the hydrogel without liposomes group. Afterward, a fast release of 81.9 ± 3.8% of the total VEGF occurred over 21 days. On the other hand, in the first 24 h, the liposomal hydrogel group had a lower initial release of 13.5 ± 4.6%, followed by a sustained delivery of 31.0 ± 5.9%, resulting in a cumulative release of 44.5 ± 3.8% VEGF after three weeks. The findings confirmed the potential use of liposomal hydrogel in the sustained release of VEGF for an extended period of at least three weeks.

2.2. *In-Vitro Characterization of Cell–Hydrogel 2.5D Culture*

2.2.1. Effect of Sustained Delivery of VEGF on Cell Morphology

The cellular morphology and interaction of t hydrogels were observed under FESEM on days 0, 7, 14, and 21 of culture. The FESEM images in Figure 5 demonstrate that cells with both spherical and spindle-shaped morphology were visible 24 h after seeding (day 0). On day 7, the cells displayed a spindle-like appearance, with cytoplasmic extensions contacting each other to form a network structure. On day 14, the cells became more polygonal and better spread, typical of differentiated osteoblastic phenotype. On day 21, calcium mineral nodules of varying sizes could be observed. Comparing the two hydrogel groups, it was revealed that the number of cells in both groups increased continuously from day 7 to day 21. The cells in the liposomal hydrogel + VEGF group spread more extensively, eventually covering the whole surface of the hydrogel, resulting in significantly higher deposits of apatite nodules by day 21 compared with the hydrogel without liposomes group, which exhibited more sparse mineral deposits. This result suggest that the liposomal hydrogel scaffold was most superior in the induction of cell differentiation and mineralization.

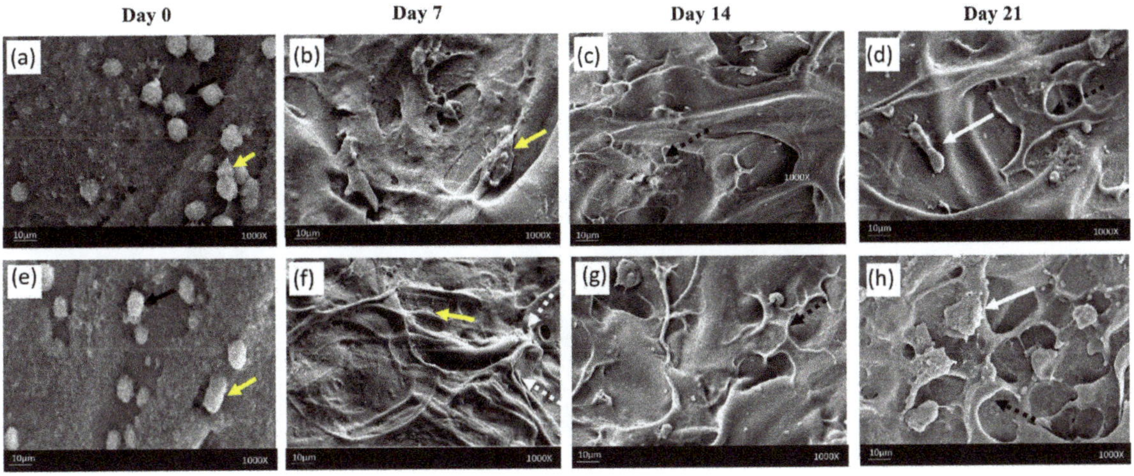

Figure 5. Surface cellular morphology viewed under FESEM at ×1000 magnification: Both spherical (black arrows) and spindle shaped cells (yellow arrows) were seen on day 0 (**a**,**e**). The presence of cytoplasmic extensions (white arrows, broken line) was observed on day 7 (**b**,**f**). Polygonal and flattened cells (black arrows, broken line) were seen on day 14 (**c**,**g**). On day 21, the liposomal hydrogel + VEGF group (**h**) showed denser cellular network and higher accumulation of mineralized nodules compared with the hydrogel without liposomes group (**d**). White arrows indicate deposition of calcium nodules.

2.2.2. Effect of Sustained Delivery of VEGF on Cell Viability

The impact of VEGF on cell viability using MTT assay is displayed in Figure 6. All the experimental groups demonstrated a time-dependent increment in the cell viability of MG-63 cells. On days 7, 14, and 21 of culture, the cell viabilities of the two hydrogel groups (hydrogel without liposomes and liposomal hydrogel) were significantly greater than the two control groups (OM and OM + VEGF) ($p < 0.05$). Between the control groups, the OM + VEGF group demonstrated significantly higher cell viability than the OM control group on day 7 of incubation ($p < 0.05$), due to the supplementation of VEGF as a bolus for one week. Nevertheless, from day 14 onwards, there were no significant differences between the control groups ($p > 0.05$). Between the two hydrogel groups, the cell viability in the liposomal hydrogel + VEGF group was significantly higher than the hydrogel without liposomes + VEGF group on days 14 and 21 of culture ($p < 0.05$). This finding indicated that the controlled delivery of VEGF via liposomal hydrogel significantly increased the cell viability of MG-63 cells over the 21-day culture period.

2.2.3. Effect of Sustained Delivery of VEGF on ALP Activity

The effect of VEGF on ALP activity is displayed in Figure 7. The ALP activity of MG-63 cells in all the experimental groups climaxed on day 14 of cultivation but reduced on day 21. Between the control groups, on day 7 of incubation, the OM + VEGF group demonstrated significantly higher ALP activity than the OM-negative control group ($p < 0.05$). Nevertheless, no significant differences were noted between the control groups from day 14 onwards ($p > 0.05$). This is because VEGF was delivered as a bolus for one week. On day 7, 14, and 21 of culture, the ALP activities of the two hydrogel groups (hydrogel without liposomes and liposomal hydrogel) were significantly higher than the two control groups (OM and OM + VEGF) ($p < 0.05$). Between the two hydrogel groups, the ALP activity in the liposomal hydrogel + VEGF group was significantly greater than the hydrogel without liposomes + VEGF group on days 14 and 21 of culture ($p < 0.05$). This finding suggested that the sustained delivery of VEGF via liposomal hydrogel significantly enhanced the ALP activity of MG-63 cells over the 21-day experimental period.

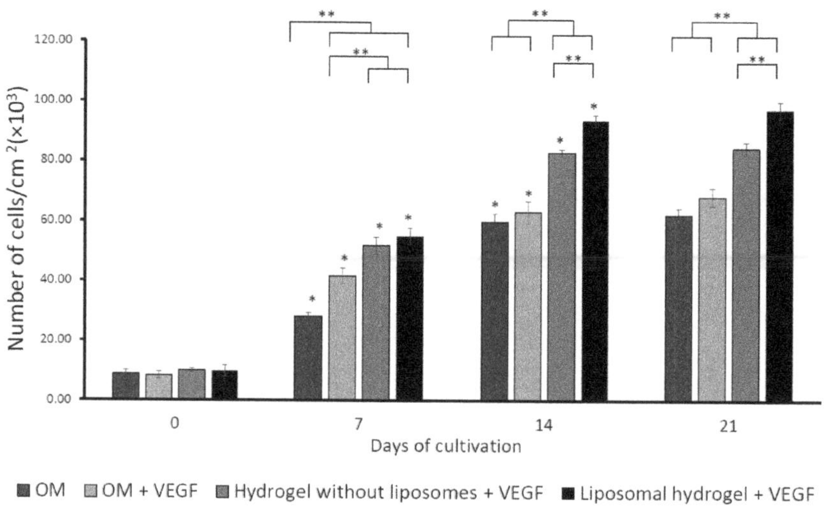

Figure 6. The effects of sustained delivery of VEGF on cell viability was determined with MTT assay on days 0, 7, 14, and 21 of cultivation. Each value was expressed as mean ± SD ($n = 3$ per group). * denotes statistical significance between different culture time points of each group ($p < 0.05$). ** denotes significant differences between groups at each time point ($p < 0.05$).

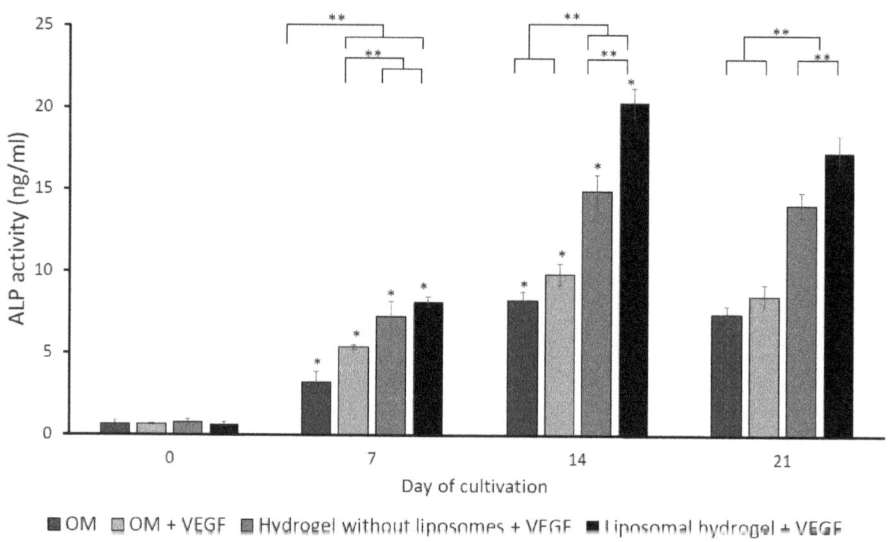

Figure 7. The effects of sustained delivery of VEGF on ALP activity assayed on day 0, 7, 14, and 21 of cultivation, and the ALP activity is expressed as nmol/ng. Each value is expressed as mean ± SD ($n = 3$ per group). * denotes statistical significance between different culture time points of each group ($p < 0.05$). ** denotes significant differences between groups at each time point ($p < 0.05$).

2.2.4. Effect of Sustained Delivery of VEGF on Matrix Mineralization

In the later stages of the matured osteoblasts, calcium ions were incorporated into ECM, resulting in matrix mineralization. The calcium deposits formed were stained by Alizarin Red S after three weeks of culture and quantitatively assayed. The quantification results are shown in Figure 8.

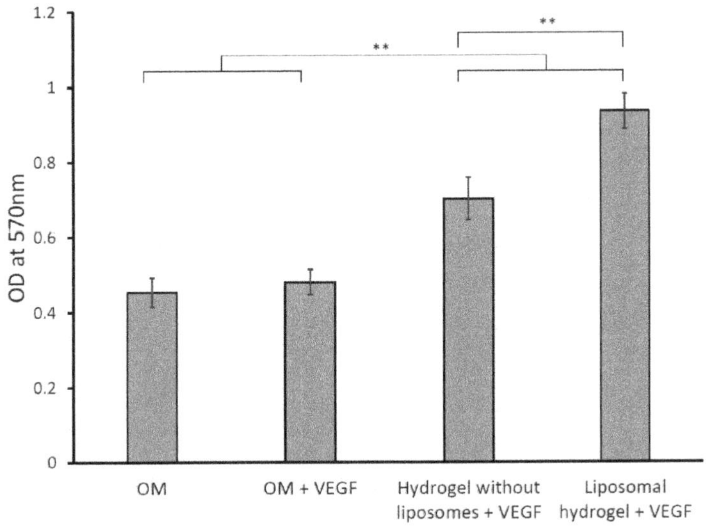

Figure 8. The effects of sustained delivery of VEGF on mineralization: Alizarin Red S staining was quantitatively assayed at 570 nm on day 21. Each value is expressed as mean ± SD (n = 3 per group). ** denotes significant differences between groups at each time point ($p < 0.05$).

The calcium deposits in the cell-seeded liposomal hydrogel + VEGF group were the highest ($p < 0.05$). The mineral deposition in the hydrogel without liposomes + VEGF group was significantly greater than the control groups but lesser than the liposomal hydrogel + VEGF group ($p < 0.05$). The calcium deposition in the positive control group (OM + VEGF) was slightly greater than the negative control (OM), but the difference was not significant ($p > 0.05$). This result indicates that the sustained release of VEGF via liposomal hydrogel significantly enhanced the matrix mineralization of MG-63 cells over the three-week experimental period.

3. Discussions

The existing literature confirmed that 21 days of continued induction was necessary for the complete osteogenic process to take place [22]. Among the different carrier systems available, Ruel-Gariépy et al. (2000) [23] reported that preloading of bioactive molecules into microparticles or liposomes is necessary to achieve sustained delivery over more than one week from thermosensitive chitosan gels. The double encapsulation strategy using liposomal hydrogel in this study successfully fulfilled the requirement of sustained delivery of VEGF for at least three weeks.

3.1. Characterization of Scaffold

The liposome formulation may influence the drug release profile. According to previous laboratory studies, the addition of cholesterol to liposome bilayers can prevent lipid exchange, thus providing an additional stabilizing effect [24]. In the current study, we followed an optimized formulation for the preparation of the liposomes, with dipalmitoylphosphatidylcholine (DPPC), dimethyldioctadecylammonium bromide (DDAB), and cholesterol (CH), at a molar ratio of 2:0.01:1 [25]. This formulation is in agreement with the proportions employed in most studies, i.e., 2:1 ratio (two parts of lipids and one part of cholesterol). However, the underlying reason for using these ratios is still unknown [24,26].

The parameters that determine the quality and suitability of a nanocarrier formulation for drug delivery include the average particle size/diameter, polydispersity index (PDI), and zeta-potential. The average particle size influences the drug release profile and bioavailability. This is because particle sizes on the order of 100 nm have a large surface-to-volume ratio, thus leading to a more controlled substance release [27,28]. The polydispersity index (PDI) measures the heterogeneity of size distribution of nanoparticles. PDI values smaller than 0.3 are deemed suitable for drug delivery purposes [29]. Zeta-potential measures the surface charge of nanoparticles in solution, and it is indicative of the colloidal stability. Zeta-potential values greater than +25 mV or less than −25 mV typically have a high level of stability [30]. Therefore, the VEGF-loaded liposomes with average particle size of 171 nm, PDI of 0.174, and zeta-potential of +38.5 mV in the current study were considered optimum for drug delivery. On average, the VEGF-loaded liposomes demonstrated a slight increase in particle size, PDI, and zeta-potential. However, the differences between liposomes with and without VEGF were statistically insignificant ($p > 0.05$). These findings showed that the loaded VEGF did not significantly alter the physical properties and morphology of the liposomes.

The presence of liposomes within the polymeric chitosan hydrogel matrix did not disturb the gelation process. However, the gelation time for the hydrogel system was shortened after the liposomes were loaded. This was due to the alteration in the rheological properties of the hydrogel in the presence of secondary carriers (liposomes) [31]. The resultant liposomal hydrogel maintained the optimal microstructure (pore size and porosity), which was appropriate for cell attachment and growth. The average pore diameter of 100–325 μm was considered ideal for bone-tissue-engineered scaffolds according to Murphy et al. (2010) [32]. Previous studies found that bigger pore diameters (100–200 μm) favored bone regeneration after implantation, while smaller pore sizes below 100 μm resulted in fibrous tissue formation [33]. Porosity within the range of 80–90% was recommended for the cell penetration and vascularization of implanted scaffolds in bone defects [34]. Highly porous scaffolds (>90%) were associated with higher permeability and surface area for cell–biomaterial interactions. However, this resulted in inferior mechanical properties and higher degradability [35]. Therefore, the liposomal hydrogel with a pore diameter of 205.8 ± 25.5 μm and porosity of 84.0 ± 2.9% in this current study was acceptable for cell culture.

The FTIR analysis is a spectroscopic technique that uses infrared light to identify the chemical properties of scanned samples [36]. The FTIR result for hydrogel without liposomes from our study was similar to the study by Qin et al. (2018) [37]. The spectra of the lyophilized hydrogels with and without liposomes showed similar trends. This showed that the addition of liposomes did not alter the chemical structure of the hydrogel. Nevertheless, an absorbance peak at 2868 cm^{-1} was observed in the liposomal hydrogel spectrum, which corresponded to the tensile vibrations of C-H bonds and carbon cyclic rings. The region between 2800 cm^{-1} and 3000 cm^{-1} is a hallmark of the presence of cholesterol in liposomes [38]. Reduced wavenumbers of amide I and amide II bands as compared with the spectrum of chitosan were observed in both the hydrogel spectra. The FTIR results indicate that the sol–gel transition was due to the electrostatic interaction between negatively charged β-glycerophosphate and positively charged chitosan [39,40]. No new peaks could be observed in both hydrogel spectra, confirming the absence of chemical bonds between chitosan and β- glycerophosphate [39].

It was perceived that owing to the presence of liposome, the liposomal hydrogel had a lower initial release of about 13.5% in the first 24 h, with a cumulative release of 44.5% VEGF over 21 days, nearly half of that observed in the hydrogel without liposomes group. These results suggest that the presence of liposomes as secondary carriers provided a continuous release of VEGF below 40 ng/mL over 21 days, as required for the complete osteogenic process to take place. This result corroborates the findings of other studies, in which the sustained delivery of bioactive molecules was accomplished using liposomes [41,42]. Due to liposome leakage, VEGF first escaped from the liposomes into the hydrogel matrix. Later, VEGF was discharged into the culture medium, first through diffusion and then through degradation of the hydrogel matrix.

3.2. Effects of Liposomal Hydrogel on Cell Viability, Differentiation, and Mineralization

Osteogenesis comprises three major stages, proliferation, differentiation, and mineralization, taking place under continuous osteogenic induction for at least 21 days. Throughout these phases, expressions of specific osteogenic markers are temporally and sequentially organized [43,44]. Previous studies only compared the effects of VEGF-containing hydrogel without any secondary nanocarrier with negative controls over 14–21 days [45,46]. Therefore, in order to further investigate the effects of VEGF on osteogenesis in this study, three methods of delivery of VEGF, namely bolus delivery (OM + VEGF group), burst release (hydrogel without liposomes + VEGF), and sustained delivery (liposomal hydrogel + VEGF), were conducted, and the effects on cell proliferation, differentiation, and mineralization were evaluated for 21 days.

The findings of the current study showed that the effect of sustained delivery of VEGF from liposomal hydrogel was the most superior compared with the hydrogel without liposomes carrier system reported by Wu et al. (2019) [45] and Elango (2023) [46], as well as the bolus delivery of VEGF without carrier [47]. This is explained by the sustained delivery of VEGF over the 21-day culture period. These results are in agreement with a previous study, in which the sustained delivery of bioactive molecules was accomplished using liposomes as secondary carriers [41].

The MG-63 cells exhibited favorable cell–material interaction on the hydrogels due to the presence of chitosan. The surface of chitosan promoted cell attachment by being positively charged, to which the negatively charged cell surfaces are naturally attached [48]. The porous and sponge-like microstructure of the hydrogels also facilitated cell attachment and proliferation [49]. While both hydrogels were proven to be non-cytotoxic and biocompatible, the liposomal hydrogel + VEGF was shown to be superior in the induction of cell differentiation and mineralization compared with hydrogel without liposomes + VEGF, as demonstrated by the more densely packed differentiated osteoblastic cells with a higher amount of calcium mineral deposits in the liposomal hydrogel + VEGF group.

VEGF stimulated the proliferative and differentiation potential of cells in a time-dependent manner, similar to other studies [47,50]. Increased proliferation and differentiation were due to the direct interaction of VEGF with the receptors expressed by the osteogenic cells, as proven by the presence of VEGFR-1 and VEGFR-2 receptors and VEGF binding proteins, Neuropilin-1 and -2 [51]. The study by Lee et al. (2012) showed that VEGF did not have a significant effect on the proliferation of human periodontal ligament stem cells, which was not in agreement with this study. However, the effects of VEGF were studied for a short period of time which did not extend beyond 5 days. In the current study, the proliferation stimulative effect of VEGF was no longer significant after the cell differentiation was initiated on around day 14. This current observation is consistent with the results in other studies also reporting that cell proliferation declines after two weeks of the osteogenic induction [46,52,53]. Both the hydrogels, with and without liposomes groups, exhibited higher ALP activities than the OM and OM + VEGF groups, consistent with the findings by Takagishi et al. (2006) [54], in which the promotion of osteogenic differentiation was reported in MG-63 cells cultured on gelatin hydrogels. The liposomal hydrogel + VEGF group expressed the highest ALP activity on day 14, because the osteoblastic differentiation was initiated in the MG-63 cells during this period, allowing the effects of persistent exposure of VEGF on osteoblastic differentiation to be fully expressed.

Following the rise in ALP activity, matrix mineralization takes place. ALP breaks down pyrophosphate and inorganic phosphate, which results in the deposition of calcium minerals in the collagen fibers of ECM to form hydroxyapatite crystals from approximately the 14th day of culture [55]. Being a weak base, BGP was able to induce thermo-sensitive gelation when it was added to the acidic chitosan solution. At the same time, BGP also served as a source of organic phosphate for mineralization [56]. A previous study found that the sustained release of organophosphates from CS/BGP hydrogel may result in significantly greater calcium deposition [43]. Calcium mineral deposition exhibited the same trend as cell differentiation assays. The liposomal hydrogel + VEGF group induced the highest level of mineralization in MG-63 cell cultures. The current result again indicated that the synergistic effect between OM and sustained delivery of VEGF was the most effective in the induction of terminal differentiation and mineralization, supported by other studies conducted on human dental pulp stem cells [47,57].

Nevertheless, the optimal concentrations of VEGF for enhanced osteogenesis are controversial. The heterogeneity between studies may be influenced by the cell type, culturing conditions, and duration of time exposure of cells to VEGF. A recent study showed enhanced osteogenesis in human bone mesenchymal stem cells cultured for 21 days on a collagen–hydrogel scaffold [46]. However, the study by Aksel et al. (2017) on dental pulp stem cells showed that the highest levels of mineralization and expression of the osteogenic gene were seen when VEGF was given only during the first week of incubation, instead of continuous supplementation over 21 days [57]. Song et al. (2011) [58] reported that higher concentrations of VEGF over a longer incubation period could induce the upregulation of the inhibitor of DNA-binding 1 protein (Id1), which may retard terminal osteoblast differentiation and mineralization.

An in vitro study on human dental pulp stem cells demonstrated the highest level of mineralization when 40 ng/mL VEGF was used [47]. Another study showed enhanced osteogenesis of human periodontal ligament stem cells in vitro and in vivo in the presence of 25 ng/mL of VEGF [59]. The study by Wu et al. (2019) on dental pulp stem cells (DPSCs) reported that the odontogenic effect of 100 ng/mL VEGF loaded in a hydrogel delivery system was significantly greater than 100 ng/mL VEGF supplementation without carriers [45]. This finding was supported by the most recent study by Elango (2023), in which the differentiation and mineralization were accelerated when human bone mesenchymal stem cells were cultured on a collagen–hydrogel carrier system loaded with 100 ng/mL VEGF [46]. On the other hand, the studies by Behr et al. (2011) [60] and Aksel et al. (2017) [57] showed that different concentrations of VEGF (5–200 ng/mL) did not appear to induce significantly different osteoinductive effects on human adipose-derived stem cells and human dental pulp stem cells, respectively. Due to the varying results of the studies, until further confirmatory studies are conducted, the cumulative release of VEGF around 40 ng/mL over 3 weeks via chitosan hydrogel loaded with 100 ng/mL VEGF in the current study appears to be justified, because it would be safer to maintain the release of VEGF at a minimal level to avoid the undesirable side effects of excessive local concentrations of VEGF, such as hypertrophic bones and defective angiogenesis [61,62].

3.3. Limitations and Recommendations for Future Research

Although the MG-63 cell line is a useful cell model in bone research, it does not fully emulate the behavior of primary human osteoblast cells or induced osteoblasts from human pluripotent stem cells. For the determination of gelation time, rheological analysis using storage and loss modulus would be more accurate than the conventional tube inversion method. In addition, the parameters that are essential to maximize bone formation, such as the loading concentration and duration of delivery of VEGF, require further optimization. The 2.5D culture design could be upgraded to a 3D culture system with the usage of stem cells to further investigate the possibility of cell encapsulation combined with liposomal delivery of VEGF in bone regeneration. The release profile of VEGF from liposomes alone without hydrogel could be added in future studies for more thorough characterization of the release kinetics of VEGF.

4. Conclusions

Besides functioning as a scaffold in BTE, liposomal hydrogel is a very promising vehicle to deliver VEGF in a sustained manner in order to enhance the osteogenesis of osteoblast-like cells, as demonstrated by the increase in cell attachment, proliferation, differentiation, and mineralization. The liposome-loaded, 2% *w*/*v* chitosan, 6% *w*/*v* BGP hydrogel also displayed a unique thermosensitive profile for use in injectable form. The potential of this double encapsulation strategy in bone regeneration can be further investigated in preclinical in-vivo studies using human MSCs.

5. Materials and Methods

5.1. Materials

VEGF165 Protein (Human Recombinant Animal Free) was obtained from Merck (Darmstadt, Germany). Chitosan (low molecular weight, 85% deacetylated) was acquired from Thermo Fisher Scientific (Waltham, MA, USA). Dipalmitoylphosphatidylcholine (DPPC), cholesterol (CH), dimethyldioctadecylammonium bromide (DDAB), beta-glycerophosphate (BGP), and other materials were bought from Sigma-Aldrich (Schnelldorf, Germany).

5.2. Preparation of Scaffold

5.2.1. Preparation of VEGF-Loaded Liposomes

Liposome nanocarriers with encapsulated VEGF were fabricated using the thin lipid film hydration method [25]. DPPC, CH, and DDAB were mixed with chloroform at a weight ratio of 5 mg:1.31 mg:0.021 mg. Rotary evaporator was used to evaporate the chloroform at 55 °C under controlled low pressure until a dry lipid film was produced. Recombinant human VEGF165 was reconstituted in phosphate-buffered saline (VEGF/PBS). Liposomes with VEGF were produced by hydrating the thin lipid films with VEGF/PBS. After that, the sonication technique was used to produce liposomes in the form of small unilamellar vesicles. The liposome samples were kept in an ice-cold bath and sonicated using a probe sonicator with the following settings: 10 s ON, 10 s OFF intervals, 40% amplitude, and 750 W, for a total period of 15 min [63].

5.2.2. Preparation of Thermoresponsive Hydrogels

An amount of 10 mL of hydrogel was prepared following the protocol mentioned in the study by Qin et al. [37]. Chitosan (CS) solution was made by mixing 200 mg CS powder in 8 mL acetic acid solution (0.75% *v*/*v*) at room temperature for 3 h. Subsequently, the dissolved CS was kept at 4 °C for 24 h to reduce the bubbles inside. BGP solution was prepared by adding 600 mg BGP in 2 mL distilled water and kept chilled at 4 °C. The chilled BGP was added drop by drop into the CS solution under continuous stirring in an ice bath for 10 min. The volume ratio of CS:BGP was 4:1. After that, liposomal hydrogel (liposomal hydrogel + VEGF) was obtained by adding the appropriate amount of VEGF loaded liposomes into the CS/BGP mixture under gentle stirring for another 10 min to attain a final concentration of 100 ng/mL (*w*/*v*) VEGF [45,46]. Hydrogel without liposomes (hydrogel without liposomes + VEGF) was obtained by adding VEGF/PBS to attain the same final concentration of 100 ng/mL (*w*/*v*) VEGF. Gelation times for both hydrogels were ascertained using the tube inversion method in 37 °C water baths. The gelation point was established when there was no flow over 30 s, with the glass bottles being turned upside down [64].

5.3. Characterization of Scaffold

5.3.1. Characterization of Physical Properties and Morphology of Liposomes

The particle sizes, polydispersity indexes (PDI), and zeta-potentials of liposomes (with and without VEGF) were ascertained in triplicates with the dynamic light scattering (DLS) method using a Zetasizer Nano ZS ZEN 3500 instrument (Malvern Instruments Ltd., Worcestershire, UK). The morphology of the liposomes with and without VEGF was observed by using a Talos-L120C transmission electron microscope (TEM, Thermo Scientific,

Waltham, USA) at an accelerating voltage of 100 kV. The liposomes were stained with 1% ammonium molybdate (pH 7) on carbon-coated copper grids and analyzed under TEM.

5.3.2. Morphological Characterization of Hydrogel Using FESEM

After gelation, the hydrogels were lyophilized in a freeze drier overnight at −80 °C. After that, the samples were sectioned and analyzed under field emission scanning electron microscope (FESEM) (Merlin VP Compact, Zeiss, Oberkochen, Germany) to obtain the surface microstructure images of the hydrogels. The mean pore diameter of each sample was measured using the software Image-J by considering 20 pores [37].

5.3.3. Porosity Determination of Hydrogels

The percentage of porosity was measured using the Archimedes method [65]. Initially, the dry weights of the lyophilized hydrogels (W_{dry}) were measured and recorded. Then, the hydrogels were prewet with dehydrated alcohol and soaked in water. The submerged weights (W_{sub}) were recorded when the hydrogels were submerged under water. After removal from the water, the weights were measured again (W_{wet}). The porosities of the hydrogels were calculated with the formula below [65]:

$$\text{Porosity (\%)} = [(W_{wet} - W_{dry})/(W_{wet} - W_{sub})] \times 100\%$$

where W_{wet} is the wet weight, W_{dry} is the dry weight, and W_{sub} represents the weight of the submerged hydrogel.

5.3.4. Characterization of Chemical Properties of Hydrogels Using FTIR Spectroscopy

Chemical properties of the lyophilized hydrogels and raw materials (CS and β-GP) were characterized using a SPECTRUM 400 FTIR spectrometer (Perkin Elmer, Waltham, USA) equipped with an attenuated total reflectance (ATR) accessory. All FTIR spectra (32 scans per spectrum) were collected within the wavelength range of 4000–500 cm^{-1} at a resolution of 4.0 cm^{-1} [37].

5.3.5. Release Profile of VEGF from the Hydrogels

An amount of 2 mL of phosphate-buffered saline (PBS) was added to the hydrogels and incubated at 37 °C for 24 h. The leachate was collected and immediately frozen at −80 °C every 2 days over 3 weeks. After each leachate collection, the same volume of PBS was replenished. The concentrations of VEGF in the leachate were quantified utilizing the human VEGF enzyme-linked immunosorbent assay (ELISA) kit (FineTest, Wuhan, China). The cumulative release of VEGF (in ng/mL) was plotted as a function of time [45].

5.4. Cell–Hydrogel 2.5D Co-Culture

The cell line used for this study was the MG-63 cell line. The cell line was procured from the American Type Culture Collection (ATCC, Bethesda, MD, USA) (ATCC No. CRL-1427™). Complete medium (CM), containing Eagle's minimum essential medium (EMEM), 10% (*v*/*v*) fetal bovine serum (FBS), and 1% (*v*/*v*) penicillin–streptomycin (Gibco, Grand Island, NY, USA), was used to culture the cells. Osteogenic medium (OM) composed of complete medium supplemented with 50 μg/mL L-ascorbic-2-phosphate (AAP) was used for osteogenic induction [66].

Hydrogels were prepared in aseptic environment according to the protocol mentioned previously. The sterile hydrogels were added into 24-well plates (500 μL/well). The hydrogels were incubated at 37 °C for 15 min for gelation to occur. MG-63 cells were seeded at $1 \times 10^4/cm^2$ seeding density on the hydrogel without liposomes + VEGF and liposomal hydrogel + VEGF scaffold groups and cultured in CM. After 24 h, the CM was discarded and replaced with OM. This marked day 0 of the analysis. In the positive control group, MG-63 cells were cultured without hydrogel in OM, initially supplemented with a bolus of 100 ng/mL (*w*/*v*) VEGF (OM + VEGF group). The VEGF was not replenished after one week to simulate the burst release effect. As a negative control, MG-63 cells were cultured without hydrogel in OM with no VEGF supplementation (OM group). The medium was replaced with fresh OM every 72 h. Subsequent experiments were performed in triplicates.

5.5. In Vitro Characterization of Cell–Hydrogel Co-Culture

5.5.1. Cell Morphology Analysis Using FESEM

The cell attachment and morphology of the hydrogels were assessed using a field emission scanning electron microscope (FESEM) on day 7, 14, and 21. The samples were first fixed overnight in 2.5% v/v glutaraldehyde. Then, the samples were washed with PBS, followed by dehydration in a freeze drier for 24 h at −80 °C. After that, the samples were examined under FESEM (Merlin VP compact, Zeiss, Oberkochen, Germany) after sputter-coating.

5.5.2. Cell Viability Analysis Using MTT Assay

MTT assay was used to investigate the cell viability on day 0, 7, 14, and 21. MTT solution and complete medium (volume ratio of 1:9) were added to each well and incubated at 37 °C for 4 h. Dimethyl sulfoxide (DMSO) in glycine buffer at pH 7.4 was used to dissolve the formazan crystal generated as a result of mitochondrial dehydrogenase activity. The solubilized formazan product was then transferred to a 96-well plate, and the optical density reading at 570 nm was recorded using an ELISA microplate reader (Varioskan Flash, Thermo Scientific, Waltham, USA). Each experiment was performed in triplicate. Number of cells/cm^2 was plotted against the day of analysis.

5.5.3. Cell Differentiation Analysis Using Alkaline Phosphatase (ALP) Assay

Alkaline phosphatase (ALP) is the key marker for determining osteoblastic phenotype. At the end of the 0-, 7-, 14-, and 21-day incubation time points, ALP activity was assayed using a Sensolyte® pNPP alkaline phosphatase assay kit (AnaSpec, California, USA), as per the manufacturer's protocol. In the presence of ALP, p-nitrophenol was released from p-nitrophenyl phosphate (PNPP) solutions. The concentration of p-nitrophenol was assayed in triplicate by recording the light absorbance at 405 nm using an ELISA reader (Varioskan Flash, Thermo Scientific, Waltham, USA). The result of the ALP assay was presented by plotting a graph of ALP activity (ng/mL) against the day of culture.

5.5.4. Mineralization Analysis Using Alizarin Red S Staining Quantification

Alizarin Red S (ARS) staining was carried out to detect the calcium mineral deposit present. On day 21, the cell layers were washed with PBS, fixed with 10% formalin, and stained with 1% ARS (Sigma-Aldrich, Schnelldorf, Germany). Quantification of the calcium mineral deposition was performed by eluting the alizarin red stain with 10% (w/v) cetylpyridinium chloride. The absorbance was recorded at 570 nm using a microplate reader (Varioskan Flash, Thermo Scientific, Waltham, USA).

5.6. Statistical Analysis

The data analysis was conducted using the SPSS version 26. Descriptive data is expressed as mean ± standard deviation (SD). Normality of the data distribution was checked with Shapiro–Wilk test. An independent t-test was used for the comparison of microstructural characteristics between liposomes with and without VEGF. A Mann–Whitney U test was used for the comparison between hydrogel with and without liposomes. One-way analysis of variance (ANOVA) with the Bonferroni's post hoc test was used to compare the differences between experimental groups in 2.5D co-culture. For the MTT and ALP results, one-way repeated measure ANOVA (Bonferroni post hoc) was used to compare the differences between experimental time points within the same group. A value of $p < 0.05$ was considered statistically significant.

Author Contributions: Conceptualization, R.M.A.W., F.A. and F.Y.; methodology, R.M.A.W., S.H.Z.A. and F.Y.; formal analysis, M.H.T. and F.Y.; investigation, M.H.T. and F.Y.; resources, F.A.; data curation, M.H.T.; writing—original draft preparation, M.H.T. and R.M.A.W.; writing—review and editing, R.M.A.W., S.H.Z.A., F.A. and F.Y.; supervision, R.M.A.W., F.A. and F.Y.; project administration, F.Y.; funding acquisition, F.Y. All authors have read and agreed to the published version of the manuscript.

Funding: This research was financially supported by the Fundamental Research Grant Scheme (FRGS), Ministry of Higher Education, Malaysia (FRGS/1/2020/SKK0/UKM/03/4).

Institutional Review Board Statement: The research protocol was reviewed and approved by the Research Ethics Committee of Universiti Kebangsaan Malaysia (UKM) before the start of the study (JEP-2021-408).

Informed Consent Statement: Not applicable.

Data Availability Statement: The data presented in this study are available on request from the corresponding author.

Acknowledgments: The authors gratefully acknowledge Universiti Kebangsaan Malaysia through Fundamental Research Grant Scheme from the Ministry of Higher Education Malaysia for providing an opportunity to work on this research.

Conflicts of Interest: The authors declare no conflict of interest.

References

1. Pacelli, S.; Basu, S.; Whitlow, J.; Chakravarti, A.; Acosta, F.; Varshney, A.; Modaresi, S.; Berkland, C.; Paul, A. Strategies to develop endogenous stem cell-recruiting bioactive materials for tissue repair and regeneration. *Adv. Drug Deliv. Rev.* **2017**, *120*, 50–70. [CrossRef] [PubMed]
2. Luque-Martín, E.; Tobella-Camps, M.-L.; Rivera-Baró, A. Alveolar graft in the cleft lip and palate patient: Review of 104 cases. *Med. Oral Patol. Oral Cir. Bucal* **2014**, *19*, e531–e537. [CrossRef] [PubMed]
3. Wahab, R.M.A.; Ng, W.; Yazid, F.; Luchman, N.; Ariffin, S. Orthodontic considerations in bone graft selection for alveolar cleft repair. *Sains Malays.* **2020**, *49*, 349–356. [CrossRef]
4. Sheikh, Z.; Najeeb, S.; Khurshid, Z.; Verma, V.; Rashid, H.; Glogauer, M. Biodegradable Materials for Bone Repair and Tissue Engineering Applications. *Materials* **2015**, *8*, 5744–5794. [CrossRef]
5. Myeroff, C.; Archdeacon, M. Autogenous bone graft: Donor sites and techniques. *J. Bone Jt. Surg. Am. Vol.* **2011**, *93*, 2227–2236. [CrossRef]
6. Mankin, H.J.; Hornicek, F.J.; Raskin, K.A. Infection in massive bone allografts. *Clin. Orthop. Relat. Res.* **2005**, *432*, 210–216. [CrossRef]
7. Wahab, R.M.A.; Abdullah, N.; Zainal Ariffin, S.H.; Che Abdullah, C.A.; Yazid, F. Effects of the sintering process on nacre-derived hydroxyapatite scaffolds for bone engineering. *Molecules* **2020**, *25*, 3129. [CrossRef]
8. Zhao, R.; Yang, R.; Cooper, P.R.; Khurshid, Z.; Shavandi, A.; Ratnayake, J. Bone Grafts and Substitutes in Dentistry: A Review of Current Trends and Developments. *Molecules* **2021**, *26*, 3007. [CrossRef]
9. Oppenheimer, A.J.; Mesa, J.; Buchman, S.R. Current and emerging basic science concepts in bone biology: Implications in craniofacial surgery. *J. Craniofacial Surg.* **2012**, *23*, 30–36. [CrossRef]
10. Stamnitz, S.; Klimczak, A. Mesenchymal Stem Cells, Bioactive Factors, and Scaffolds in Bone Repair: From Research Perspectives to Clinical Practice. *Cells* **2021**, *10*, 1925. [CrossRef]
11. Gromolak, S.; Krawczenko, A.; Antończyk, A.; Buczak, K.; Kiełbowicz, Z.; Klimczak, A. Biological Characteristics and Osteogenic Differentiation of Ovine Bone Marrow Derived Mesenchymal Stem Cells Stimulated with FGF-2 and BMP-2. *Int. J. Mol. Sci.* **2020**, *21*, 9726. [CrossRef] [PubMed]
12. Schipani, E.; Maes, C.; Carmeliet, G.; Semenza, G.L. Regulation of osteogenesis-angiogenesis coupling by HIFs and VEGF. *J. Bone Miner. Res.* **2009**, *24*, 1347–1353. [CrossRef] [PubMed]
13. Grosso, A.; Burger, M.G.; Lunger, A.; Schaefer, D.J.; Banfi, A.; Di Maggio, N. It takes two to tango: Coupling of angiogenesis and osteogenesis for bone regeneration. *Front. Bioeng. Biotechnol.* **2017**, *5*, 68. [CrossRef] [PubMed]
14. Baldo, B.A. Side effects of cytokines approved for therapy. *Drug Saf.* **2014**, *37*, 921–943. [CrossRef]
15. Hassani, A.; Avci Ç, B.; Kerdar, S.N.; Amini, H.; Amini, M.; Ahmadi, M.; Sakai, S.; Bagca, B.G.; Ozates, N.P.; Rahbarghazi, R.; et al. Interaction of alginate with nano-hydroxyapatite-collagen using strontium provides suitable osteogenic platform. *J. Nanobiotechnol.* **2022**, *20*, 310. [CrossRef]
16. Oliveira, M.Z.; Albano, M.B.; Namba, M.M.; da Cunha, L.A.; de Lima Gonçalves, R.R.; Trindade, E.S.; Andrade, L.F.; Vidigal, L. Effect of hyaluronic acids as chondroprotective in experimental model of osteoarthrosis. *Rev. Bras. Ortop.* **2014**, *49*, 62–68. [CrossRef]
17. Vieira, S.; Strymecka, P.; Stanaszek, L.; Silva-Correia, J.; Drela, K.; Fiedorowicz, M.; Malysz-Cymborska, I.; Janowski, M.; Reis, R.L.; Łukomska, B.; et al. Mn-Based Methacrylated Gellan Gum Hydrogels for MRI-Guided Cell Delivery and Imaging. *Bioengineering* **2023**, *10*, 427. [CrossRef]

18. Alavi, S.; Haeri, A.; Dadashzadeh, S. Utilization of chitosan-caged liposomes to push the boundaries of therapeutic delivery. *Carbohydr. Polym.* **2017**, *157*, 991–1012. [CrossRef]
19. Elieh-Ali-Komi, D.; Hamblin, M.R. Chitin and chitosan: Production and application of versatile biomedical nanomaterials. *Int. J. Adv. Res.* **2016**, *4*, 411–427.
20. Saravanan, S.; Vimalraj, S.; Thanikaivelan, P.; Banudevi, S.; Manivasagam, G. A review on injectable chitosan/beta glycerophosphate hydrogels for bone tissue regeneration. *Int. J. Biol. Macromol.* **2019**, *121*, 38–54. [CrossRef]
21. Thirumaleshwar, S.; Kulkarni, P.; Gowda, D. Liposomal hydrogels: A novel drug delivery system for wound dressing. *Curr. Drug Ther.* **2012**, *7*, 212–218. [CrossRef]
22. Robert, A.W.; Angulski, A.B.B.; Spangenberg, L.; Shigunov, P.; Pereira, I.T.; Bettes, P.S.L.; Naya, H.; Correa, A.; Dallagiovanna, B.; Stimamiglio, M.A. Gene expression analysis of human adipose tissue-derived stem cells during the initial steps of in vitro osteogenesis. *Sci. Rep.* **2018**, *8*, 4739. [CrossRef]
23. Ruel-Gariépy, E.; Chenite, A.; Chaput, C.; Guirguis, S.; Leroux, J. Characterization of thermosensitive chitosan gels for the sustained delivery of drugs. *Int. J. Pharm.* **2000**, *203*, 89–98. [CrossRef] [PubMed]
24. Nakhaei, P.; Margiana, R.; Bokov, D.O.; Abdelbasset, W.K.; Jadidi Kouhbanani, M.A.; Varma, R.S.; Marofi, F.; Jarahian, M.; Beheshtkhoo, N. Liposomes: Structure, Biomedical Applications, and Stability Parameters with Emphasis on Cholesterol. *Front. Bioeng. Biotechnol.* **2021**, *9*, 705886. [CrossRef]
25. Marasini, N.; Giddam, A.K.; Ghaffar, K.A.; Batzloff, M.R.; Good, M.F.; Skwarczynski, M.; Toth, I. Multilayer engineered nanoliposomes as a novel tool for oral delivery of lipopeptide-based vaccines against group A Streptococcus. *Nanomedicine* **2016**, *11*, 1223–1236. [CrossRef]
26. Briuglia, M.L.; Rotella, C.; McFarlane, A.; Lamprou, D.A. Influence of cholesterol on liposome stability and on in vitro drug release. *Drug Deliv. Transl. Res.* **2015**, *5*, 231–242. [CrossRef]
27. Bahari, L.A.S.; Hamishehkar, H. The impact of variables on particle size of solid lipid nanoparticles and nanostructured lipid carriers; A comparative literature review. *Adv. Pharm. Bull.* **2016**, *6*, 143–151. [CrossRef] [PubMed]
28. Singh, R.; Lillard, J.W., Jr. Nanoparticle-based targeted drug delivery. *Exp. Mol. Pathol.* **2009**, *86*, 215–223. [CrossRef]
29. Danaei, M.; Dehghankhold, M.; Ataei, S.; Hasanzadeh Davarani, F.; Javanmard, R.; Dokhani, A.; Khorasani, S.; Mozafari, M.R. Impact of particle size and polydispersity index on the clinical applications of lipidic nanocarrier systems. *Pharmaceutics* **2018**, *10*, 57. [CrossRef] [PubMed]
30. Honary, S.; Zahir, F. Effect of zeta potential on the properties of nano-drug delivery systems—A review (Part 1). *Trop. J. Pharm. Res.* **2013**, *12*, 255–264. [CrossRef]
31. Li, Y.; Liu, C.; Tan, Y.; Xu, K.; Lu, C.; Wang, P. In situ hydrogel constructed by starch-based nanoparticles via a Schiff base reaction. *Carbohydr. Polym.* **2014**, *110*, 87–94. [CrossRef] [PubMed]
32. Murphy, C.M.; Haugh, M.G.; O'Brien, F.J. The effect of mean pore size on cell attachment, proliferation and migration in collagen-glycosaminoglycan scaffolds for bone tissue engineering. *Biomaterials* **2010**, *31*, 461–466. [CrossRef] [PubMed]
33. Iviglia, G.; Kargozar, S.; Baino, F. Biomaterials, current strategies, and novel nano-technological approaches for periodontal regeneration. *J. Funct. Biomater.* **2019**, *10*, 3. [CrossRef] [PubMed]
34. Tripathi, A.; Melo, J.S. Preparation of a sponge-like biocomposite agarose–chitosan scaffold with primary hepatocytes for establishing an in vitro 3D liver tissue model. *RSC Adv.* **2015**, *5*, 30701–30710. [CrossRef]
35. Abbasi, N.; Hamlet, S.; Love, R.M.; Nguyen, N.-T. Porous scaffolds for bone regeneration. *J. Sci. Adv. Mater. Devices* **2020**, *5*, 1–9. [CrossRef]
36. Kowalczuk, D.; Pitucha, M. Application of FTIR method for the assessment of immobilization of active substances in the matrix of biomedical materials. *Materials* **2019**, *12*, 2972. [CrossRef]
37. Qin, H.; Wang, J.; Wang, T.; Gao, X.; Wan, Q.; Pei, X. Preparation and characterization of chitosan/β-glycerophosphate thermal-sensitive hydrogel reinforced by graphene oxide. *Front. Chem.* **2018**, *6*, 565. [CrossRef]
38. Pilch, E.; Musiał, W. Selected Physicochemical Properties of Lyophilized Hydrogel with Liposomal Fraction of Calcium Dobesilate. *Materials* **2018**, *11*, 2143. [CrossRef]
39. Boido, M.; Ghibaudi, M.; Gentile, P.; Favaro, E.; Fusaro, R.; Tonda-Turo, C. Chitosan-based hydrogel to support the paracrine activity of mesenchymal stem cells in spinal cord injury treatment. *Sci. Rep.* **2019**, *9*, 6402. [CrossRef]
40. Assaad, E.; Maire, M.; Lerouge, S. Injectable thermosensitive chitosan hydrogels with controlled gelation kinetics and enhanced mechanical resistance. *Carbohydr. Polym.* **2015**, *130*, 87–96. [CrossRef]
41. Cheng, R.; Yan, Y.; Liu, H.; Chen, H.; Pan, G.; Deng, L.; Cui, W. Mechanically enhanced lipo-hydrogel with controlled release of multi-type drugs for bone regeneration. *Appl. Mater. Today* **2018**, *12*, 294–308. [CrossRef]
42. Melling, G.E.; Colombo, J.S.; Avery, S.J.; Ayre, W.N.; Evans, S.L.; Waddington, R.J.; Sloan, A.J. Liposomal delivery of demineralized dentin matrix for dental tissue regeneration. *Tissue Eng. Part A* **2018**, *24*, 1057–1065. [CrossRef] [PubMed]
43. Kim, S.; Tsao, H.; Kang, Y.; Young, D.A.; Sen, M.; Wenke, J.C.; Yang, Y. In vitro evaluation of an injectable chitosan gel for sustained local delivery of BMP-2 for osteoblastic differentiation. *J. Biomed. Mater. Res. Part A* **2011**, *99B*, 380–390. [CrossRef] [PubMed]
44. Rutkovskiy, A.; Stensløkken, K.-O.; Vaage, I.J. Osteoblast Differentiation at a Glance. *Med. Sci. Monit. Basic Res.* **2016**, *22*, 95–106. [CrossRef] [PubMed]
45. Wu, S.; Zhou, Y.; Yu, Y.; Zhou, X.; Du, W.; Wan, M.; Fan, Y.; Zhou, X.; Xu, X.; Zheng, L. Evaluation of chitosan hydrogel for sustained delivery of VEGF for odontogenic differentiation of dental pulp stem cells. *Stem Cells Int.* **2019**, *2019*, 1515040. [CrossRef]

46. Elango, J. Proliferative and Osteogenic Supportive Effect of VEGF-Loaded Collagen-Chitosan Hydrogel System in Bone Marrow Derived Mesenchymal Stem Cells. *Pharmaceutics* **2023**, *15*, 1297. [CrossRef]
47. D' Alimonte, I.; Nargi, E.; Mastrangelo, F.; Falco, G.; Lanuti, P.; Marchisio, M.; Miscia, S.; Robuffo, I.; Capogreco, M.; Buccella, S.; et al. Vascular endothelial growth factor enhances in vitro proliferation and osteogenic differentiation of human dental pulp stem cells. *J. Biol. Regul. Homeost. Agents* **2011**, *25*, 57–69.
48. Aranaz, I.; Mengíbar, M.; Harris, R.; Paños, I.; Miralles, B.; Acosta, N.; Galed, G.; Heras, Á. Functional characterization of chitin and chitosan. *Curr. Chem. Biol.* **2009**, *3*, 203–230.
49. Loh, E.Y.X.; Mohamad, N.; Fauzi, M.B.; Ng, M.H.; Ng, S.F.; Mohd Amin, M.C.I. Development of a bacterial cellulose-based hydrogel cell carrier containing keratinocytes and fibroblasts for full-thickness wound healing. *Sci. Rep.* **2018**, *8*, 2875. [CrossRef]
50. Kozhukharova, I.; Minkevich, N.; Alekseenko, L.; Domnina, A.; Lyublinskaya, O. Paracrine and autocrine effects of VEGF are enhanced in human eMSC spheroids. *Int. J. Mol. Sci.* **2022**, *23*, 14324. [CrossRef]
51. Jacobsen, K.A.; Al-Aql, Z.S.; Wan, C.; Fitch, J.L.; Stapleton, S.N.; Mason, Z.D.; Cole, R.M.; Gilbert, S.R.; Clemens, T.L.; Morgan, E.F.; et al. Bone formation during distraction osteogenesis is dependent on both VEGFR1 and VEGFR2 signaling. *J. Bone Miner. Res.* **2008**, *23*, 596–609. [CrossRef] [PubMed]
52. Persson, M.; Lehenkari, P.P.; Berglin, L.; Turunen, S.; Finnilä, M.A.J.; Risteli, J.; Skrifvars, M.; Tuukkanen, J. Osteogenic differentiation of human mesenchymal stem cells in a 3D woven scaffold. *Sci. Rep.* **2018**, *8*, 10457. [CrossRef]
53. Hanna, H.; Mir, L.M.; Andre, F.M. In vitro osteoblastic differentiation of mesenchymal stem cells generates cell layers with distinct properties. *Stem Cell Res. Ther.* **2018**, *9*, 203. [CrossRef] [PubMed]
54. Takagishi, Y.; Kawakami, T.; Hara, Y.; Shinkai, M.; Takezawa, T.; Nagamune, T. Bone-like tissue formation by three-dimensional culture of MG63 osteosarcoma cells in gelatin hydrogels using calcium-enriched medium. *Tissue Eng.* **2006**, *12*, 927–937. [CrossRef] [PubMed]
55. Orimo, H. The mechanism of mineralization and the role of alkaline phosphatase in health and disease. *J. Nippon. Med. Sch.* **2010**, *77*, 4–12. [CrossRef] [PubMed]
56. Chang, Y.L.; Stanford, C.M.; Keller, J.C. Calcium and phosphate supplementation promotes bone cell mineralization: Implications for hydroxyapatite (HA)-enhanced bone formation. *J. Biomed. Mater. Res.* **2000**, *52*, 270–278. [CrossRef] [PubMed]
57. Aksel, H.; Huang, G.T. Combined effects of vascular endothelial growth factor and bone morphogenetic protein 2 on odonto/osteogenic differentiation of human dental pulp stem cells in vitro. *J. Endod.* **2017**, *43*, 930–935. [CrossRef]
58. Song, X.; Liu, S.; Qu, X.; Hu, Y.; Zhang, X.; Wang, T.; Wei, F. BMP2 and VEGF promote angiogenesis but retard terminal differentiation of osteoblasts in bone regeneration by up-regulating Id1. *Acta Biochim. Et Biophys. Sin.* **2011**, *43*, 796–804. [CrossRef]
59. Lee, J.H.; Um, S.; Jang, J.; Seo, B.M. Effects of VEGF and FGF-2 on proliferation and differentiation of human periodontal ligament stem cells. *Cell Tissue Res.* **2012**, *348*, 475–484. [CrossRef]
60. Behr, B.; Tang, C.; Germann, G.; Longaker, M.T.; Quarto, N. Locally applied vascular endothelial growth factor A increases the osteogenic healing capacity of human adipose-derived stem cells by promoting osteogenic and endothelial differentiation. *Stem Cells* **2011**, *29*, 286–296. [CrossRef]
61. Dreyer, C.H.; Kjaergaard, K.; Ding, M.; Qin, L. Vascular endothelial growth factor for in vivo bone formation: A systematic review. *J. Orthop. Transl.* **2020**, *24*, 46–57. [CrossRef] [PubMed]
62. Maes, C.; Goossens, S.; Bartunkova, S.; Drogat, B.; Coenegrachts, L.; Stockmans, I.; Moermans, K.; Nyabi, O.; Haigh, K.; Naessens, M.; et al. Increased skeletal VEGF enhances beta-catenin activity and results in excessively ossified bones. *EMBO J.* **2010**, *29*, 424–441. [CrossRef] [PubMed]
63. Šturm, L.; Poklar Ulrih, N. Basic methods for preparation of liposomes and studying their interactions with different compounds, with the emphasis on polyphenols. *Int. J. Mol. Sci.* **2021**, *22*, 6547. [CrossRef] [PubMed]
64. Thiele, J.; Ma, Y.; Bruekers, S.M.C.; Ma, S.; Huck, W.T.S. 25th Anniversary Article: Designer Hydrogels for Cell Cultures: A Materials Selection Guide. *Adv. Mater.* **2014**, *26*, 125–148. [CrossRef] [PubMed]
65. Ho, S.T.; Hutmacher, D.W. A comparison of micro CT with other techniques used in the characterization of scaffolds. *Biomaterials* **2006**, *27*, 1362–1376. [CrossRef]
66. Hadzir, S.N.; Ibrahim, S.N.; Abdul Wahab, R.M.; Zainol Abidin, I.Z.; Senafi, S.; Ariffin, Z.Z.; Abdul Razak, M.; Zainal Ariffin, S.H. Ascorbic acid induces osteoblast differentiation of human suspension mononuclear cells. *Cytotherapy* **2014**, *16*, 674–682. [CrossRef]

gels

MDPI

Article

Numerical Optimization of Prednisolone–Tacrolimus Loaded Ultraflexible Transethosomes for Transdermal Delivery Enhancement; Box–Behnken Design, Evaluation, Optimization, and Pharmacokinetic Study

Munerah M. Alfadhel [1], Randa Mohammed Zaki [1,2,*], Basmah Nasser Aldosari [3] and Ossama M. Sayed [4]

[1] Department of Pharmaceutics, College of Pharmacy, Prince Sattam Bin Abdulaziz University, P.O. Box 173, Al-Kharj 11942, Saudi Arabia; m.alfadhel@psau.edu.sa
[2] Department of Pharmaceutics and Industrial Pharmacy, Faculty of Pharmacy, Beni-Suef University, Beni-Suef 62514, Egypt
[3] Department of Pharmaceutics, College of Pharmacy, King Saud University, P.O. Box 2457, Riyadh 11451, Saudi Arabia; baldosari@ksu.edu.sa
[4] Department of Pharmaceutics, Faculty of Pharmacy, Sinai University-Kantara Branch, Ismailia 41612, Egypt; osama.sayed@su.edu.eg
* Correspondence: r.abdelrahman@psau.edu.sa; Tel.: +966-540617870

Abstract: The aim of the present study is to formulate highly permeable carriers (i.e., transethosomes) for enhancing the delivery of prednisolone combined with tacrolimus for both topical and systemic pathological conditions. A Box–Behnken experimental design was implemented in this research. Three independent variables: surfactant concentration (X1), ethanol concentration (X2), and tacrolimus concentration (X3) were adopted in the design while three responses: entrapment efficiency (Y1), vesicle size (Y2), and zeta potential (Y3) were investigated. By applying design analysis, one optimum formulation was chosen to be incorporated into topical gel formulation. The optimized transethosomal gel formula was characterized in terms of pH, drug content, and spreadability. The gel formula was challenged in terms of its anti-inflammatory effect and pharmacokinetics against oral prednisolone suspension and topical prednisolone–tacrolimus gel. The optimized transethosomal gel achieved the highest rate of rat hind paw edema reduction (98.34%) and highest pharmacokinetics parameters (Cmax 133.266 ± 6.469 μg/mL; $AUC_{0-\infty}$ 538.922 ± 49.052 μg·h/mL), which indicated better performance of the formulated gel.

Keywords: prednisolone; tacrolimus; Box–Behnken design; transdermal delivery; transethosomes

Citation: Alfadhel, M.M.; Zaki, R.M.; Aldosari, B.N.; Sayed, O.M. Numerical Optimization of Prednisolone–Tacrolimus Loaded Ultraflexible Transethosomes for Transdermal Delivery Enhancement; Box–Behnken Design, Evaluation, Optimization, and Pharmacokinetic Study. *Gels* **2023**, *9*, 400. https://doi.org/10.3390/gels9050400

Academic Editors: Ying Huang, Zhengwei Huang and Xuanjuan Zhang

Received: 26 March 2023
Revised: 27 April 2023
Accepted: 3 May 2023
Published: 10 May 2023

1. Introduction

For many years, prednisolone (PRED) has been utilized extensively in the treatment of inflammatory illnesses (both acute and chronic). For many rheumatic illnesses, a long-term remedy with these medications is frequently required to manage the symptoms. Long-term use of PRED in therapy has a number of negative consequences on the heart and metabolism of bones. One of the most common adverse effects of long-term PRED use is bone loss. The anti-inflammatory effect of PRED is underpinned by very intricate processes that obstruct the operation of numerous systems. In addition to the risks connected to the prolonged use of PRED at a super-physiological dose, there are issues connected to stopping the steroid medication. Reduced unfavorable side effects would be extremely beneficial for therapeutic usage of PRED, which is the greatest anti-inflammatory medicine currently on the market. For patients, worry regarding their safety has always been the main barrier to the use of oral PRED [1,2]. One of the means to reduce the side effects of PRED is to shift the route of application to the transdermal route.

Recently, the name "trans-ethosomes" was given to lipid-based nanovesicles that include an edge activator and ethanol. The benefits of both ethosomes and transfersomes are present in transethosomes (TETSMs). TETSMs provide a number of advantages over other drug delivery methods because of the inclusion of these components. Ethanol improves medication penetration through the minute holes created in the stratum corneum as a result of fluidization by enhancing the lipid's fluidity and decreasing the density of the lipid bilayer [3]. These vesicles' edge activator weakens the phospholipid bilayer, making them ultradeformable and extremely elastic [4]. Therefore, it is anticipated that these drug-loaded nanovesicles will have a positive effect on therapeutic activity.

Tacrolimus (TAC) alters the humoral and cell-mediated immune reactions linked to inflammation in a number of different ways. The key method of action involves reducing the immunophilin FK506 binding protein 12's ability to activate calcineurin as a phosphatase, preventing the generation of interleukin (IL)-2, and inhibiting the signal transduction pathway that activates T cells. Although cyclosporin also inhibits calcineurin by forming a compound with a different immunophilin, in vitro and in vivo research shows that TAC has 10–100 times greater immunosuppressive activity than cyclosporin. Additionally, nitric oxide synthase activation and apoptosis may be inhibited by TAC, which may also enhance the effects of corticosteroids in these processes [5].

Many sources in the literature confirmed that combining corticosteroids therapy with TAC leads to lower required doses of corticosteroids in renal transplant operations [6]. This could be beneficial in terms of fewer side effects from steroids.

The aim of this research is to examine the possibility of combining PRED and TAC in TETSMs carriers and utilizing the transdermal route to increase the anti-inflammatory effect and pharmacokinetics of PRED.

2. Results and Discussion

2.1. *Effect of Formulation Variables on EE%, Vesicle Size, and Zeta Potential*

The chief objective of this study was to assess the effectiveness of TETSMs in improving the penetration of PRED through the skin for the management of inflammatory disorders. TETSMs are capable of effectively penetrating the stratum corneum due to their high alcohol content and surfactant properties [7]. Additionally, the study also examined the impact of TAC on the delivery of PRED to the skin.

An analysis was conducted to examine the impact of various formulation variables, including concentration of surfactant (X1), concentration of ethanol (X2), and concentration of TAC (X3) on EE% (Y1), vesicular size (Y2), and zeta potential (Y3). The results of the regression analysis, which were used to determine the best fitting model for each response, are summarized in Table 1.

Table 1. Composition of PRED-TAC TETSMs Formulations and the results of different responses (mean ± SD).

Formula Code	Independent Variables			Dependent Variables			
	Surfactant Concentration *w/v* % (X1)	Ethanol Concentration *v/v* % (X2)	Tacrolimus Concentration *w/v* % (X3)	EE% (Y1)	Vesicles Size (nm) (Y2)	Zeta Potential (mv) (Y3)	PDI
1	1	20	0.065	87.6 ± 2.14	330.5 ± 10.12	−32.2 ± 0.92	0.132 ± 0.08
2	1	40	0.065	80.4 ± 1.87	307.8 ± 8.75	−36.1 ± 0.87	0.315 ± 0.09
3	1	30	0.1	84.4 ± 1.46	320.7 ± 7.35	−33.8 ± 1.01	0.326 ± 0.11
4	0.2	20	0.065	62.4 ± 2.01	253.6 ± 6.54	−19.3 ± 1.17	0.321 ± 0.12
5	0.6	30	0.065	72.3 ± 3.27	283.5 ± 5.43	−30.4 ± 1.28	0.276 ± 0.03
6	0.6	30	0.065	72.1 ± 2.98	283.2 ± 6.98	−30.1 ± 0.84	0.376 ± 0.11
7	0.2	30	0.1	58.4 ± 1.54	242.1 ± 4.33	−22.1 ± 0.58	0.123 ± 0.02
8	0.6	30	0.065	72.4 ± 1.62	282.8 ± 7.21	−30.3 ± 0.23	0.265 ± 0.06
9	0.2	30	0.03	56.7 ± 3.21	231.5 ± 3.75	−21.8 ± 0.67	0.225 ± 0.14
10	0.6	20	0.03	75.6 ± 2.19	290.4 ± 7.25	−27.3 ± 0.95	0.431 ± 0.10
11	0.6	30	0.065	72.6 ± 1.92	283.7 ± 3.98	−30.5 ± 0.26	0.259 ± 0.07
12	0.6	20	0.1	76.5 ± 1.64	302.6 ± 9.43	−28.1 ± 1.11	0.239 ± 0.13
13	0.6	40	0.1	68.7 ± 2.54	278.5 ± 5.38	−32.7 ± 0.83	0.185 ± 0.05
14	0.2	40	0.065	51.3 ± 2.73	230.4 ± 6.27	−24.8 ± 0.63	0.173 ± 0.16
15	0.6	30	0.065	71.9 ± 3.81	283.3 ± 4.79	−30.6 ± 0.27	0.265 ± 0.08
16	1	30	0.03	83.2 ± 1.26	307.6 ± 10.3	−33.1 ± 1.03	0.439 ± 0.14
17	0.6	40	0.03	66.5 ± 1.87	266.7 ± 7.83	−31.8 ± 0.68	0.286 ± 0.12

2.1.1. Effect of Formulation Variables on EE%

EE% data of the formulated TETSMs are shown in Table 1. Model fitting showed a fairly good fit with a quadratic interaction model, with a correlation coefficient R^2 of 0.9994, adjusted R^2 of 0.9986, and predicted R^2 of 0.9932. The small difference (less than 0.2) between the adjusted and predicted R^2 confirms the validity of the model [8]. Additionally, the model showed high adequate precision of 125.84 (greater than 4), indicating the ability of the model to navigate the design space [9], as represented in Table 2. ANOVA in Table 3 showed that EE% was significantly affected by all independent variables ($p < 0.05$) with the following equation representing the combined effect of independent variables on EE% of TETSMs:

$$\begin{aligned} EE\% = &+72.26 + 13.35X1 - 4.40X2 + 0.7500\ X3 + 0.9750X1X2 - 0.1250X1X3 \\ &+0.3250X2X3 - 1.49X12 - 0.3425\ X22 - 0.0925X32 \end{aligned} \quad (1)$$

where X1 is the concentration of surfactant, X2 is the concentration of ethanol, and X3 is the concentration of TAC. The response surface curves of the EE% values from the interaction of different independent variables are shown in Figures 1A and 2A.

Table 2. Output data of Box–Behnken design for optimization of PRED–TAC-loaded TETSMs.

Dependent Variables	R^2	Adjusted R^2	Predicted R^2	Adequate Precision
Y1: % EE	0.9994	0.9986	0.9932	125.8424
Y2: Vesicle size (nm)	0.9999	0.9998	0.9990	325.6015
Y3: Zeta potential (mV)	0.9988	0.9974	0.9877	89.5319

Table 3. ANOVA for Box–Behnken design of PRED–TAC-loaded TETSMs.

Dependent Variable	Source	SS	Df	Mean Square	F Value	p Value	
Y1	Model	1599.74	9	177.75	1313.87	<0.0001	significant
	A-Surfactant concentration	1425.78	1	1425.78	10,539.03	<0.0001	
	B-Ethanol concentration	154.88	1	154.88	1144.84	<0.0001	
	C-Tacrolimus concentration	4.50	1	4.50	33.26	0.0007	
	AB	3.80	1	3.80	28.11	0.0011	
	AC	0.0625	1	0.0625	0.4620	0.5185	
	BC	0.4225	1	0.4225	3.12	0.1205	
	A^2	9.38	1	9.38	69.33	<0.0001	
	B^2	0.4939	1	0.4939	3.65	0.0977	
	C^2	0.0360	1	0.0360	0.2663	0.6217	
Y2	Model	13,537.78	9	1504.20	9232.25	<0.0001	significant
	A-Surfactant concentration	11,942.85	1	11,942.85	73,301.15	<0.0001	
	B-Ethanol concentration	1106.85	1	1106.85	6793.48	<0.0001	
	C-Tacrolimus concentration	288.00	1	288.00	1767.65	<0.0001	
	AB	0.0625	1	0.0625	0.3836	0.5553	
	AC	1.69	1	1.69	10.37	0.0146	
	BC	0.0000	1	0.0000	0.0000	1.0000	
	A^2	145.21	1	145.21	891.22	<0.0001	
	B^2	41.18	1	41.18	252.77	<0.0001	
	C^2	16.80	1	16.80	103.11	<0.0001	

Table 3. *Cont.*

Dependent Variable	Source	SS	Df	Mean Square	F Value	*p* Value	
Y3	Model	345.94	9	38.44	671.83	<0.0001	significant
	A-Surfactant concentration	278.48	1	278.48	4867.32	<0.0001	
	B-Ethanol concentration	42.78	1	42.78	747.74	<0.0001	
	C-Tacrolimus concentration	0.9112	1	0.9112	15.93	0.0053	
	AB	0.6400	1	0.6400	11.19	0.0123	
	AC	0.0400	1	0.0400	0.6991	0.4307	
	BC	0.0025	1	0.0025	0.0437	0.8404	
	A^2	21.84	1	21.84	381.72	<0.0001	
	B^2	0.0000	1	0.0000	0.0005	0.9835	
	C^2	0.6821	1	0.6821	11.92	0.0106	

Y1: % EE, Y2: Vesicle size (nm), Y3: Zeta potential (mV), SS: sum of squares, Df: degree of freedom.

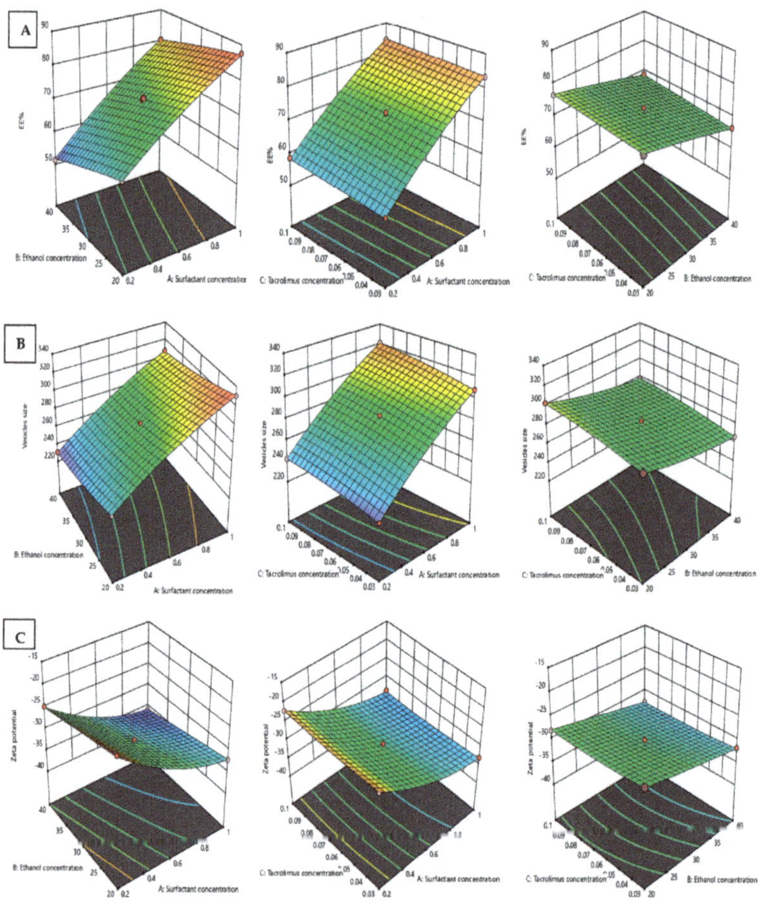

Figure 1. A 3D response surface graph for the effect of independent factors: surfactant concentration, Ethanol concentration, and Tacrolimus concentration on the dependent responses, EE% (**A**), vesicle size (**B**), and zeta potential (**C**) of PRED–TAC-loaded TETSMs. Red dots indicate the replicates in our design.

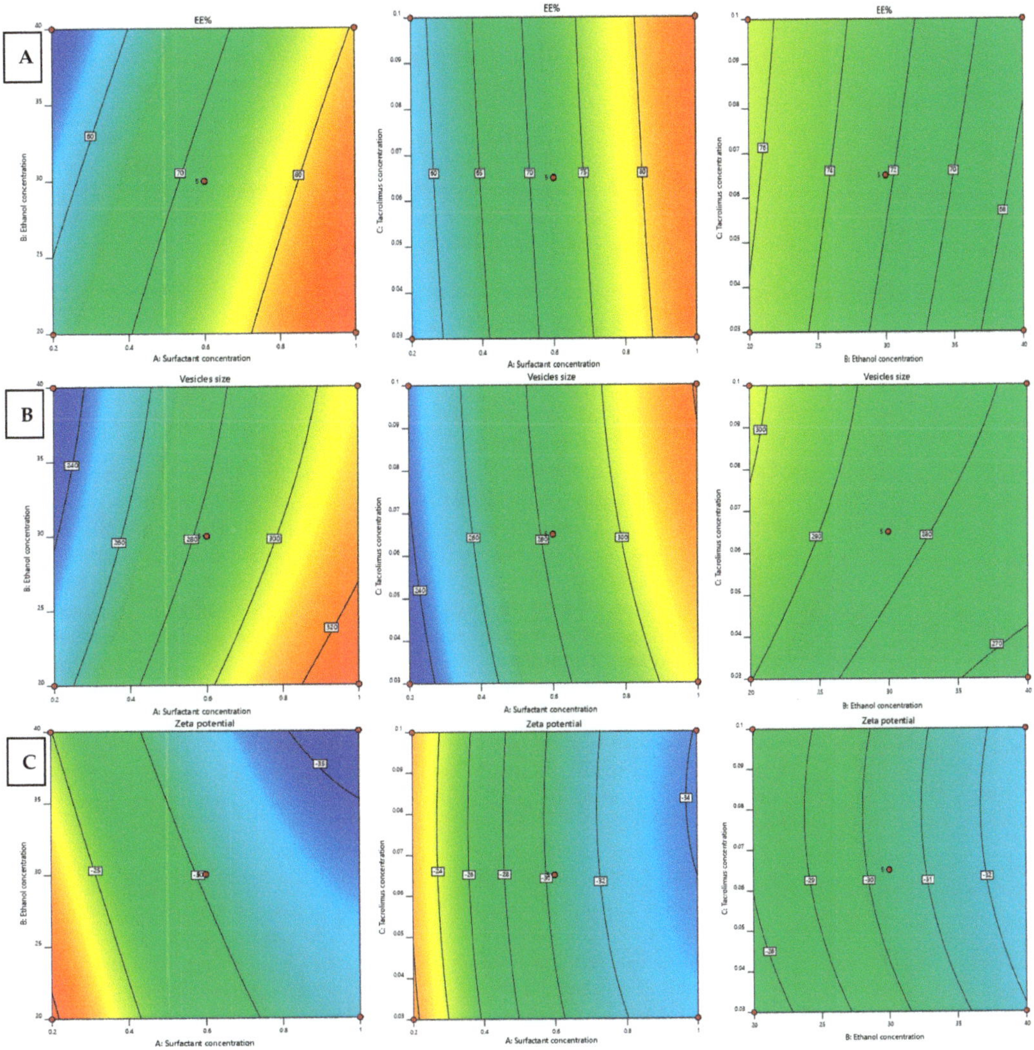

Figure 2. Contour graph for the effect of independent factors on different responses, EE% (**A**), vesicle size (**B**), and zeta potential (**C**) of PRED–TAC-loaded TETSMs. Red dots indicate the replicates in our design.

Increasing the surfactant concentration led to an increase in EE% of PRED as confirmed by the positive sign of the correlation coefficient in Equation (1). This could be explained based on the low HLB value of span 60, which increases the lipophilic domain of the lipid bilayer and hence increases the entrapped PRED in this hydrophobic domain [10,11].

Ethanol concentrations have a negative effect individually on the EE% of PRED in TETSMs, as represented by the negative sign of the correlation coefficient in Equation (1). The decrease in the EE% may be due to the increase in the fluidity and the leakage of the vesicles [12–15]. This result is in contrast to the previous research, which reported that enhancing the concentration of ethanol from 20 to 40% will have a positive impact on the EE% [16]. In addition, TAC showed an improvement in EE% by increasing its concentration from 0.03 to 0.1%, which aids the solubilization of drugs in the lipid mixture. Including lipid soluble TAC could increase the solubility of PRED.

2.1.2. Effect of Formulation Variables on TETSMs Vesicle Size

The vesicle sizes of different formulated TETSMs are shown in Table 1. Model fitting revealed a quadratic interaction between independent variables, with R^2 of 0.9999, adjusted R^2 of 0.9998, and predicted R^2 of 0.9990. There was a small difference (less than 0.2) between the adjusted and predicted R^2. Additionally, the model showed high adequate precision of 325.6, as represented in Table 2. The data from ANOVA in Table 3 revealed that vesicle size of different formulations was significantly affected by all independent variables ($p < 0.05$) with the following equation representing the combined effect of independent variables on vesicle size of TETSMs:

$$\begin{aligned}\text{Vesicles size} = {} & +283.32 + 38.64\,X1 - 11.76X2 + 6.00X3 + 0.1250\,X1X2 + 0.6500\,X1X3 \\ & +0.0000X2X3 - 5.87X12 + 3.13X22 - 2.00X32\end{aligned} \quad (2)$$

where X1 is the concentration of surfactant, X2 is the concentration of ethanol, and X3 is the concentration of TAC. Figures 1B and 2B show the surface response curves of the combined effects of prepared independent variables on vesicle size. Generally, the surfactant concentration range in ethosomal formulations is 0.2–1% [17]. It was observed that increasing the surfactant (span 60) ratio resulted in a slight or moderate increase in vesicle size. The positive effect of surfactant on the vesicle size agrees with the earlier outcomes and can be due to the reduction in the hydrophilic portion of the surfactant in the presence of low HLB surfactant at high concentrations [10,11]. In addition, TAC had a positive impact on the vesicle size values, meaning that increasing TAC concentration from 0.03 to 0.1% resulted in a simultaneous increase in vesicle size. This condition was observed parallel to the increase in span 60. In contrast, the concentration of ethanol was found to have a negative impact on the vesicle size values and this is confirmed by the negative sign of the correlation coefficient (X2) in Equation (2). This observation agrees with the earlier literature [18].

Ethanol has been found to be a potent penetration enhancer. Ethanol concentration has been reported in the literature to be in the range of 10–50% [7,19]. Enhancing the concentration of ethanol resulted in a decrease in vesicle size, as stated in the previous works [7,15,20–32].

2.1.3. Effect of Preparation Variables on Zeta Potential

Zeta potential is an indicator for the stability of nanosystems as it gives information about the magnitude of attraction and repulsion between nanoparticles. As zeta potential values increase, the repulsion between nanoparticles increases, and therefore there is improvement in the system stability. As shown in Table 1, the zeta potential values of all TETSMs formulations ranged from -19.3 ± 1.17 to -36.1 ± 0.87, indicating high stability for the fabricated formulations.

The model fitting of the combined effects of TETSMs variables on zeta potential proposed a quadratic model with R^2 of 0.9988, adjusted R^2 of 0.9974, and predicted R^2 of 0.9877. There was a small difference (less than 0.2) between the adjusted and predicted R^2. In addition, the model showed high adequate precision of 89.53 as represented in Table 2.

The data from ANOVA in Table 3 revealed that the zeta potential of different formulations was considerably affected by all independent variables ($p < 0.05$) with the following equation representing the effect of independent variables on zeta potential of TETSMs:

$$\begin{aligned}\text{Zeta potential} = {} & +30.38 + 5.90X1 + 2.31X2 + 0.3375X3 - 0.4000X1X2 + 0.1000X1 + 0.0250\,X2X3 \\ & -2.28X1^2 - 0.0025X2^2 - 0.4025X32\end{aligned} \quad (3)$$

where X1 is the concentration of surfactant, X2 is the concentration of ethanol, and X3 is the concentration of TAC. As clear from the above equation, all independent variables caused an increase in the zeta potential values. Increasing the zeta potential is so beneficial. The increase in zeta potential values with the increase in span 60 concentration may be attributed to the low HLB value (hydrophilic lipophilic balance) of span 60 resulting in high adsorption of OH ions from the hydration medium on the nanoparticles causing an increase in zeta potential values [15].

The positive impact of ethanol concentration on zeta potential values may be due to the negative charge imparted by ethanol on the particles' surfaces. Ethanol's effect on zeta potential is well documented in the literature [17,33–35].

The effect of different independent variables on zeta potential is shown in Figures 1C and 2C.

2.2. Formulation Optimization and Validation

A numeric optimization was performed by Design Expert software to select the optimum formula based on the highest desirability value. The optimum formula was found to consist of 0.999 (*w*/*v* %) span 60, 39.99 (*v*/*v* %) ethanol, and 0.03 (*w/v* %) TAC. The predicted responses of EE%, vesicle size and zeta potential were 79.308%, 298.929 nm, and −35.047 mV, respectively, with a desirability of 0.704 as represented in Table 4 and Figure 3. Validation of the optimum formula resulted in a % relative error less than 5% for all predicted responses, confirming the fitness of the model [36].

Table 4. The composition and validation of the optimized formula with its expected responses.

The Optimized Formula	Independent Variables			Predicted Responses			Desirability
	Surfactant Concentration *w/v* % (X1)	Ethanol Concentration *v/v* % (X2)	Tacrolimus Concentration *w/v* % (X3)	EE%	Vesicle Size (nm)	Zeta Potential (mv)	0.704
	0.9999	39.998	0.03000	79.3083	298.929	−35.0471	

Validation of the Optimum Formula

Responses	Predicted value	Experimental value	% Relative error
EE%	79.3083	81.892	3.258
Vesicle size (nm)	298.929	305.325	2.139
Zeta potential (mv)	−35.0471	−34.46	1.675

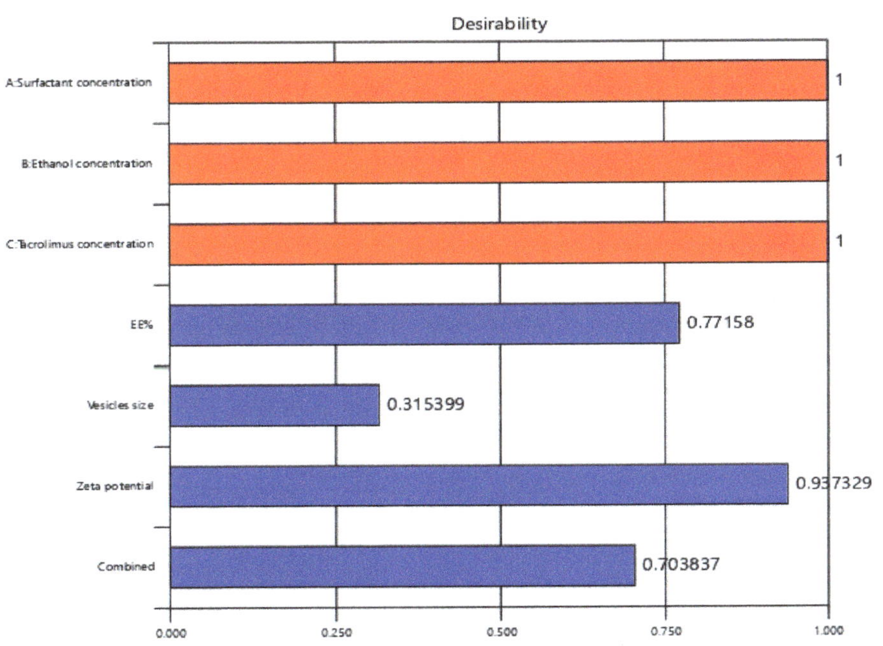

Figure 3. *Cont.*

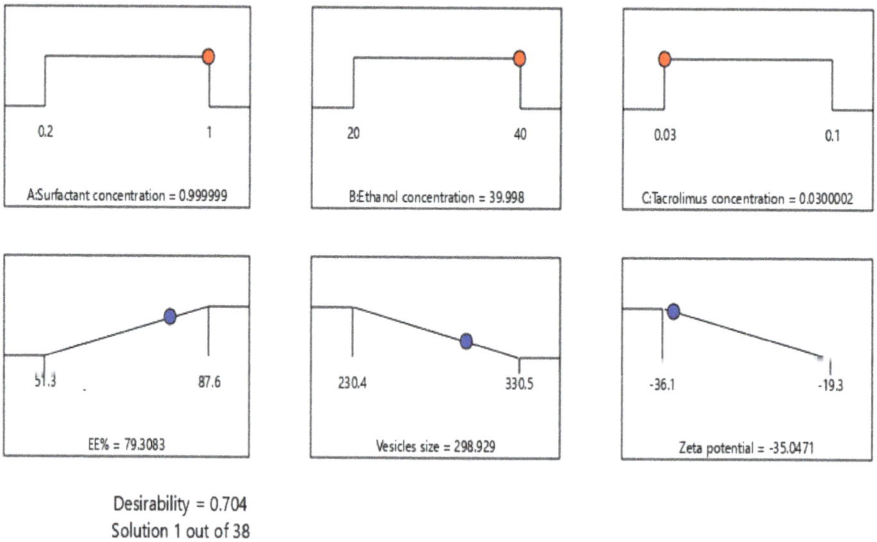

Figure 3. Desirability plot of the numerical optimization of PRED–TAC-loaded TETSMs.

2.3. *In Vitro Release of Prednisolone from Optimized TETSMs Containing Gel*

The release profile of PRED is illustrated in Figure 4. The release profile showed that the PRED–TAC-loaded TETSMs gel achieved higher released PRED (82.93% ± 2.75) than the PRED-loaded gel (42.56% ± 3.11) and PRED suspension (50.45% ± 2.12). The higher release achieved could be attributed to the increased thermodynamic activity of PRED solubilized in the TETSMs lipid bilayers, and to the nanosize of the vesicles that led to higher release [37]. All gel formulations exhibited Higuchi diffusion model release.

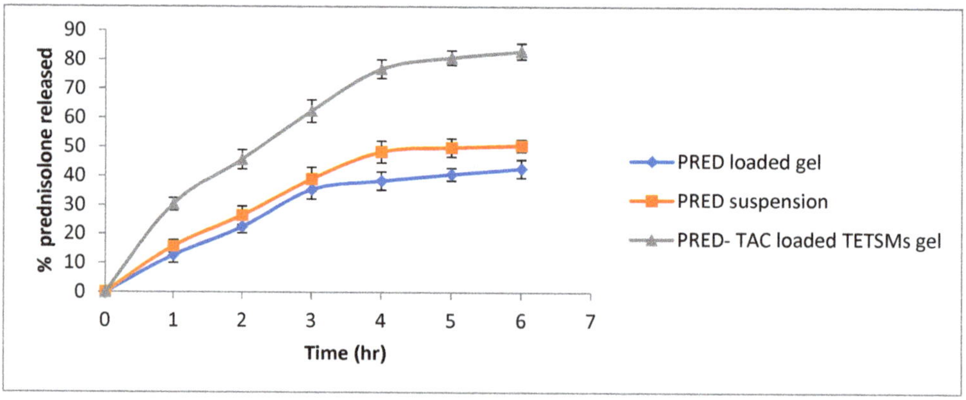

Figure 4. In vitro release profile of PRED from the optimum PRED–TAC-loaded TETSMs gel compared to PRED suspension and the PRED-loaded gel.

2.4. *PRED Permeation from Optimized Formula*

The permeation profile of PRED from the optimized-formula-loaded gel compared to PRED-loaded gel and PRED suspension is illustrated in Figure 5, while permeation parameters are presented in Table 5. Preparations of PRED–TAC-loaded TETSMs gels showed the highest percent of PRED (62.65% ± 2.87) permeated through the skin, compared to PRED-loaded gels and PRED suspension (26.5% ± 2.21 and 36.76% ± 1.98, respectively), and this is due to the lower rigidity of the vesicles, which permits their squashing between cells. The combined effect of the presence of both ethanol and edge activator (i.e., surfactant) enhanced the transepidermal flux of the gel formulation (45.27± 2.87 × 10^4 µg/cm^2·h^{-1}) compared to the PRED-loaded gels and PRED suspension.

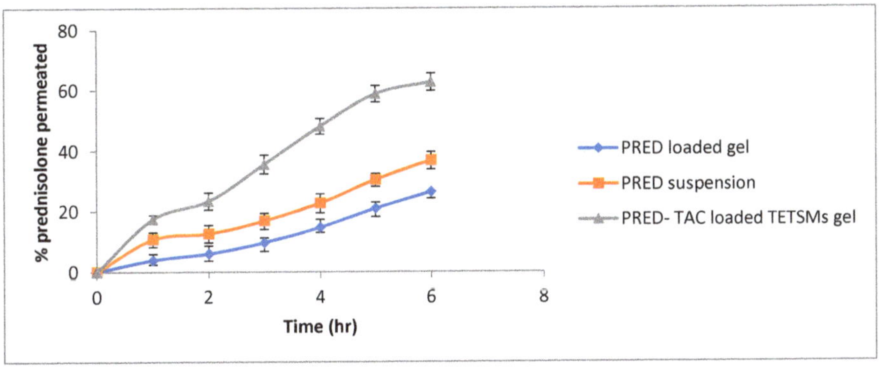

Figure 5. Ex vivo permeation of prednisolone from the optimum-formula-loaded gel compared to PRED suspension and the PRED-loaded gel.

Table 5. Ex vivo permeation parameters of PRED from the optimum-formula-loaded gel compared to PRED suspension and PRED-loaded gel (n = 3, mean ± SD).

Formulation Code	Flux (J_{ss}) (µg/cm^2 h^{-1}) × 10^4	Permeability Coefficient (P) (cm/h) × 10^{-6}	Partition Coefficient (K_P) × 10^4
PRED–TAC-loaded TETSMs gel	45.27 ± 2.87	9.21 ± 2.87	242 ± 2.87
PRED suspension	27.87 ± 2.43	5.11 ± 2.43	95.41 ± 2.43
PRED-loaded gel	19.32 ± 3.12	3.42 ± 3.12	38.54 ± 3.12

2.5. Morphological Examination of PRED–TAC-Loaded TETSMs

The transmission electron image of the optimum formula is represented in Figure 6. The image shows spherical vesicles with large aqueous cores. The vesicles are smooth, well defined, and non-aggregated, confirming their stability. The image is also in agreement with the size analysis data.

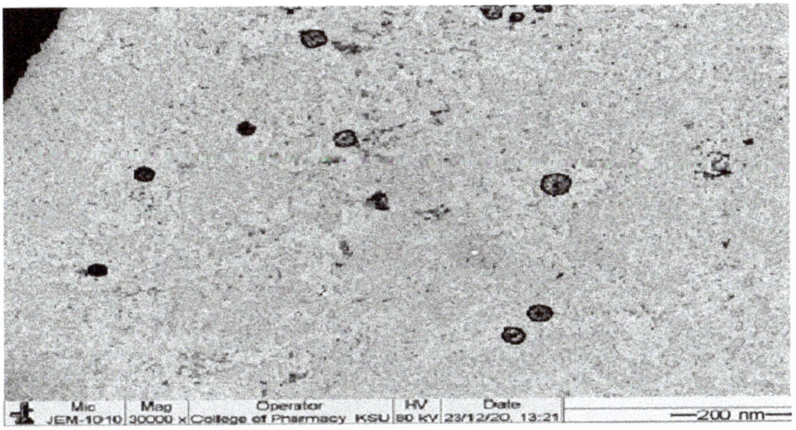

Figure 6. TEM image of the optimum formula.

2.6. Characterization of Gel Formulations

Gel formulations of PRED were characterized regarding pH, % drug content, and spreadability. The data of these characterizations are shown in Table 6. Regarding pH, all formulations showed a pH around 6.5 which is dermatologically acceptable according to the literature [38]. All formulations kept a good % drug content meaning good compatibility of the mixed ingredients.

Table 6. Characterization of PRED containing gel formulations.

Formula	pH	%Drug Content	Spreadability
PRED–TAC-loaded TETSMs gel	6.92 ± 0.014	96.53 ± 1.34	4.23 ± 0.12
PRED-loaded gel	6.54 ± 0.015	98.23 ± 0.42	3.87 ± 0.23
PRED–TAC-loaded gel	6.73 ± 0.009	97.48 ± 1.51	3.92 ± 0.15

The spreadability has a substantial role in patient agreement and aids in uniform gel administration to the skin. An acceptable gel needs less time to spread over the skin and will have great spreadability. The values of spreadability (4.23–3.87 (g cm)/s) confirm that the gels spread without difficulty by application of a sheer minimum and have satisfactory bioadhesion [39].

2.7. In Vivo Evaluation of Anti-Inflammatory Effect of PRED Containing TETSMs Gels

The rat hind paw edema model is one of the most used anti-inflammatory models to study the effect of topically applied anti-inflammatory agents. The percent edema inhibition of PRED-loaded gel, PRED–TAC-loaded gel, and PRED–TAC-loaded TETSMs gel are shown in Table 7 and Figure 7. It was noted that PRED–TAC-loaded TETSMs gel achieved the highest rate of percent inhibition ($p < 0.05$) among the other gel formulations (98.34% in 180 min), followed by PRED–TAC-loaded gel (63.26%), and PRED-loaded gel (41.25%).

Table 7. % Edema inhibition of PRED gel formulations.

Time (min)	% Edema Inhibition		
	PRED-Loaded Gel	PRED–TAC-Loaded Gel	PRED–TAC-Loaded TETSMs Gel
30	5.34	10.12	21.56
60	9.34	16.24	33.25
90	14.67	31.27	62.87
120	22.11	43.24	78.43
150	34.32	55.32	91.2
180	41.25	63.26	98.34

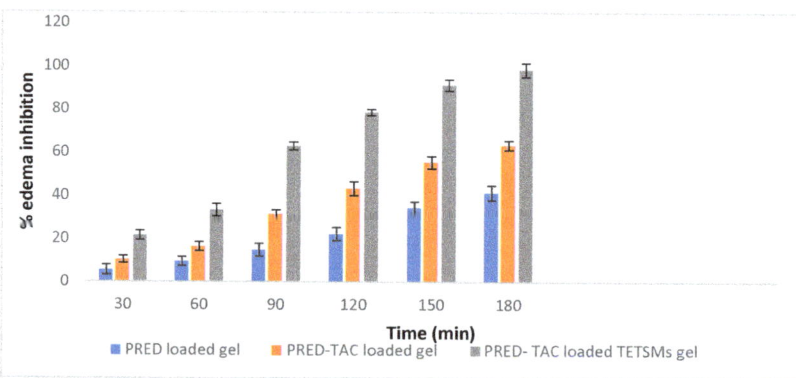

Figure 7. % Edema inhibition of PRED gel formulations.

PRED, as one of the corticosteroids, has the ability to suppress every step of the inflammatory cascade including synthesis of inflammatory mediators and cell-mediated immunity [40]. The TETSMs gel formulation enhanced the permeation of PRED and its localization in skin layers due to both ethanol and edge activator (surfactant) effects. Additionally, TAC has a synergistic effect on the inflammatory process due to its effect on inhibiting the production of cytokines involved in the inflammatory cascade [41,42].

2.8. Comparative Pharmacokinetic Study of Prednisolone Gel Formulations against Oral PRED Suspension

Pharmacokinetic parameters of PRED-containing gel formulations and oral PRED suspension are shown in Table 8. Plasma PRED profiles are illustrated in Figure 8. It is noticed that PRED–TAC-loaded TETSMs gel achieved higher pharmacokinetic parameters ($p < 0.05$) compared to PRED-loaded gel (topical) and PRED oral suspension in terms of Cmax and AUC 0-∞ (133.266 ± 6.469 μg/mL, 538.922 ± 49.052 μg·h/mL), (1.3 and 2.2-fold increase in Cmax and 1.47 and 1.88 fold increase in AUC 0-∞ compared to oral suspension and PRED gel, respectively), with significance ($p < 0.05$) according to ANOVA results in Table 9. The higher systemic concentration of PRED after application of PRED–TAC-loaded TETSMs gel than after oral administration could be attributed to the presence of edge activators in the TETSMs composition which increase skin penetration by enhancing the fluidity of the TETSMs phospholipid bilayer, making them ultradeformable and extremely elastic, and therefore facilitating their squeezing into the skin pores [43,44]. Additionally, the presence of ethanol in TETSMs improves drug penetration through the minute holes created in the stratum corneum as a result of fluidization by enhancing the lipids fluidity and decreasing the density of the lipid bilayer [11]. Moreover, the addition of TAC to TETSMs gel formulations resulted in higher pharmacokinetic parameters of PRED, as TAC is previously reported to increase corticosteroids accumulation [45]. These findings support the fact that transdermal delivery bypasses first-pass metabolism for many active ingredients including corticosteroids.

Table 8. Pharmacokinetic parameters of PRED formulations (mean ± SD).

Pharmacokinetic Parameters	PRED Suspension (Oral)	PRED-Loaded Gel (Topical)	PRED–TAC-Loaded TETSMs Gel
Cmax	103.333 ± 5.686	61.7 ± 6.564	133.266 ± 6.469
Tmax	1.00 ± 0.000	2.00 ± 0.000	1.96 ± 0.057
t1/2	3.449 ± 0.413	4.926 ± 0.344	3.665 ± 0.428
AUC 0-t	341.080 ± 33.666	240.056 ± 36.342	490.233 ± 32.855
AUC 0-inf_obs	365.769 ± 42.054	285.776 ± 45.123	538.922 ± 49.052
MRT 0-inf_obs	4.120 ± 0.480	6.293 ± 0.348	4.852 ± 0.466

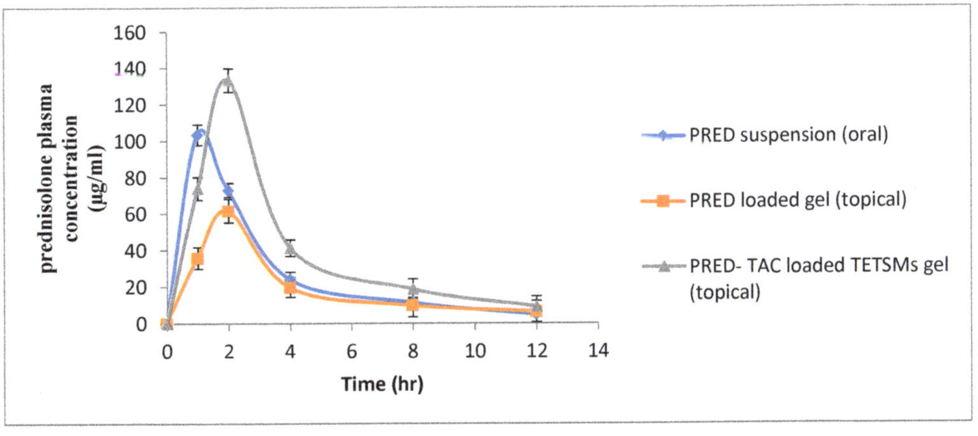

Figure 8. Plasma concentration time profiles of PRED formulations.

Table 9. One way ANOVA of pharmacokinetic parameters of PRED formulations.

Pharmacokinetic Parameters	SS	Df	MS	F	p-Value	F Crit
Cmax	7751.127	2	3875.563	99.13899	0.017047	5.143253
Tmax	1.935556	2	0.967778	871	0.001355	5.143253
t1/2	3.816098	2	1.908049	12.09148	0.007855	5.143253
AUC 0-t	95040.8	2	47520.4	40.34386	0.000332	5.143253
AUC 0-inf_obs	100463.8	2	50231.89	24.26347	0.001332	5.143253
MRT 0-inf_obs	7.33255	2	3.666275	19.28791	0.002439	5.143253

3. Conclusions

The present research proved that incorporating TAC with PRED in TETSMs and transdermal application increased the efficacy and optimized the pharmacokinetics of PRED, thus improving the efficacy of the therapy for both topical and systemic pathological conditions. These findings focus the spotlight on a way to decrease the proposed doses of corticosteroids, utilizing the benefits of both changing the route of administration and the incorporation of TAC. This will lead to better management of inflammatory conditions and fewer side effects linked to the use of corticosteroids.

4. Material and Methods

4.1. Materials

Prednisolone (PRED) and tacrolimus (TAC) were kindly gifted by Aljazeera Pharmaceuticals Company. Phosphatidylcholine from soy lecithin (Phospholipon 90G, pure phosphatidylcholine stabilized with 0.1% ascorbyl palmitate and max 0.3% tocopherol) was a gift from LIPOID GmbH, Germany. Ethanol, span 60, and hydroxypropyl methyl cellulose were purchased from Sigma-Aldrich.

4.2. Experimental Design Setup

A Box–Behnken design was utilized to optimize the formulation variables for creating PRED-loaded TETSMs with the goal of achieving high encapsulation efficiency, small vesicle size, and high zeta potential. Experiments were conducted using Design Expert software, with 17 total experiments conducted, including 13 at the midpoint of each edge of a multidimensional cube and 4 replicates of the cube's center point. Three independent variables were estimated, including surfactant concentration, ethanol concentration, and tacrolimus concentration. The dependent variables were encapsulation efficiency, vesicle size, and zeta potential. The study design and composition of the prepared PRED-loaded TETSMs are presented in Tables 1 and 10. The optimized formula was selected by Design Expert software based on the highest value of desirability for achieving our goal of high encapsulation efficiency, high zeta potential, and small particle size. The selected optimized formula was fabricated and assessed again for different responses to detect the accuracy of the model.

Table 10. Dependent and independent variables in Box–Behnken design.

Independent Variables	Levels	
	Low	High
Surfactant concentration w/v % (X1)	0.2	1
Ethanol concentration v/v % (X2)	20	40
Tacrolimus (TAC) concentration (X3)	0.03	0.1
Dependent values (Responses)	Desirability	
EE% (Y1)	maximize	
Vesicle size (Y2)	minimize	
Zeta potential (Y3)	maximize	

4.3. Preparation of PRED-Loaded TETSMs

Fabrication of PRED-loaded TETSMs was performed according to the thin film hydration method [46]. Briefly, the accurate weight of phosphatidylcholine, PRED (100 mg), TAC, and span 60 were dissolved in 10 mL organic solvent, which consisted of 2:1 chloroform: methanol in a rounded flask. Then the organic solvent was evaporated under vacuum using a rotary evaporator (Buchi Rotavapor R-200, Switzerland) at 60 °C and 100 rpm. The formed film was then hydrated with 10 mL of a certain ethanol concentration as specified in Table 2. The hydration step was carried out at 60 °C to be higher than the lipid transition temperature [11]. Finally, the formulated dispersions were kept in a refrigerator for further analysis.

4.4. Characterization of Formulated PRED-Loaded TETSMs

4.4.1. Entrapment Efficiency Estimation

The entrapment efficiency (EE%) of PRED in the fabricated preparations was calculated by determining the percentage of PRED found in the fabricated preparations. To separate unentrapped drug, cooling centrifugation was performed at 16,000 rpm and 4 °C using a Sigma cooling centrifuge from Sigma Laborzentrifugen GmbH. The clear supernatant was then analyzed for PRED content at λ max 254 nm using a Jasco UV–Vis spectrophotometer from Jasco, Japan. The EE% can be computed by subtraction of the quantity of PRED present in the supernatant from the initial quantity of PRED added [47].

4.4.2. Vesicle Size and Zeta Potential Analysis

The zeta potential, polydispersity index (PDI), and average vesicle size of all fabricated preparations were measured using photon correlation spectroscopy with a Zetasizer 2000 (Malvern Instruments Ltd., Malvern, UK) [48]. The nanodispersions were exposed to dilution before analysis and experiments were conducted in triplicate at a 90-degree scattering angle and a temperature of 25 degrees Celsius prior to taking measurements.

4.4.3. Formulation Optimization

The optimized formulation for PRED was selected using Design Expert® software (Ver. 12, Stat-Ease, Minneapolis, MN, USA), with the goal of achieving the highest EE% and zeta potential, while minimizing vesicle size. The selected optimized formulation was then fabricated and tested in triplicate to validate the accuracy of the model by comparing the predicted responses from the software with actual measurements, using the equation for relative error [49,50].

$$\%\ \text{Relative error} = ((\text{predicted value} - \text{actual value})/\text{predicted value}) \times 100 \qquad (4)$$

4.5. Preparation of Gels Containing Optimized TETSMs:

The optimized formula was amalgamated into a gel base using hydroxypropyl methylcellulose (HPMC, K4M) as a gelling agent at 2.5% (w/v). Briefly, the accurate weight of HPMC (0.25 g) was dispersed in 10 mL distilled water while stirring at 1000 rpm to obtain homogenous dispersion. Then the optimum TETSMs formula was dispersed into the gel base with continuous stirring to obtain a final gel formulation with 5 mg PRED per 1 g gel [36].

4.6. Evaluation of Gel Formulations

4.6.1. pH Evaluation

A 100 mg sample of the TETSMs-loaded gel formulation was weighed and added to a 50 mL volumetric flask. The volume was then filled to 50 mL with double-distilled water. The pH of the HPMC gel loaded with TETSMs was noted down using a glass microelectrode (Mettler Instruments, Giessen, Germany) by measuring the pH after 1 min of equilibration. The experiments were repeated three times to confirm the neutralization of the gel from different batches. The pH was measured on the first, 15th, and 30th day after preparation to check for any changes in pH over time.

4.6.2. Spreadability

A modified instrument called the spreadability apparatus was used to test spreadability. It consisted of two glass slides with gel in the middle, with the upper slide attached to a balance by a hook and the lower slide fixed to a wooden plate. On the basis of the gel's slip and drag characteristics, spreadability was measured [51]. Spreadability was computed using the following equation:

$$s = m \times l/t \quad (5)$$

where s represents the spreadability, m is the weight in pan (g), l is the fixed distance moved by the slide and t is the time.

4.6.3. Drug Content Determination

The created gels were given 48 h to rest before 1 g of each was taken in a 10 mL volumetric flask, dissolved in methanol, and the remaining 10 mL was filled with water. For PRED, maximum absorbance values were determined spectrophotometrically using the Jasco V530, Tokyo, Japan at 254 nm. From a standard calibration curve in methanol, concentrations of PRED were determined [52].

4.6.4. In Vitro Drug Release Studies

The release of PRED from the PRED–TAC-loaded TETSMs gel, PRED-loaded gel, and PRED suspension was tested in vitro using a dialysis bag method. An amount of 3 mg of PRED was placed in a dialysis bag with a molecular weight cut-off of 12,000 Da and released into 50 mL of Sorensen's phosphate buffer at pH 7.4. Samples were taken from the medium at various intervals and replaced with fresh medium over a period of 6 h [53]. The samples were analyzed using spectrophotometry at a wavelength of 254 nm (using a Jasco V530 from Tokyo, Japan). The release kinetics were determined by applying Korsmeyer–Peppas, Zero-order, First-order, Higuchi, and Hixson models to the data obtained.

4.6.5. In Vitro Permeation Studies

The barrier membrane samples of hairless rat skin were used to conduct the study after approval by the Institutional Animal Ethical Committee (IAEC) number SCBR-055-2022, Prince Sattam Bin Abdulaziz University. Firstly, the animals were anesthetized by ether and sacrificed; then the abdominal hair was cautiously shaved off. Secondly, skin samples were excised, and the subcutaneous tissues were detached carefully and finally washed with saline and stored at −20 °C until use. The membrane samples were placed in vertical Franz diffusion cells, consisting of a receptor compartment with a capacity of 5 mL and a diffusion surface area of 1 cm^2. Isotonic phosphate buffer saline (1.55 M, pH 7.4) was placed in the receptor compartments and stirred at 600 rpm in a temperature-controlled water bath. Before starting the experiment, phosphate buffer saline was added to the donor compartment at 37 °C to equilibrate the biological membranes for thirty minutes. After this time, the phosphate buffer saline was cautiously detached from the donor compartment. Afterward, a suitable quantity of TETSMs gel equivalent to 3 mg PRED was added to the donor compartment and spread over the membrane. To maintain occlusive conditions in the skin, glass slips were put on the donor compartments during the experiment. The in vitro permeation studies were conducted for 6 h and, at different time intervals, 1 mL samples were withdrawn from the receptor compartment and substituted with the same quantity of buffer. All experiments were repeated five times [53,54].

The cumulative amount permeated was computed using Equation (3) by taking into account the volume of the receptor phase (V_R), the volume of the sample at each time point (V_s), and the quantified concentration of the sample taken at the nth time point (C_n). This cumulative amount calculated at each time point was then divided by the diffusion area of the Franz cells (measured in cm^2) to obtain the final result.

$$\text{Cumulative amount} = V_R \times C_n + \lfloor V_s(\sum C_1 + \cdots + C_{n-1}) \rfloor \quad (6)$$

4.6.6. Data Analysis for Permeation Studies

The different parameters of the skin permeation test were studied and assessed. The steady state flux (J_{ss}) of the drug in $\mu g \cdot cm^{-2} \cdot h^{-1}$ at time t was computed from the slope of the linear portion of the plot representing the cumulative amount of PRED permeated per unit area over time. The cumulative amount of PRED in the receptor compartment after 6 h was known as Qcum ($\mu g \cdot cm^{-2}$). The permeability coefficient (k_p) of PRED in each formulation was computed by dividing the steady state flux (J_{ss}) by the initial concentration of PRED in the donor compartment (which is considered the saturated solubility of PRED in each formulation) (C_o) as presented in Equation (4). The results were represented as mean ± S.D. [55].

$$k_p = \frac{J_{ss}}{C_o} \quad (7)$$

ANOVA was used to analyze the data of flux and to compare the data of flux with the cumulative amount of the drug in the receptor compartment of different formulations, followed by a Tukey's honestly significant difference test using IBM SPSS® version 23 for Windows®. A probability of less than 5% ($p < 0.05$) was considered significant.

4.7. In Vivo Estimation of Anti-Inflammatory Effect of PRED-Loaded TETSMs Gels

Using adult male albino rats, the anti-inflammatory effectiveness of the best formulation of PRED–TAC-loaded TETSMs gel was compared to PRED–TAC-loaded gel and PRED-loaded gel. The study was conducted according to the guidelines of the Declaration of Helsinki and approved by the Institutional Animal Ethical Committee (IAEC) number SCBR-055-2022 of CPCSEA (Committee for Control and Supervision of Experiments on Animals), Prince Sattam Bin Abdulaziz University. Rats ranged in weight from 100 to 185 g. Before beginning the tests, they were housed for a week at the animal house at 25 ± 2 °C and a light: dark cycle of 12:12 h. Rats were fed a conventional rat pellet diet, which was removed 12 h before the experiment, although water was still available. All studies were conducted in accordance with the ethical standards for research using experimental animals as well as the guidelines for the care of laboratory animals. In the present investigation, 24 rats were separated into four groups of six rats each (n = 6), and the following study protocols were used.

First group: served as a positive control and received PRED-loaded gel.
Second group: received PRED–TAC-loaded gel.
Third group: received PRED–TAC-loaded TETSMs gel.
Fourth group: served as a negative control group.

4.7.1. Carrageenan-Induced Acute Inflammation

By using carrageenan-induced paw edema, the anti-inflammatory effect of the best formulation of PRED–TAC-loaded TETSMs gel was compared to PRED–TAC-loaded gel and PRED-loaded gel. The right paw of each animal was massaged 50 times with 0.5 g of each treatment's optimal formulation of PRED–TAC-loaded TETSMs gel, PRED–TAC-loaded gel, and PRED-loaded gel an hour before carrageenan administration to allow formulation penetration through the skin. Acute inflammation was generated in all groups an hour after treatment by injecting 0.1 mL of 1% *w/v* carrageenan saline solution into each rat's right hind paw's sub-plantar tissue. Following carrageenan administration, the rats were monitored for three hours [56,57].

4.7.2. Comparative Pharmacokinetics

Blood samples of 0.5 mL were collected at 1, 2, 4, 8, and 12 h from the abdominal aorta after application of oral PRED suspension, topical PRED-loaded gel, and topical PRED–TAC-loaded TETSMs gel at a dose of 50 mg/kg. To separate the plasma from different blood samples, all blood samples were centrifuged at 2000× *g* for 15 min. Then plasma was instantaneously conveyed into clean tubes and stored at −20 °C for further analysis.

4.8. HPLC Conditions

HPLC assay of Prednisolone in Plasma

To prepare the plasma samples for the HPLC assay, all samples were vortexed; then aliquots of 300 μL of plasma were put into falcon tubes. After that, 40 μL of 1 μg/mL dexamethasone (internal standard) was added to the plasma samples. For Prednisolone extraction, 1 mL of ethyl acetate was added and vortexed for 10 min then centrifuged at 2500× *g* for 10 min (4 °C) to allow phase separation. Finally, the upper layer was transferred to a glass tube and evaporated at 45 °C to dryness then reconstituted with 300 μL of mobile phase and injected into the HPLC system.

For quantitative determination of prednisolone in plasma samples, 100 μL aliquots were injected in a Shimadzu HPLC system (SHIMADZU 1200 series HPLC system (Kyoto, Japan) equipped with a quaternary pump, an online degasser, and an autosampler (SHIMADZU1200, Kyoto, Japan). a Ther-mosil® C-18 column (250 mm × 4.6 mm i.d., 5 μm particle size) was used and operated at 30 °C. The system was equipped with UV–Vis detectors set at 254 nm. Isocratic elution was performed using acetonitrile and water (36:64) as the mobile phase (1) with a flow rate of 1.2 mL/min and a total time of 12 min. The liquid chromatography instrument was interfaced with a computer software using Microsoft Windows 7 [58].

4.9. Pharmacokinetic Analysis

Pharmacokinetic parameters were calculated using WinNonlin software (version 1.5, Scientific Consulting, Inc., Rockville, MD, USA). Pharmacokinetic parameters include C max (maximum plasma concentration), Tmax (time required for peak concentration), AUC (area under the curve), $T_{1\backslash 2}$ (half-life), and MRT (mean residence time). The obtained parameters were subjected to analysis by ANOVA and Fisher's PSLD test for multiple evaluations among groups. Results were considered significant if the *p* value was less than 0.05. All results were reported as the mean ± SD.

Author Contributions: Conceptualization, M.M.A. and O.M.S.; Methodology, R.M.Z. and B.N.A.; Software, R.M.Z.; Validation, M.M.A. and R.M.Z.; Formal analysis, M.M.A. and O.M.S.; Investigation, B.N.A. and O.M.S.; Resources, M.M.A., R.M.Z., B.N.A. and O.M.S.; Data curation, B.N.A. and O.M.S.; Writing—original draft, O.M.S. and R.M.Z.; Writing—review & editing, R.M.Z.; Visualization, M.M.A. and B.N.A.; Project administration, B.N.A. and M.M.A.; Funding acquisition, R.M.Z. All authors have read and agreed to the published version of the manuscript.

Funding: The authors extend their appreciation to the Deputyship for Research and Innovation, Ministry of Education in Saudi Arabia, for funding this research work through the project number (IF-PSAU-2021/03/18810).

Institutional Review Board Statement: The study was conducted according with the Declaration of Helsinki, and approved by the Institutional Animal Ethical Committee (IAEC) number SCBR-055-2022 of CPCSEA (Committee for Control and Supervision of Experiments on Animals), Prince Sattam Bin Abdulaziz University.

Informed Consent Statement: Not applicable.

Data Availability Statement: The data is contained in the manuscript.

Acknowledgments: The authors extend their appreciation to the Deputyship for Research and Innovation, and Prince Sattam Bin Abdulaziz University, Ministry of Education, Saudi Arabia.

Conflicts of Interest: The authors declare no conflict of interest.

References

1. Nagarsenker, M.S.; Meshram, R.N.; Ramprakash, G. Solid dispersion of hydroxypropyl β-cyclodextrin and ketorolac: Enhancement of in-vitro dissolution rates, improvement in anti-inflammatory activity and reduction in ulcerogenicity in rats. *J. Pharm. Pharmacol.* **2000**, *52*, 949–956. [CrossRef] [PubMed]
2. Paul-Clark, M.J.; Mancini, L.; Del Soldato, P.; Flower, R.J.; Perretti, M. Potent antiarthritic properties of a glucocorticoid derivative, NCX-1015, in an experimental model of arthritis. *Proc. Natl. Acad. Sci. USA* **2002**, *99*, 1677–1682. [CrossRef] [PubMed]
3. Rao, Y.; Zheng, F.; Zhang, X.; Gao, J.; Liang, W. In vitro percutaneous permeation and skin accumulation of finasteride using vesicular ethosomal carriers. *Aaps Pharmscitech* **2008**, *9*, 860–865. [CrossRef] [PubMed]
4. Gupta, P.N.; Mishra, V.; Rawat, A.; Dubey, P.; Mahor, S.; Jain, S.; Chatterji, D.; Vyas, S.P. Non-invasive vaccine delivery in transfersomes, niosomes and liposomes: A comparative study. *Int. J. Pharm.* **2005**, *293*, 73–82. [CrossRef] [PubMed]
5. Scott, L.J.; McKeage, K.; Keam, S.J.; Plosker, G.L. Tacrolimus. *Drugs* **2003**, *63*, 1247–1297. [CrossRef] [PubMed]
6. Anglicheau, D.; Flamant, M.; Schlageter, M.H.; Martinez, F.; Cassinat, B.; Beaune, P.; Legendre, C.; Thervet, E. Pharmacokinetic interaction between corticosteroids and tacrolimus after renal transplantation. *Nephrol. Dial. Transplant.* **2003**, *18*, 2409–2414. [CrossRef] [PubMed]
7. Touitou, E.; Dayan, N.; Bergelson, L.; Godin, B.; Eliaz, M. Ethosomes—Novel vesicular carriers for enhanced delivery: Characterization and skin penetration properties. *J. Control. Release* **2000**, *65*, 403–418. [CrossRef] [PubMed]
8. Zaki, R.M.; Alfadhel, M.M.; Alshahrani, S.M.; Alsaqr, A.; Al-Kharashi, L.A.; Anwer, M.K. Formulation of Chitosan-Coated Brigatinib Nanospanlastics: Optimization, Characterization, Stability Assessment and In-Vitro Cytotoxicity Activity against H-1975 Cell Lines. *Pharmaceuticals* **2022**, *15*, 348. [CrossRef]
9. Said, M.; Aboelwafa, A.A.; Elshafeey, A.H.; Elsayed, I. Central composite optimization of ocular mucoadhesive cubosomes for enhanced bioavailability and controlled delivery of voriconazole. *J. Drug Deliv. Sci. Technol.* **2021**, *61*, 102075. [CrossRef]
10. Yeo, L.K.; Olusanya, T.O.; Chaw, C.S.; Elkordy, A.A. Brief effect of a small hydrophobic drug (cinnarizine) on the physicochemical characterisation of niosomes produced by thin-film hydration and microfluidic methods. *Pharmaceutics* **2018**, *10*, 185. [CrossRef]
11. Albash, R.; Abdelbary, A.A.; Refai, H.; El-Nabarawi, M.A. Use of transethosomes for enhancing the transdermal delivery of olmesartan medoxomil: In vitro, ex vivo, and in vivo evaluation. *Int. J. Nanomed.* **2019**, *14*, 1953. [CrossRef] [PubMed]
12. Paolino, D.; Lucania, G.; Mardente, D.; Alhaique, F.; Fresta, M. Ethosomes for skin delivery of ammonium glycyrrhizinate: In vitro percutaneous permeation through human skin and in vivo anti-inflammatory activity on human volunteers. *J. Control. Release* **2005**, *106*, 99–110. [CrossRef] [PubMed]
13. Faisal, W.; Soliman, G.M.; Hamdan, A.M. Enhanced skin deposition and delivery of voriconazole using ethosomal preparations. *J. Liposome Res.* **2018**, *28*, 14–21. [CrossRef] [PubMed]
14. Abdulbaqi, I.M.; Darwis, Y.; Khan, N.A.K.; Abou Assi, R.; Khan, A.A. Ethosomal nanocarriers: The impact of constituents and formulation techniques on ethosomal properties, in vivo studies, and clinical trials. *Int. J. Nanomed.* **2016**, *11*, 2279. [CrossRef] [PubMed]
15. Nayak, D.; Tawale, R.M.; Aranjani, J.M.; Tippavajhala, V.K. Formulation, optimization and evaluation of novel ultra-deformable vesicular drug delivery system for an anti-fungal drug. *AAPS PharmSciTech* **2020**, *21*, 140. [CrossRef] [PubMed]
16. Prasanthi, D.; Lakshmi, P. Development of ethosomes with taguchi robust design-based studies for transdermal delivery of alfuzosin hydrochloride. *Int. Curr. Pharm. J.* **2012**, *1*, 370–375. [CrossRef]
17. Limsuwan, T.; Amnuaikit, T. Development of ethosomes containing mycophenolic acid. *Procedia Chem.* **2012**, *4*, 328–335. [CrossRef]
18. Zhang, J.-P.; Wei, Y.-H.; Zhou, Y.; Li, Y.-Q.; Wu, X.-A. Ethosomes, binary ethosomes and transfersomes of terbinafine hydrochloride: A comparative study. *Arch. Pharmacal Res.* **2012**, *35*, 109–117. [CrossRef]
19. Puri, R.; Jain, S. Ethogel topical formulation for increasing the local bioavailability of 5-fluorouracil: A mechanistic study. *Anti-Cancer Drugs* **2012**, *23*, 923–934. [CrossRef]
20. Li, G.; Fan, Y.; Fan, C.; Li, X.; Wang, X.; Li, M.; Liu, Y. Tacrolimus-loaded ethosomes: Physicochemical characterization and in vivo evaluation. *Eur. J. Pharm. Biopharm.* **2012**, *82*, 49–57. [CrossRef]
21. Sarwa, K.K.; Suresh, P.K.; Rudrapal, M.; Verma, V.K. Penetration of tamoxifen citrate loaded ethosomes and liposomes across human skin: A comparative study with confocal laser scanning microscopy. *Curr. Drug Deliv.* **2014**, *11*, 332–337. [CrossRef] [PubMed]
22. Jain, S.; Patel, N.; Madan, P.; Lin, S. Quality by design approach for formulation, evaluation and statistical optimization of diclofenac-loaded ethosomes via transdermal route. *Pharm. Dev. Technol.* **2015**, *20*, 473–489. [CrossRef] [PubMed]
23. Zhang, Z.; Wo, Y.; Zhang, Y.; Wang, D.; He, R.; Chen, H.; Cui, D. In vitro study of ethosome penetration in human skin and hypertrophic scar tissue. *Nanomed. Nanotechnol. Biol. Med.* **2012**, *8*, 1026–1033. [CrossRef]
24. Mishra, D.; Mishra, P.K.; Dabadghao, S.; Dubey, V.; Nahar, M.; Jain, N.K. Comparative evaluation of hepatitis B surface antigen–loaded elastic liposomes and ethosomes for human dendritic cell uptake and immune response. *Nanomed. Nanotechnol. Biol. Med.* **2010**, *6*, 110–118. [CrossRef]
25. Zhou, Y.; Wei, Y.; Liu, H.; Zhang, G.; Wu, X.a. Preparation and in vitro evaluation of ethosomal total alkaloids of Sophora alopecuroides loaded by a transmembrane pH-gradient method. *Aaps Pharmscitech* **2010**, *11*, 1350–1358. [CrossRef] [PubMed]

26. Lopez-Pinto, J.; Gonzalez-Rodriguez, M.; Rabasco, A. Effect of cholesterol and ethanol on dermal delivery from DPPC liposomes. *Int. J. Pharm.* **2005**, *298*, 1–12. [CrossRef]
27. Liu, X.; Liu, H.; Liu, J.; He, Z.; Ding, C.; Huang, G.; Zhou, W.; Zhou, L. Preparation of a ligustrazine ethosome patch and its evaluation in vitro and in vivo. *Int. J. Nanomed.* **2011**, *6*, 241. [CrossRef]
28. Patel, K.K.; Kumar, P.; Thakkar, H.P. Formulation of niosomal gel for enhanced transdermal lopinavir delivery and its comparative evaluation with ethosomal gel. *AAPS PharmSciTech* **2012**, *13*, 1502–1510. [CrossRef]
29. Rakesh, R.; Anoop, K. Formulation and optimization of nano-sized ethosomes for enhanced transdermal delivery of cromolyn sodium. *J. Pharm. Bioallied Sci.* **2012**, *4*, 333.
30. Ahad, A.; Aqil, M.; Kohli, K.; Sultana, Y.; Mujeeb, M. Enhanced transdermal delivery of an anti-hypertensive agent via nanoethosomes: Statistical optimization, characterization and pharmacokinetic assessment. *Int. J. Pharm.* **2013**, *443*, 26–38. [CrossRef]
31. Zhaowu, Z.; Xiaoli, W.; Yangde, Z.; Nianfeng, L. Preparation of matrine ethosome, its percutaneous permeation in vitro and anti-inflammatory activity in vivo in rats. *J. Liposome Res.* **2009**, *19*, 155–162. [CrossRef] [PubMed]
32. Abdellatif, M.M.; Khalil, I.A.; Khalil, M.A. Sertaconazole nitrate loaded nanovesicular systems for targeting skin fungal infection: In-vitro, ex-vivo and in-vivo evaluation. *Int. J. Pharm.* **2017**, *527*, 1–11. [CrossRef] [PubMed]
33. Aute, P.P.; Kamble, M.S.; Chaudhari, P.D.; Bhosale, A.V. A comprehensive review on ethosomes. *Int. J. Res. Dev. Pharm. Life Sci.* **2012**, *2*, 218–224.
34. Kumar, N.; Dubey, A.; Mishra, A.; Tiwari, P. Ethosomes: A Novel Approach in Transdermal Drug Delivery System. *Int. J. Pharm. Life Sci.* **2020**, *11*, 6598–6608.
35. Pathan, I.B.; Jaware, B.P.; Shelke, S.; Ambekar, W. Curcumin loaded ethosomes for transdermal application: Formulation, optimization, in-vitro and in-vivo study. *J. Drug Deliv. Sci. Technol.* **2018**, *44*, 49–57. [CrossRef]
36. Zaki, R.M.; Seshadri, V.D.; Mutayran, A.S.; Elsawaf, L.A.; Hamad, A.M.; Almurshedi, A.S.; Yusif, R.M.; Said, M. Wound Healing Efficacy of Rosuvastatin Transethosomal Gel, I Optimal Optimization, Histological and In Vivo Evaluation. *Pharmaceutics* **2022**, *14*, 2521. [CrossRef] [PubMed]
37. Salem, H.F.; Kharshoum, R.M.; Sayed, O.M.; Abdel Hakim, L.F. Formulation design and optimization of novel soft glycerosomes for enhanced topical delivery of celecoxib and cupferron by Box–Behnken statistical design. *Drug Dev. Ind. Pharm.* **2018**, *44*, 1871–1884. [CrossRef]
38. Lukić, M.; Pantelić, I.; Savić, S.D. Towards optimal ph of the skin and topical formulations: From the current state of the art to tailored products. *Cosmetics* **2021**, *8*, 69. [CrossRef]
39. Salem, H.F.; Kharshoum, R.M.; Abou-Taleb, H.A.; Farouk, H.O.; Zaki, R.M. Fabrication and appraisal of simvastatin via tailored niosomal nanovesicles for transdermal delivery enhancement: In vitro and in vivo assessment. *Pharmaceutics* **2021**, *13*, 138. [CrossRef]
40. McEwen, B.S.; Biron, C.A.; Brunson, K.W.; Bulloch, K.; Chambers, W.H.; Dhabhar, F.S.; Goldfarb, R.H.; Kitson, R.P.; Miller, A.H.; Spencer, R.L. The role of adrenocorticoids as modulators of immune function in health and disease: Neural, endocrine and immune interactions. *Brain Res. Rev.* **1997**, *23*, 79–133. [CrossRef]
41. Thapa, R.K.; Yoo, B.K. Evaluation of the effect of tacrolimus-loaded liquid crystalline nanoparticles on psoriasis-like skin inflammation. *J. Dermatol. Treat.* **2014**, *25*, 22–25. [CrossRef] [PubMed]
42. Umar, B.U.; Rahman, S.; Dutta, S.; Islam, T.; Nusrat, N.; Chowdhury, K.; Ahmad, W.F.S.B.W.; Haque, M. Management of Atopic Dermatitis: The Role of Tacrolimus. *Cureus* **2022**, *14*, e28130. [CrossRef] [PubMed]
43. Ramadon, D.; McCrudden, M.T.; Courtenay, A.J.; Donnelly, R.F. Enhancement strategies for transdermal drug delivery systems: Current trends and applications. *Drug Deliv. Transl. Res.* **2021**, *12*, 758–791. [CrossRef]
44. Dhopavkar, S.; Kadu, P. Transfersomes-a Boon for Transdermal Delivery. *Indo Am. J. Pharm. Sci.* **2017**, *4*, 2908–2919.
45. Teng, F.; Zhang, W.; Wang, W.; Chen, J.; Liu, S.; Li, M.; Li, L.; Guo, W.; Wei, H. Population pharmacokinetics of tacrolimus in Chinese adult liver transplant patients. *Biopharm. Drug Dispos.* **2022**, *43*, 76–85. [CrossRef] [PubMed]
46. Manca, M.L.; Zaru, M.; Manconi, M.; Lai, F.; Valenti, D.; Sinico, C.; Fadda, A.M. Glycerosomes: A new tool for effective dermal and transdermal drug delivery. *Int. J. Pharm.* **2013**, *455*, 66–74. [CrossRef]
47. Divakar, P.; Kumar, D.; Praveen, C.; Sowmya, C.; Reddy, C.S. Formulation and in vitro evaluation of liposomes containing metformin hydrochloride. *Int. J. Res. Pharm. Biomed. Sci.* **2013**, *4*, 479–485.
48. Vezočnik, V.; Rebolj, K.; Sitar, S.; Ota, K.; Tušek-Žnidarič, M.; Štrus, J.; Sepčić, K.; Pahovnik, D.; Maček, P.; Žagar, E. Size fractionation and size characterization of nanoemulsions of lipid droplets and large unilamellar lipid vesicles by asymmetric-flow field-flow fractionation/multi-angle light scattering and dynamic light scattering. *J. Chromatogr. A* **2015**, *1418*, 185–191. [CrossRef]
49. Zaki, R.M.; Alfadhel, M.M.; Alossaimi, M.A.; Elsawaf, L.A.; Devanathadesikan Seshadri, V.; Almurshedi, A.S.; Yusif, R.M.; Said, M. Central Composite Optimization of Glycerosomes for the Enhanced Oral Bioavailability and Brain Delivery of Quetiapine Fumarate. *Pharmaceuticals* **2022**, *15*, 940. [CrossRef]
50. Mazyed, E.A.; Abdelaziz, A.E. Fabrication of transgelosomes for enhancing the ocular delivery of acetazolamide: Statistical optimization, in vitro characterization, and in vivo study. *Pharmaceutics* **2020**, *12*, 465. [CrossRef]
51. Kute, S.; Saudagar, R. Emulsified gel A Novel approach for delivery of hydrophobic drugs: An overview. *J. Adv. Pharm. Educ. Res.* **2013**, *3*, 368–376.

52. Rawat, S.; Warade, S.; Lahoti, S. In situ gel formulation of ornidazole for the treatment of periodontal disease. *J. Curr. Pharma Res.* **2010**, *1*, 60.
53. Zaki, N.M.; Awad, G.A.; Mortada, N.D.; Abd ElHady, S.S. Enhanced bioavailability of metoclopramide HCl by intranasal administration of a mucoadhesive in situ gel with modulated rheological and mucociliary transport properties. *Eur. J. Pharm. Sci.* **2007**, *32*, 296–307. [CrossRef] [PubMed]
54. Qian, S.; Wong, Y.C.; Zuo, Z. Development, characterization and application of in situ gel systems for intranasal delivery of tacrine. *Int. J. Pharm.* **2014**, *468*, 272–282. [CrossRef]
55. Dias, M.; Hadgraft, J.; Lane, M.E. Influence of membrane–solvent–solute interactions on solute permeation in model membranes. *Int. J. Pharm.* **2007**, *336*, 108–114. [CrossRef]
56. Kurakula, M.; Srinivas, C.; Kasturi, N.; Diwan, P.V. Formulation and evaluation of prednisolone proliposomal gel for effective topical pharmacotherapy. *Int. J. Pharm. Sci. Drug Res.* **2012**, *4*, 35.
57. Eroğlu, İ.; Azizoğlu, E.; Özyazıcı, M.; Nenni, M.; Gürer Orhan, H.; Özbal, S.; Tekmen, I.; Ertam, I.; Ünal, İ.; Özer, Ö. Effective topical delivery systems for corticosteroids: Dermatological and histological evaluations. *Drug Deliv.* **2016**, *23*, 1502–1513. [CrossRef]
58. Del Sole, M.J.; Schaiquevich, P.; Aba, M.A.; Lanusse, C.E.; Moreno, L. Plasma and ocular prednisolone disposition after oral treatment in cats. *BioMed Res. Int.* **2013**, *2013*, 209439. [CrossRef]

gels

MDPI

Article

Emulsion-Based Gel Loaded with Ibuprofen and Its Derivatives

Adebukola Abiola Agboola [1], Anna Nowak [2], Wiktoria Duchnik [2], Łukasz Kucharski [2], Anna Story [3], Grzegorz Story [3], Łukasz Struk [4], Adrian Krzysztof Antosik [1] and Paula Ossowicz-Rupniewska [1,*]

1 Department of Chemical Organic Technology and Polymeric Materials, Faculty of Chemical Technology and Engineering, West Pomeranian University of Technology in Szczecin, Piastów Ave. 42, 71-065 Szczecin, Poland; adebukolaabiola856@gmail.com (A.A.A.); adriankrzysztofantosik@gmail.com (A.K.A.)

2 Department of Cosmetic and Pharmaceutical Chemistry, Pomeranian Medical University in Szczecin, Powstańców Wielkopolskich Ave. 72, 70-111 Szczecin, Poland; anowak@pum.edu.pl (A.N.); wiktoria.duchnik@pum.edu.pl (W.D.); lukasz.kucharski@pum.edu.pl (Ł.K.)

3 Department of Chemical and Process Engineering, Faculty of Chemical Technology and Engineering, West Pomeranian University of Technology in Szczecin, Piastów Ave. 42, 71-065 Szczecin, Poland; anna.story@zut.edu.pl (A.S.); gstory@zut.edu.pl (G.S.)

4 Department of Organic and Physical Chemistry, Faculty of Chemical Technology and Engineering, West Pomeranian University of Technology in Szczecin, Piastów Ave. 42, 71-065 Szczecin, Poland; lstruk@zut.edu.pl

* Correspondence: possowicz@zut.edu.pl; Tel.: +48-4494801

Abstract: The aim of this study was to evaluate the effect of vehicle and chemical modifications of the structure of active compounds on the skin permeation and accumulation of ibuprofen (IBU). As a result, semi-solid formulations in the form of an emulsion-based gel loaded with ibuprofen and its derivatives, such as sodium ibuprofenate (IBUNa) and L-phenylalanine ethyl ester ibuprofenate ([PheOEt][IBU]), were developed. The properties of the obtained formulations were examined, including density, refractive index, viscosity, and particle size distribution. The parameters of release and permeability through the pig skin of the active substances contained in the obtained semi-solid formulations were determined. The results indicate that an emulsion-based gel enhanced the skin penetration of IBU and its derivatives compared to two commercial preparations in the form of a gel and a cream. The average cumulative mass of IBU after a 24 h permeation test from an emulsion-based gel formulation through human skin was 1.6–4.0 times higher than for the commercial products. Ibuprofen derivatives were evaluated as chemical penetration enhancers. After 24 h of penetration, the cumulative mass was 1086.6 ± 245.8 for IBUNa and 948.6 ± 87.5 μg IBU/cm^2 for [PheOEt][IBU], respectively. This study demonstrates the perspective of the transdermal emulsion-based gel vehicle in conjunction with the modification of the drug as a potentially faster drug delivery system.

Keywords: carbomer; increasing drug permeability; nonsteroidal anti-inflammatory drugs; ibuprofen; transdermal drug delivery; structural modification of ibuprofen

Citation: Agboola, A.A.; Nowak, A.; Duchnik, W.; Kucharski, Ł.; Story, A.; Story, G.; Struk, Ł.; Antosik, A.K.; Ossowicz-Rupniewska, P. Emulsion-Based Gel Loaded with Ibuprofen and Its Derivatives. *Gels* **2023**, *9*, 391. https://doi.org/10.3390/gels9050391

Academic Editors: Ying Huang, Zhengwei Huang and Xuanjuan Zhang

Received: 21 April 2023
Revised: 4 May 2023
Accepted: 6 May 2023
Published: 8 May 2023

1. Introduction

Administering drugs through the skin, or transdermal drug delivery, has many advantages over other routes of administration, such as oral or injection routes [1–6]. In particular, it is a non-invasive method of drug administration that does not require punctures or incisions, which is especially important for patients who are afraid of needles [7,8]. It can also be used for patients who have difficulty taking oral medications. Transdermal drug delivery can provide continuous drug delivery over a longer period of time, which can improve therapeutic efficacy and reduce the need for frequent dosing. In addition, transdermal administration of the drug avoids the first-pass effect, which allows for higher concentrations of the drug in the blood. This route of administration allows for continuous drug delivery, which may improve patient compliance as it reduces the need for frequent dosing and may prevent missed doses. In this case, we can target the delivery of the drug to specific areas of the body, such as joints, providing local therapeutic effects. Overall, transdermal drug delivery offers a convenient and effective method of drug administration with the potential for improved therapeutic efficacy and patient compliance [1,7,9–11].

One of the main challenges in developing topical drug products is achieving adequate penetration of the drug through the skin. The skin is a complex barrier that provides protection against external threats but also hinders the penetration of many therapeutic agents. The outermost layer of the skin, called the *stratum corneum*, is composed of dead skin cells that are tightly packed together [8,12–14]. This layer acts as a barrier to the penetration of molecules larger than a few hundred Daltons, preventing the absorption of many drugs. In addition to the *stratum corneum*, other factors such as skin thickness, hydration, and the presence of hair follicles and sweat glands can also impact the penetration of drugs through the skin [15–17].

Various approaches have been developed to overcome these challenges and enhance drug penetration through the skin. In general, various methods of increasing the permeability of the *stratum corneum* are distinguished, including methods related to the modification of drug molecules, selection of the appropriate carrier and drug form, and incorporation of the drug into the carrier; methods related to the modification of the *stratum corneum* properties—hydration, agents that increase chemical penetration; electrically assisted methods and devices—iontophoresis, electroporation, ultrasounds; methods related to bypassing or removing the *stratum corneum*—microneedles, jet injections, and ablations [1,17–19].

In this publication, we will use two parallel methods related to the modification of the drug molecule and the development of the emulsion-based gel formulation.

Emulsion-based gels allow for greater stability of the resulting emulsion due to the presence of an emulsifying agent, which allows for a longer shelf life and better stability of the drug. Furthermore, this form of the drug allows for increased bioavailability, in particular for hydrophobic drugs, by increasing their solubility in the oil phase. In addition, it can prevent drug degradation in the aqueous phase. In addition, these types of formulations can increase drug penetration through the skin, allowing for a more even distribution of the drug over the skin surface and extending the time the drug is in contact with the skin. The gelling agent contained in the formulation can also control drug release over a longer period of time, providing a lasting therapeutic effect and reducing the need for frequent dosing. In addition, emulsion-based gels are easy to apply and spread evenly on the skin, providing a convenient way to administer medications [20–25].

The aim of this research will be the development of the composition of the emulsion-based gel as a ready-to-use pharmaceutical formulation and an assessment of its properties. Furthermore, these studies aim to develop a delivery system for a poorly water-soluble model drug, ibuprofen, to increase its solubility by developing a microemulsion system as a matrix and then introducing it into the gel phase. Additionally, changes in the structure of ibuprofen were also introduced in order to increase the solubility of the drug in water, and the obtained compounds were introduced into the preparations obtained for application to the skin.

2. Results and Discussion

The *stratum corneum*, which consists primarily of lipids, inhibits the penetration of topically applied drugs [26]. Consequently, in the penetration study, the selection of a suitable substrate may be the key to increasing penetration and achieving a rapid therapeutic effect in the subcutaneous tissues. The physicochemical properties of the compound, such as its lipophilicity, solubility, and molar mass, and the type of pharmaceutical base have a significant impact on its ability to penetrate the *stratum corneum* [27]. The ingredients of the formulation are responsible for the physicochemical properties of the pharmaceutical form and affect the penetrating ability of a given substance. The modification of the active ingredient, consisting primarily of an increase in its lipophilicity, is an additional method for enhancing penetration [28,29].

In our previous publication, we showed that amino acid alkyl ester ibuprofenates have increased skin permeability compared to ibuprofen [28,30]. The possibility of creating formulations with two commercially available, modern pharmaceutical bases—Pentravan® [31] and Celugel® [32]—were also tested.

As is known, the selection and use of the appropriate vehicle are crucial for the degree of absorption and therapeutic effectiveness of the drug. An ideal pharmaceutical vehicle should meet the following conditions: have the right consistency, be easy to mix with therapeutic agents, keep the active substance in constant dispersion, not react with the drug and other excipients or external factors, not cause allergies or irritations of the skin or mucous membranes, even if they are of a short-term nature, and not decompose rapidly.

2.1. *Ibuprofenate of Ethyl L-Phenylalanine Ester*

As a result of the 3-stage reaction shown in Scheme 1, an ibuprofen derivative—L-phenylalanine ethyl ester ibuprofenate—was obtained with a total yield of 91%. The identity of the obtained compound was confirmed based on the analysis of ^{1}H NMR, ^{13}C NMR, and FT-IR spectra. Signal assignments, their interpretation, and all spectra are shown in Figures S1–S3. In addition, a full characterization of the obtained compound was performed.

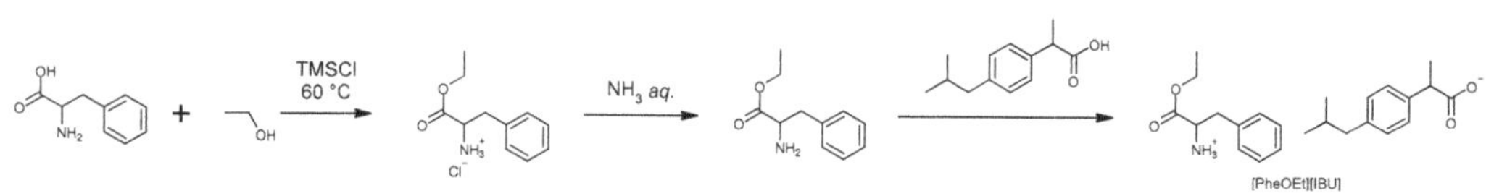

Scheme 1. Scheme of synthesis of L-phenylalanine ethyl ester ibuprofenate used in studies.

The signal for the amino group (NH_3^+), confirming the ionic structure of the obtained compound, is 5.09 ppm. Furthermore, the ionic structure of L-phenylalanine ethyl ester ibuprofenate was confirmed according to the presence of two distinct absorption bands—asymmetric $v(COO^-)_{as}$ at 1609.07 cm^{-1} and symmetric stretching vibrations $v(COO^-)_{sym}$ at 1382.20 cm^{-1}. The difference in frequency values between these two bands was greater than 200 cm^{-1}, confirming the presence of the carboxylate anion COO^- and the ionic structure of the obtained ibuprofenate [30,33].

The properties of the obtained structural modification of ibuprofen ([PheOEt][IBU]) were compared with the properties of unmodified ibuprofen (IBU) and sodium salt (IBUNa).

As can be seen in Figure 1, the diffractograms of the derivatives obtained are different from those of the starting ibuprofen. The obtained results confirmed the crystalline nature of both unmodified ibuprofen and its derivatives.

The thermal stability of drugs is an essential property necessary to control the formulation method of the finished drug. Ibuprofen, sodium ibuprofen, and L-phenylalanine ethyl ester ibuprofenate thermal stabilities were assessed and contrasted. Figure S4 displays the TG, DTG, and c-DTA curves of [PheOEt][IBU]. Both the temperature at which decomposition begins (T_{onset}) and the temperature at which mass loss accelerates to its maximum (T_{max}) were identified. As seen in Table 1, [PheOEt][IBU] is the least stable (Tonset = 160.8 °C), while the [IBUNa] compound is the most stable (T_{onset} = 265.4 °C). The temperatures of the decompositions of ibuprofen and ibuprofen salt are similar. However, differences in the stages of decomposition are visible. Ibuprofen decomposes in one step, [PheOEt][IBU]—in two stages, and IBUNa—in several stages.

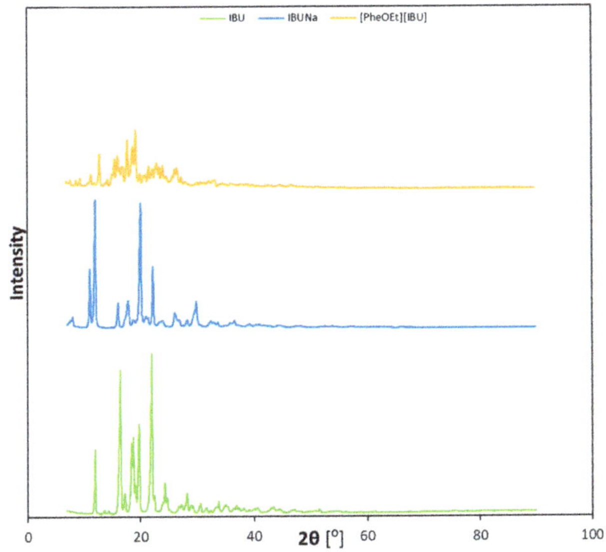

Figure 1. Diffractograms of IBU, IBUNa, and [PheOEt][IBU].

Table 1. Thermal stability results of the tested APIs.

Compound	T_m (°C)	T_{onset} (°C)	T_{max} (°C)	$T_{5\%}$ (°C)	$T_{50\%}$ (°C)
IBU	79.6	193.2	223.8	162.1	210.3
IBUNa	194.6	265.4	444.4	285.4	445.1
[PheOEt][IBU]	82.3	160.8	215.6	141.6	221.7

T_m, melting point; T_{onset}, the onset temperature of the thermal degradation; T_{max}, the temperature of the maximum mass loss rate; $T_{5\%}$, decomposition temperature, corresponds to 5% weight loss of the initial mass; $T_{50\%}$, decomposition temperature, corresponds to 50% weight loss of the initial mass.

The melting point was determined from the DSC curves. The melting point for IBU is 79.6 °C, and for [PheOEt][IBU] it is 82.3 °C. It demonstrates that the obtained [PheOEt][IBU] has a higher melting point than IBU. Since L-phenylalanine ethyl ester ibuprofenate has an ionic structure, which has already been confirmed, and a melting point below 100 °C, it can be classified as an ionic liquid. As is well known, obtaining ionic liquids has several advantages, including avoiding the phenomenon of polymorphism.

The solubility of a substance in water is used to assess the bioavailability and effectiveness of a drug. Therefore, the solubility of IBUNa (9.297 ± 0.672 g/L; 8.402 ± 0.607 g IBU/L) and [PheOEt][IBU] (0.716 ± 0.080 g/L; 0.370 ± 0.045 g IBU/L) was determined and compared to the solubility of ibuprofen (0.076 g/L). As can be seen, converting the drug to a salt form increases the solubility of the drug in water, which is a common method.

Lipophilicity is essential for predicting the hydrophobicity and partitioning of drugs in biological systems. Therefore, the test was prepared using the shake-flask method. IBU and [PheOEt][IBU] showed a positive log p of 2.758 ± 0.014 and 1.336 ± 0.046, respectively. On the other hand, IBUNa's negative log p-value (−0.341 ± 0.176) indicates it is the most hydrophilic. Therefore, [PheOEt][IBU] is more hydrophilic than IBU but less hydrophilic than IBUNa.

2.2. Pre-Formulation Studies: Selection of Formulation Components

Pre-formulation studies are essential to selecting the best emulsion composition and the ingredient in which ibuprofen is most soluble. The results of ibuprofen solubility in various oils and surfactants are summarized in Table 2. It was shown that among the tested oils, ibuprofen was the most soluble in hemp oil (95.44 mg/mL), while among hydrophilic surfactants, Tween 60 (519.63 mg/mL). Furthermore, Span 80 was selected as the lipophilic surfactant, and the solubility of ibuprofen was 56.89 mg/mL. Furthermore, no interaction was observed between ibuprofen and the formulation ingredients, i.e., unplanned and unintended changes in the physicochemical properties (e.g., appearance, smell, and colour). This means that the chosen surfactant and auxiliary substances are the most appropriate for the formulation of an emulsion-based gel.

Table 2. Solubility of ibuprofen in various oils and surfactants at 25 °C (mean ± S.D., n = 3).

	Components	Solubility (mg/mL)
Oils	Sunflower oil	65.10 ± 1.92
	Hemp oil	95.44 ± 0.03
	Primrose oil	76.74 ± 0.35
Surfactants	Span 80	56.89 ± 0.15
	Tween 40	466.13 ± 1.71
	Tween 60	519.63 ± 0.69
	Tween 80	411.10 ± 4.18

2.3. Preparation of Base Formulations

As a result, four emulsions with different ratios of the water phase to the oil phase were obtained (Table 3), which resulted in different amounts of surfactants being used. The ratio of hydrophilic and lipophilic surfactants was determined for the required HLB of the oil.

Table 3. Amount of ingredients in basic emulsions.

Formulation	Amount of Water (g)	Mass of Surfactant 1 (Tween 60) (g)	Mass of Surfactant 2 (Span 80) (g)	Mass of Oil (Hemp Oil) (g)
F1	25.32	2.69	2.38	25.02
F2	30.02	2.16	1.86	20.02
F3	35.02	1.62	1.40	15.01
F4	40.40	1.25	0.92	10.03

The stability of each of the four obtained formulations was studied at an elevated temperature of 40 °C and in a freeze-thaw stability test. There were no physical changes in color, and the appearance of F1, F2, and F3 then changed by showing separate physical phases after seven days of the testing period, leaving F4 to be the emulsion without any physical change in appearance. In the freeze-thaw test, the remaining emulsions stratified after the first cycle. In the freeze-thaw test, the emulsions F1–F3 stratified after the first cycle. The F4 emulsion was stable in all five cycles. Therefore, F4 was the most suitable formulation.

2.4. Preparation of Formulation with Carbomer

Since all the base formulations had inadequate consistency and were low-viscosity liquids, it was decided to refine the composition of the formulation. In order to prepare a suitable formulation, a formulation containing carbomer as a rheology-improving agent and triethanolamine as a pH regulator was developed. As a result of the work, a formulation with the desired consistency was obtained, similar to commercial pharmaceutical gel preparations for use on the skin.

2.5. *API-Loaded Emulsion-Based Gel Formulation*

According to the preparation of the emulsion, the active pharmaceutical ingredient (IBU, IBUNa, and the obtained [PheOEt][IBU]) was added, and there was no physical change in color or appearance. However, it was noticed that the addition of IBUNa significantly reduced the viscosity of the preparation. In the case of other active compounds, no differences were noticed compared to the formulation without API.

2.6. *Determination of Physicochemical Properties of Obtained Emulsion-Based Gel Formulations*

2.6.1. Microscopic Examinations

The prepared formulations were observed under an optical microscope (Delta Optical, with a MC500-W3, 5 MP camera). The images presented below in Figure 2 are for the formulation without API (MEG) and with API, which contained IBU (MEG_IBU), IBUNa (MEG_IBUNa), and [PheOEt][IBU] (MEG_[PheOEt][IBU]). The result shows that the preparation was perfectly homogenous, and there are no separate phases in the formulations with different APIs.

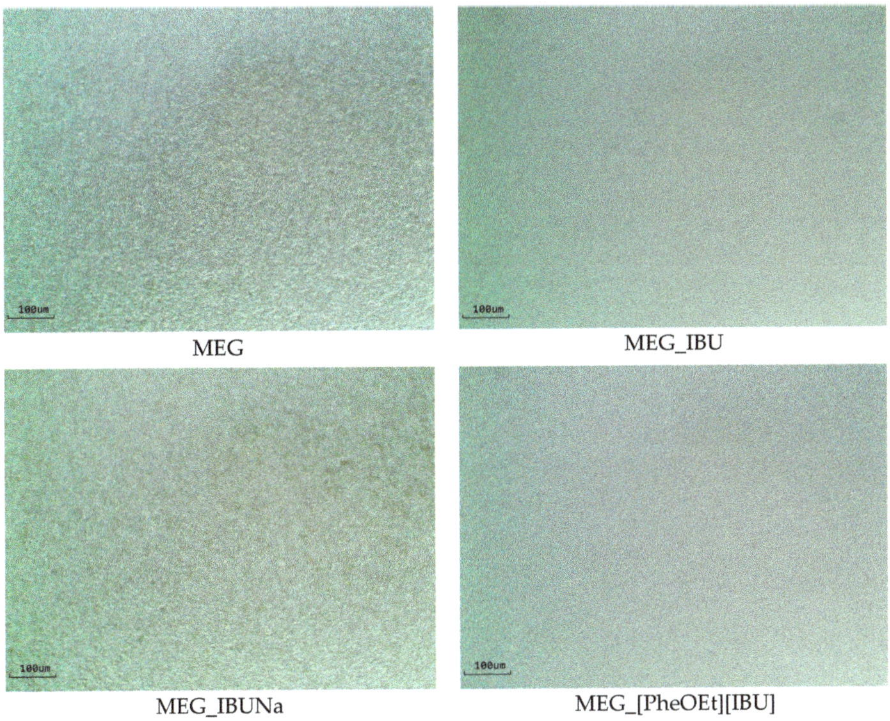

Figure 2. Microscopic images of (micro)emulsion-based gel loaded without and with ibuprofen and its derivatives.

2.6.2. Density and Refractive Index

The densities of all the resulting formulations ranged from 0.9610 to 0.9855 g/cm^3, as seen in Table 4. This result shows that the gel has a relatively lower density than water. This is due to the fact that the oil droplets in the gel are less dense than the continuous water phase, which results in a lower overall density. Compared to MEG_IBUNa, which had the highest density value, MEG_IBU displayed the lowest density. The density values are similar overall.

Table 4. Density and refractive index results of tested formulations without and with different APIs.

Formulation	Density (g/cm^3)	Refractive Index
MEG	0.9826 ± 0.0008	1.3952 ± 0.0003
MEG_IBU	0.9610 ± 0.0006	1.4226 ± 0.0005
MEG_IBUNa	0.9920 ± 0.0009	1.3910 ± 0.0010
MEG_PheOEtIBU	0.9855 ± 0.0007	1.4129 ± 0.0006

The refractive indices for all the resulting formulations ranged from 1.3952 to 1.4129. Compared to MEG_IBU, which has the highest refractive index value, MEG_IBUNa has the lowest refractive index. There is a negligible difference between the refractive index values of each MEG with and without API. The obtained formulations were transparent, indicating very small particle sizes. The refractive index of the formulations was just a little bit higher than that of water (1.3325).

2.6.3. Viscosity

Figure 3 shows that the emulsion-based gel formulations exhibited non-Newtonian shear-thinning pseudoplastic flow. From the utility point of view, the most important are the values of the flow limit at 20 °C, which can be associated with the administration of the formulation; the viscosity at low (10–100 s^{-1}) shear rates accompanying the application of the formulation on the skin; and the viscosity at shear rates in the range of 10^2–10^3 s^{-1} at body temperature, which characterize the rubbing of the formulation. The highest viscosity is found in MEG. However, the viscosity of the MEG decreases when API is added. The viscosities of the obtained formulations for MEG_IBU and MEG_PheOEtIBU are similar but two times lower than MEG. When IBUNa is added, the viscosity of the formulation decreases ten times more. The same relationships were found for the two temperatures used (20.0 and 36.6 °C). Was chosen as the application temperature of 20.0 °C because it is the typical storage temperature, and 36.6 °C is the typical body temperature. It can also be seen that the viscosity decreases slightly as the temperature rises.

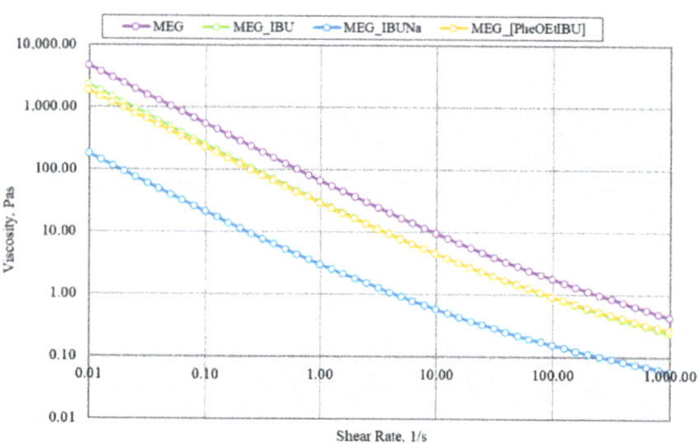

(**a**)

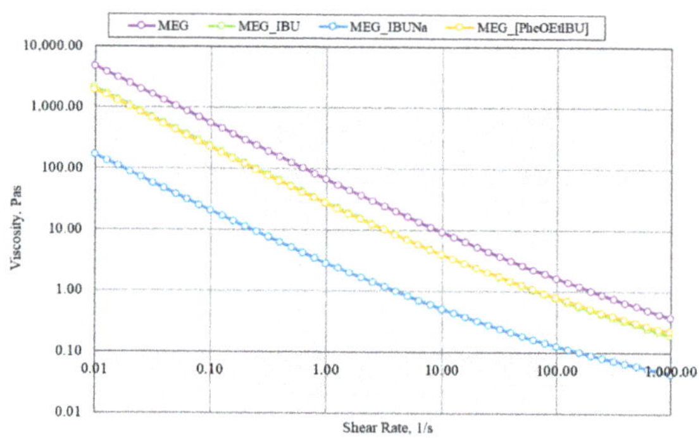

(**b**)

Figure 3. The viscosity profile of MEG_IBU, MEG_IBUNa, and MEG_[PheOEt][IBU] at temperatures (**a**) 20 °C and (**b**) 36.6 °C.

2.6.4. Particle Size Distribution

The particle size distribution of the emulsion-based gel was determined by a particle size analyzer. Figure 4 shows the particle size distribution of MEG with and without API expressed as a volume fraction (%). The emulsion droplets range from 0.3 µm to 19.89 µm for MEG, 0.3 µm to 6.3 µm for MEG_IBU, 0.3 µm to 5.6 µm for IBUNa, and 0.2 µm to 4.9 µm for [PheOEt][IBU]. The highest volume fraction for particle sizes of 2.0 µm was for MEG, MEG_IBU, and MEG_IBUNa, compared to MEG_[PheOEt][IBU], for which it was 1.8 µm. This range indicates that the MEG with and without API is of very small globule size. The addition of IBU, IBUNa, or [PheOEt][IBU] affects the size and distribution of the droplets or particles in the emulsion-based gel. This results in a narrower size distribution, as the droplets or particles are less likely to aggregate and form larger sizes.

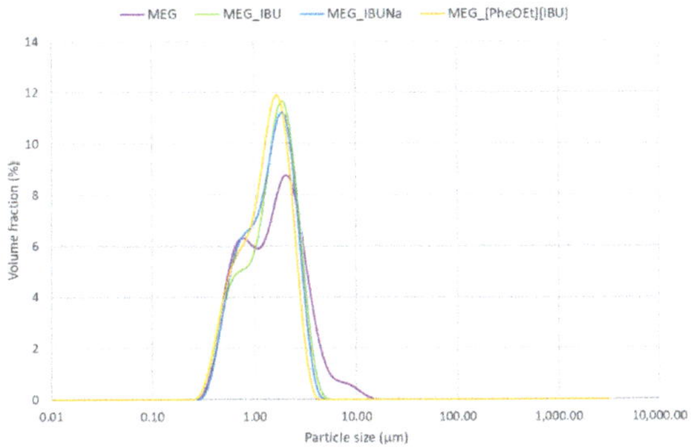

Figure 4. The particle size profile of MEG_IBU, MEG_IBUNa, and MEG_[PheOEt][IBU].

2.6.5. Release of the Active Substance from the Drug Form

The release of the API from the emulsion-based gels was also determined. For comparison, the release of ibuprofen from two commercial formulations in the form of gel and cream was also carried out. Figure 5 depicts the cumulative release in µg IBU/cm^2 as a function of time and the permeation rates of ibuprofen and its derivatives from the emulsion-based gel through pig skin. The first 180 min show the greatest release of API, which is then inhibited. As a result, the MEG formulations obtained limit the amount of active ingredient released over time. The highest statistically significant release was observed for the two ibuprofen derivatives placed in MEG formulations (Figure 6). In the release study, a statistically significant difference was observed between IBU derivatives (MEG_IBUNa and MEG_[PheOEt][IBU]), pure IBU (MEG_IBU), and two commercial preparations. No statistically significant difference was observed between MEG_IBU and CP gel. The lowest statistically significant release was observed for the commercial preparation of CP gel.

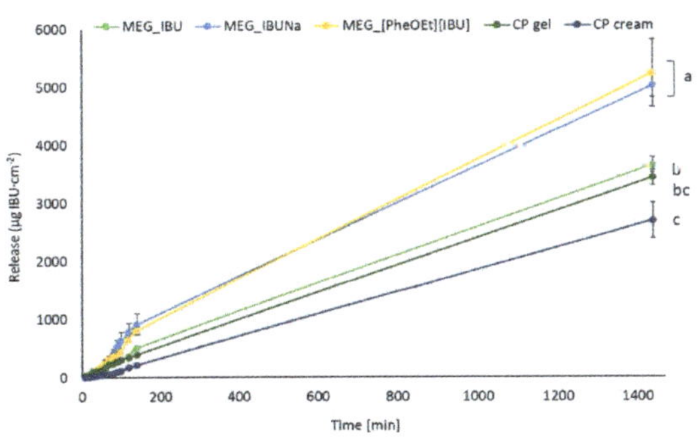

(a)

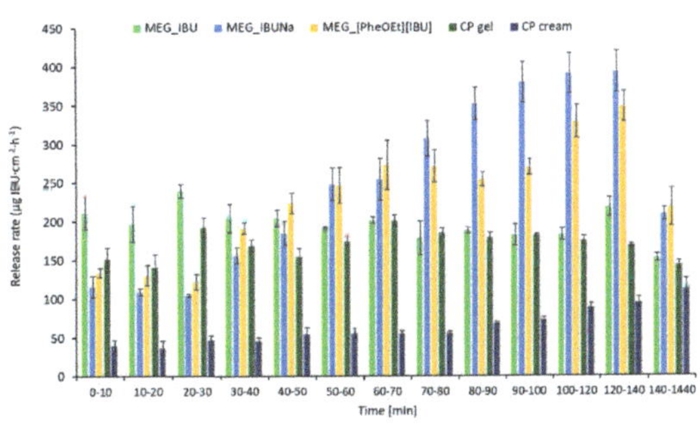

(b)

Figure 5. Time course of the cumulative mass of IBU and its derivatives during the 24 h release (**a**) and permeation rates of IBU and its derivatives from the emulsion-based gel through pig skin (**b**). Values are the means with a standard deviation; n = 3. The statistically significant difference was estimated using the ANOVA test.

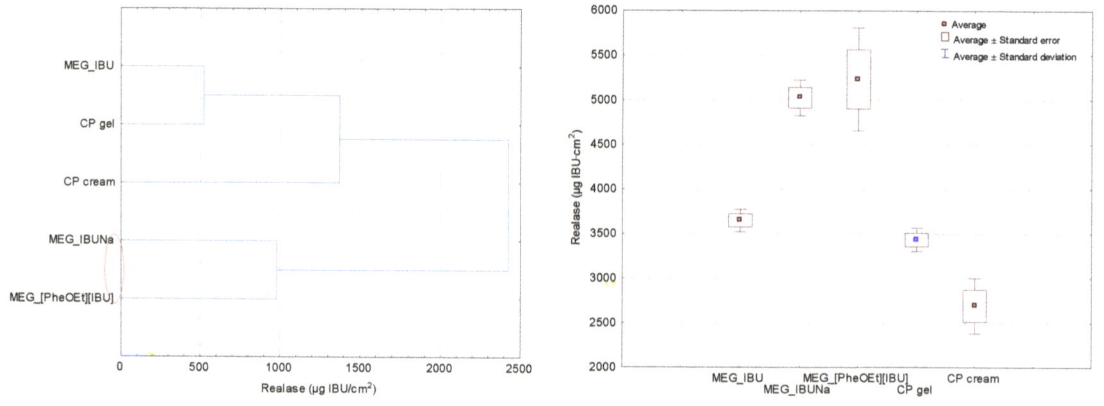

(a) (b)

Figure 6. The cluster analysis graph (**a**) and the box plot (**b**) for the cumulative mass of ibuprofen and its derivatives after a 24-h study of release. The preparations with the highest release are marked with a red circle.

2.6.6. Permeation of the Active Substance through the Skin

Figure 7 depicts individual permeation profiles for ibuprofen and its derivatives. Table 5 summarizes the cumulative mass of ibuprofen after 24 h of permeation measurement. MEG_IBUNa had a higher permeability than MEG_IBU and MEG_[PheOEt][IBU]. The total amount of substance that permeated over the course of the 24-h study was, in this case, 1086.6 ± 245.8 μg IBU/cm^2 (Table 5). The formation of structural modifications of ibuprofen in charged compounds with lower lipophilicity has an influence on better permeability compared to more lipophilic ibuprofen. After 24 h of permeation, the cumulative mass of ibuprofen derivatives in the acceptor phase was significantly higher than that of ibuprofen and significantly higher compared to commercial preparations (Figure 8). In the pig skin penetration study, a statistically significant difference was found in the μg IBU/cm^2 of testing for IBU derivatives (MEG_IBUNa and MEG_[PheOEt][IBU]) in comparison with commercial preparations. However, there was no significant difference between pure IBU (MEG_IBU) and CP gel. The lowest penetration, statistically significantly different, was observed for the commercial formulation of CP cream.

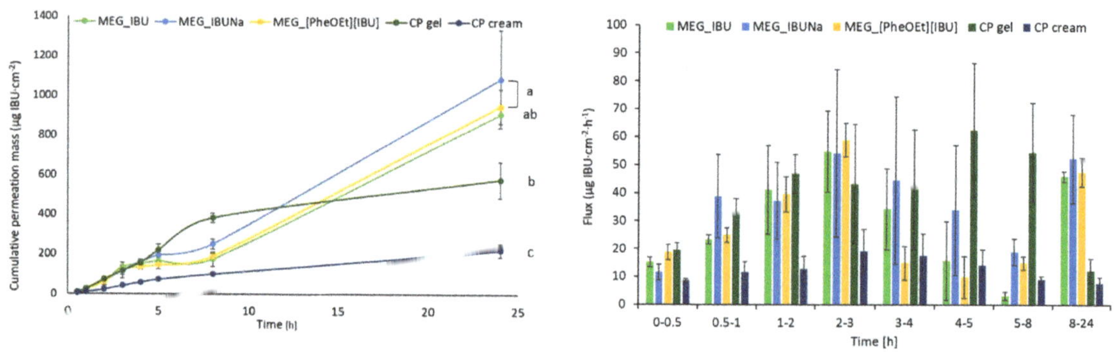

(a) (b)

Figure 7. Permeation profiles (**a**) and permeation rates (**b**) of ibuprofen and its derivatives from the emulsion-based gel through pig skin. The values are the means with standard deviation; n = 3. A statistically significant difference was estimated using the ANOVA test.

Table 5. Skin permeation parameters for ibuprofen and its derivatives from obtained formulations; n = 3.

Formulation	Cumulated Mass, µg IBU/cm^2	J_{ss}, µg IBU/cm^2·h	K_p·10^4, cm/h	L_T, h	D·10^4, cm^2/h	K_m	$Q\%_{24\,h}$	EF
MEG_IBU	909.6 ± 45.3	43.439 ± 2.262	8.688 ± 0.452	0.461 ± 0.099	9.371 ± 2.294	0.048 ± 0.012	1.82 ± 0.09	4.09
MEG_IBUNa	1086.6 ± 245.8	43.741 ± 3.258	8.749 ± 0.652	0.422 ± 0.036	9.929 ± 0.808	0.045 ± 0.016	2.17 ± 0.49	4.89
MEG_ [PheOEt][IBU]	948.6 ± 87.5	44.445 ± 3.501	8.889 ± 0.700	0.427 ± 0.081	10.009 ± 1.938	0.046 ± 0.012	1.90 ± 0.18	4.27
CP gel	577.9 ± 90.2	43.339 ± 7.893	8.668 ± 0.579	0.324 ± 0.044	12.834 ± 1.048	0.033 ± 0.008	1.15 ± 0.18	2.60
CP cream	222.2 ± 33.3	14.897 ± 1.236	2.979 ± 0.247	0.298 ± 0.024	13.926 ± 1.111	0.011 ± 0.002	0.44 ± 0.07	1.00

J_{ss}—steady-state flux; K_p—permeability coefficient; L_T—lag time; D—diffusion coefficient; K_m—skin partition coefficient; Q—the percentage of the applied dose; EF—enhancement factor.

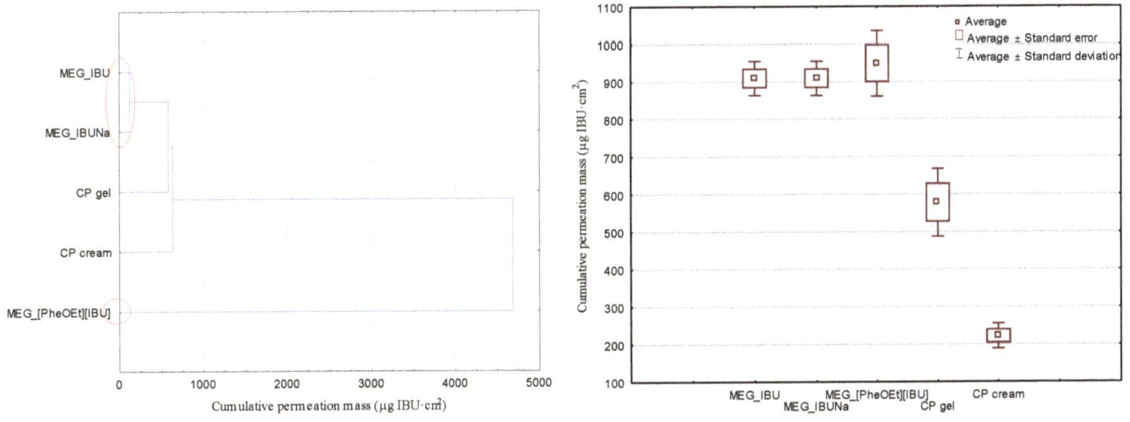

(**a**) (**b**)

Figure 8. The cluster analysis graph (**a**) and the box plot (**b**) for the cumulative mass of ibuprofen and its derivatives after a 24 h study of permeation. The preparations with the highest penetration are marked with a red circle.

The permeation parameters were obtained from the typically J-shaped profiles using Equations (1) and (2) and are listed in Table 5. The lowest ibuprofen cumulated mass was determined for a commercial product in the form of cream, and it was 222.2 ± 33.3 µg IBU/cm^2. The commercial preparation in the form of a cream used in the study was based on triglycerides of saturated fatty acids, glycerol monostearate, macrogol-30-glycerol monostearate, macrogol-100-glycerol monostearate, and propylene glycol. While for the commercial preparation in the form of a gel, the cumulative mass value was higher and amounted to 577.9 ± 90.2 µg IBU/cm^2. In this case, the commercial preparation used is a gel based on isopropyl alcohol, 2,2-dimethyl-4-hydroxymethyl-1,3-dioxalan, poloxamer 407, and triglycerides of saturated fatty acids. In the case of the obtained emulsion-based gel formulation, a higher value was obtained for ibuprofen, and it was 909.6 ± 45.3 µg IBU/cm^2. As can be seen, the use of sodium salt and phenylalanine ethyl ester had a beneficial effect on the amount of ibuprofen permeated. In this case, it was 1086.6 ± 245.8 and 948.6 ± 87.5 µg IBU/cm^2, respectively. The remaining permeability parameters were comparable, and only the cream-based commercial sample showed very low values for all parameters. The highest percentage of the dosed dose was obtained using preparations containing ibuprofen salts: MEG_IBUNa and MEG_[PheOEt][IBU], which resulted in an almost fivefold increase in the permeability of the active substance. In both commercial preparations used in the comparative studies, the producers used additional promoters for the penetration of the active substance through the skin. In both cases, they were essential oils—lavender and orange oil, due to the high content of terpenes. Compared to the obtained formulations, in which no additional penetration promoters were used, drug transport through the skin was enhanced due to changes in the drug structure and/or appropriate selection of formulation ingredients.

2.6.7. Stability of Obtained Formulations

The stability of each formulation obtained with different APIs (IBU, IBUNa, [PheOEt][IBU]) was studied at an elevated temperature of 40 °C in a freeze-thaw stability test. There were no physical changes in color or appearance of each formulation with a different API during the seven days of the testing period and the five freeze-thaw cycles.

3. Conclusions

The selection and use of the correct vehicle are crucial for the degree of absorption and therapeutic effectiveness of the drug. An ideal vehicle should meet the following conditions: have the right consistency, be easy to mix with therapeutic agents, keep the active substance in constant dispersion, not react with the drug or other excipients or external factors, not cause allergies or irritations of the skin or mucous membranes, even if they are of a short-term nature, and not decompose rapidly. It was found that the use of transdermal carriers with a modified drug molecule can be an excellent method of increasing the permeability and achieving a faster therapeutic effect, as well as a method of supplementing the action of ibuprofen with the action attributed to amino acids. As a result of the work, emulsion-based gel formulations with increased permeability of the active substance through the skin were obtained, both in comparison to commercial preparations and analogous formulations containing unmodified ibuprofen. It has been shown that the permeability of the active compound through the skin can be controlled by structural changes in the active substance and by the appropriate selection of the formulation base. It was established that to obtain drugs with increased permeability through the skin, it is most advantageous to combine various methods of increasing permeability, including drug molecule modifications and the selection of the appropriate drug form.

4. Materials and Methods

4.1. Materials

All reagents used in the study were commercially available and used without prior purification. L-Phenylalanine (99%) was provided by Alfa Aesar (Ward Hill, MA, USA). Chlorotrimethylsilane (TMSCl) (99%), ibuprofen sodium salt (98%), and acetonitrile for HPLC (99%) were provided by Sigma-Aldrich (Darmstadt, Germany). Pol-Aura (Morąg, Poland) provided ethanol (96%). The ammonium hydroxide (25%) was provided by P.P.H. Stanlab (Lublin, Polska) sp. j. Diethyl ether (99.5%) and anhydrous sodium sulfate (Na_2SO_4) (99%) were provided by Chempur (Piekary Śląskie, Polska). Ibuprofen (98%) was supplied by AmBeed (Arlington Heights, IL, USA), chloroform ($\geq$99.9%) and triethanolamine ($\geq$99.9%) were supplied by Eurochem BGD sp. z.o.o (Tarnów, Poland), and Efavit provided hemp oil. Span 80 and Tween 60 were purchased by Croda. Carbomer (Carbopol® 940 NF Polymer) was purchased from Lubrizol (Wickliffe, OH, USA). The commercial products (in gel and cream form) were purchased (Dolorgiet Pharmaceuticals, Sankt Augustin, Germany).

4.2. L-Phenylalanine Ethyl Ester Ibuprofenate Synthesis

L-Phenylalanine ethyl ester ibuprofenate [PheOEt][IBU] used for the research was synthesized using the methodology described in our previous publication [30]. In the studies, an Ibuprofen derivative in the form of an ion pair in which the anion was ibuprofenate and the cation was the ethyl ester of L-phenylalanine. The synthesis and structure of the compound used in the research and its acronym are presented in Scheme 1.

4.3. Pre-Formulation Studies: Selection of Formulation Components

Based on ibuprofen solubility studies in individual components, the compounds used to produce the formulation were selected. The following oils were used in the research: hemp oil, evening primrose oil, and sunflower oil, as well as the following surfactants: Tween 40, Tween 60, Span 80, and Tween 80. For this purpose, 1 mL of the appropriate ingredient was added in portions of 100 mg ibuprofen, obtaining a supersaturated solution. The study was carried out in screw-cap vials with continuous magnetic stirring at 100 rpm at 25 °C for 48 h. A supersaturated solution was obtained and filtered through a syringe filter with MCE (Mixed Cellulose Ester) membrane (0.45 μm, ⌀ 47 mm). The content of the compound in the solution was determined by HPLC. The HPLC analyses were performed using a Shimadzu (Kyoto, Japan) Nexera-i LC-2040C 3D High Plus liquid chromatograph equipped with a DAD/FLD detector and Kinetex® F5 column 100 Å (2.6 mm; 150 × 4.6 mm; Phenomenex, Torrance, CA, USA) maintained at 35 °C. The mixture of acetonitrile and water 60/40 (*v*/*v*) was used as the mobile phase under isocratic conditions, with a flow rate of 0.5 mL/min. The detection wavelength was 210 nm. The collected data were acquired and processed using LabSolutions/LC Solution System. Each measurement was performed in triplicate, and the results were averaged. The concentrations of ibuprofen and its salts were calculated based on peak area measurements using the calibration curve method.

4.4. Preparation of Base Formulations

A total of four emulsions were formulated, and a stability test was carried out to choose the best formulation out of the four emulsions. Table 3 shows the amount and mass of each ingredient. The amounts of surfactants were selected based on the HLB of the oil layer (HLB = 9) and the amount of oil phase used.

Hemp oil was combined with Span 80 to make the oil phase, while water and Tween 60 were combined to form the water phase; both mixtures were heated to 80 °C. The water phase was stirred with a homogenizer provided by Janke and Kunkel IKA Labortechnik at 8000 rpm. Then the oil phase was slowly added to the water phase, and both phases were vigorously mixed until the temperature of the mixture dropped to 30 °C and an emulsion was formed.

4.5. Preparation of Formulations with Carbomer

According to the result of the stability test, formulation 4 (F4) was determined to be the most stable and was therefore selected for further study. Due to the very low viscosity of the formulation (the resulting formulation was pourable at room temperature, like water), it was decided to change the rheology of the sample by adding an appropriate rheological regulator, carbomer 940. For this purpose, the oil layer and water layer of the emulsion were prepared analogously to the base preparation (F4), with the difference that 0.17 g of carbomer 940 was added to the aqueous layer. The amounts of ingredients used are shown in Table 6 (MEG). After both phases of the emulsion had been heated to 80 °C, the oil phase was added in portions to the water phase while stirring. Stirring was continued until the resulting emulsion cooled to below 30 °C. Then, 0.2 g of triethanolamine was added to the resulting formulation as a pH regulator.

Table 6. Amount of ingredients in an emulsion-based gel loaded with APIs.

Formulation	Amount of Water (g)	Mass of Surfactant 1 (Tween 60) (g)	Mass of Surfactant 2 (Span 80) (g)	Mass of Oil (Hemp Oil) (g)	API	Mass of API (g)
MEG	23.05	0.62	0.32	5.77	-	-
MEG_IBU	23.05	0.62	0.32	5.77	IBU	0.6280
MEG_IBUNa	23.05	0.62	0.32	5.77	IBUNa	0.6639
MEG_[PheOEt][IBU]	23.05	0.62	0.32	5.77	[PheOEt][IBU]	1.1760

4.6. API-Loaded Emulsion-Based Gel Formulation

Formulations containing pharmaceutically active ingredients (IBU, IBUNa, and [PheOEt][IBU]) were then prepared. The formulations were prepared analogously, as described in Section 4.5. API was added to the finished formulation, and then the system was mixed using a homogenizer. Table 6 shows the composition of formulations loaded with different APIs. The active substance was used at 5%, calculated as ibuprofen, comparable to commercially available preparations. Ibuprofen, sodium ibuprofenate, and L-phenylalanine ethyl ester ibuprofenate were used as active substances.

4.7. The Microscopic Examinations

The prepared formulations were observed under an optical microscope (Delta Optical, with a MC500-W3 5 MP camera). The camera attached to the microscope captured images at magnifications of 25×. The measurement was performed at room temperature.

4.8. Stability of Base Formulations

A stability test was conducted to select the most stable formulation. The first test was carried out by placing a sample of the preparation on a magnetic stirrer with a speed of rotation of 250 r/min at a temperature of 40 °C for 7 days. During the measurement, the behavior of the sample was monitored organoleptically, including color change and the separation of individual phases of the emulsion.

Another stability test was the freeze-thaw stability test. This stability test was performed by freezing the sample at about −4 °C for 24 h and then allowing the sample to thaw at room temperature for 24 h. The freezing-thawing process was repeated five times. During the measurement, the behavior of the sample was monitored organoleptically, including color change and the separation of individual phases of the emulsion.

4.9. Density

The density of formulations with and without API was measured using a densimeter (Densito 30px from Mettler Toledo). The test was performed at ambient temperature, repeated five times, and the value was averaged.

4.10. Refractive Index

Refractive index measurements were used to evaluate the four emulsions: those without API and those with API. One drop of each emulsion was placed on a slide, and refractive indices were measured using a refractometer (Refracto 30GS, Mettler Toledo). The test was repeated five times, and the value was averaged.

4.11. Viscosity

The rheological measurements of flow curves and apparent viscosity were made using a rotational rheometer, MCR 102 (Anton Paar), equipped with a cone-plate measuring system (CP50-1/S), with a sandblasted surface and roughness ranging from 4 to 7 μm. The sandblasted measuring spindle was used to avoid slipping the sample on its surface. The spindle diameter was 50 mm, the cone angle was 0.994°, and the taper was 103 μm. The test sample size was 0.57 mL.

During the rheological tests, the temperature of the samples was kept constant using the Peltier temperature module and was 20.0 or 36.6 °C, respectively. Measurements were carried out for a logarithmically increasing range of shear rates, γ = 0.01–1000 s^{-1}, using 10 measurement points for each decade. In total, 51 measurement points were obtained for one measurement.

Measurements for each sample were made twice, and the presented results are the average of the two measurements. Due to the repeatability of the measurement results, a larger number of repetitions was not necessary.

4.12. Particle Size Distribution

A Mastersizer 3000 particle size analyzer was used for the measurements, enabling particle size measurement by laser diffraction (Malvern Instruments Ltd., Malvern, UK), equipped with a Hydro EV adapter for wet dispersion of samples. Measurements were carried out with the assumption of sphericity in accordance with the Mie theory. As a dispersant, demineralized water at a temperature of 20 °C with a refractive index of 1.33 was used, which was constantly stirred with a propeller stirrer at a constant speed of 1200 rpm.

Prior to the measurements, a standard cleaning of the apparatus based on three cycles was performed. Then, 600 mL of demineralized water was poured into a glass beaker placed in the Hydro EV attachment, automatic alignment of the system was performed, and the measurement background was recorded. The next step was to add the test sample with a known light scattering coefficient to the dispersant in a small amount, enabling the determination of obscure in the range of 5–10% (according to the manufacturer's recommendations for such particle size). Then the measurement was performed by the red laser (632.8 nm) for 20 s and by the blue laser (470 nm) for the next 20 s. The number of repetitions for one sample was 160.

The results are presented as single curves as volume fraction (%) in relation to particle size (μm) and tabulated data for successive repetitions with the calculated mean, standard deviation, and relative standard deviation.

4.13. Release of the Active Substance from the Drug Form

The release was carried out utilizing the Franz diffusion cell (Phoenix DB-6, ABL&E-JASCO, Vienna, Austria) with diffusion regions of 1 cm^2 in accordance with the modified method [34]. The diffusion areas were covered with a dialysis tubing cellulose membrane (D 9777-100FT, Sigma Aldrich), and the prepared formulation was mounted over the membrane.

The acceptor chamber, with a capacity of 10 mL, was filled with PBS (pH 7.4). In the donor chamber, 1 g of each formulation was dosed directly onto the cellulose membrane. The release medium was maintained at 37 °C, and the experiment was carried out for 24 h. After 10, 20, 30, 40, 50, 60, 70, 80, 100, 120 min, and 24 h of stirring, the samples were reported. For this purpose, aliquots of the acceptor fluid (0.5 mL) were removed, and freshly buffered samples with the same pH were added. The concentration of the compounds in the acceptor fluid was determined using HPLC.

The Smartline model 2600 UV detector, model 1050 pump, and model 3950 autosampler with ClarityChrom 2009 software made up the HPLC system (Knauer, Berlin, Germany). At 264 nm, the detector was in operation. A chromatographic column measuring 125 × 4 mm loaded with Hyperisil ODS (C18), particle size 5 μm, was employed. With a 1 mL/min flow rate, the mobile phase contained 0.02 M potassium dihydrogen phosphate, acetonitrile, and methanol (45/45/10 *v*/*v*/*v*).

4.14. Permeation of the Active Substance through the Skin

The permeation tests were performed using the same procedure as the release tests, except that a different membrane was used. Pig skin was used as the membrane in these studies since porcine skin has a similar level of permeability to human skin [35,36]. Fresh abdominal pig skin from the neighborhood slaughterhouse was repeatedly washed in PBS buffer, pH 7.4. The 0.5 mm of skin thickness was dermatomed, and the pieces were each 2 cm × 2 cm in size. Analogously, assessing the skin's impedance, as previously described, was used to assess the skin's integrity [37,38]. Only skin samples with impedances greater than >3 kΩ, which roughly correspond to the electrical resistance of human skin, were used [38]. On the skin, 1 g of each formulation was dosed. The experiment was conducted over 24 h. After 0.5 h, 1 h, 2 h, 3 h, 4 h, 5 h, 8 h, and 24 h from the start of the permeation experiment, 0.5 mL samples of the acceptor fluid were taken out, and the acceptor chamber was refilled with fresh PBS solution. The amounts of IBU and its salts in the acceptor fluid were determined using HPLC. Based on this concentration, the cumulative mass (μg IBU/cm^2) was computed [39].

The kinetic profiling of in vitro infinite dose steady-state percutaneous absorption has been most often characterized by Fick's Laws of Diffusion [40–45], as shown in Equations (1)–(3). The flux (in μg/cm^2·h) through the pig skin into acceptor fluid was determined as the slope of the plot of cumulative mass in the acceptor fluid versus time.

The permeation parameters—fluxes of ibuprofen and its derivatives through the skin (J_{ss}), the diffusion coefficient (K_p), and the time required to reach steady-state permeation (lag time–L_T)—were obtained from typically J-shaped profiles by using Equation (1):

$$A = J_{ss}(t - L_T) \tag{1}$$

where:

A—cumulative amount of active pharmaceutical ingredient (API) of IBU and its salts permeating into the receptor compartment [μg IBU·cm^{-2}],

J_{ss}—steady-state flux [μg/cm^2·h],

t—time [h],

L_T—lag time [h].

The steady-state flux was estimated from the slope of the linear portion of the plot of cumulative mass in the acceptor phase over time. The lag time (L_T) was determined from the x-intercept of the linear part of the plot of cumulative permeation mass in the acceptor phase over time and was used to calculate the permeability coefficient (K_p) by using Equation (2):

$$K_p = \frac{J_{ss}}{C} \tag{2}$$

where:

C—concentration in the donor phase.

The diffusion coefficient (D) was calculated according to Equation (3):

$$D = \frac{I^2}{6 \cdot L_T} \tag{3}$$

where:

I—diffusional pathway length as a skin thickness [mm].

The skin partition coefficient (K_m) was calculated according to Equation (4):

$$K_m = \frac{K_p \cdot I}{D} \tag{4}$$

The enhancement factor (EF) was determined using Equation (5):

$$EF = \frac{m_{c\_formulation}}{m_{c\_CP\ cream}} \tag{5}$$

where:

$m_{c\_formulation}$—cumulated mass for ibuprofen and its derivatives from obtained formulations [μg IBU/cm^2],

$m_{c\_CP\ cream}$—cumulated mass for ibuprofen from CP cream [μg IBU/cm^2].

4.15. Statistical Analysis

This study used a one-way variance analysis (ANOVA) analysis. In the case of the cumulative mass, the significance of differences between individual formulations was evaluated with Tukey's test ($\alpha < 0.05$), where each derivative was compared to the control (MEG_IBU). In addition, a cluster analysis was carried out to determine similarities between all formulations tested, considering all time points. On this basis, the formulations presented a similar release and permeation. Statistical calculations were done using Statistica 13 PL software (StatSoft, Kraków, Poland).

Supplementary Materials: The following supporting information can be downloaded at: https://www.mdpi.com/article/10.3390/gels9050391/s1, Identity and physicochemical properties of the obtained compounds; Figure S1: ^{1}H NMR spectra for [PheOEt][IBU]; Figure S2: ^{13}C NMR spectra for [PheOEt][IBU]; Figure S3: The FT-IR spectra for [PheOEt][IBU]; Figure S4: XRD pattern of [PheOEt][IBU]; Figure S5: The DSC curve for [PheOEt][IBU].

Author Contributions: Conceptualization, P.O.-R.; methodology, A.N., A.S., G.S. and P.O.-R.; formal analysis, A.A.A., A.N., W.D., Ł.K., A.S., G.S., Ł.S. and A.K.A.; investigation, A.A.A., A.K.A. and P.O.-R.; data curation, A.N. and P.O.-R.; writing—original draft preparation, A.A.A. and P.O.-R.; writing—review and editing, P.O.-R.; visualization, A.A.A., A.N. and P.O.-R.; supervision, P.O.-R.; project administration, P.O.-R.; funding acquisition, P.O.-R. All authors have read and agreed to the published version of the manuscript.

Funding: This research was funded by the National Centre for Research and Development, grant number LIDER/53/0225/L-11/19/NCBR/2020.

Institutional Review Board Statement: Not applicable.

Informed Consent Statement: Not applicable.

Data Availability Statement: The data presented in this study are available on request from the corresponding author.

Conflicts of Interest: The authors declare no conflict of interest. The funders had no role in the design of the study, in the collection, analysis, or interpretation of data, in the writing of the manuscript, or in the decision to publish the results.

References

1. Alkilani, A.; McCrudden, M.T.; Donnelly, R. Transdermal Drug Delivery: Innovative Pharmaceutical Developments Based on Disruption of the Barrier Properties of the Stratum Corneum. *Pharmaceutics* **2015**, *7*, 438–470. [CrossRef]
2. Donnelly, R.F.; Singh, T.R.R.; Garland, M.J.; Migalska, K.; Majithiya, R.; McCrudden, C.M.; Kole, P.L.; Mahmood, T.M.T.; McCarthy, H.O.; Woolfson, A.D. Hydrogel-Forming Microneedle Arrays for Enhanced Transdermal Drug Delivery. *Adv. Funct. Mater.* **2012**, *22*, 4879–4890. [CrossRef] [PubMed]
3. Arora, A.; Prausnitz, M.R.; Mitragotri, S. Micro-scale devices for transdermal drug delivery. *Int. J. Pharm.* **2008**, *364*, 227–236. [CrossRef] [PubMed]
4. Tuan-Mahmood, T.-M.; McCrudden, M.T.C.; Torrisi, B.M.; McAlister, E.; Garland, M.J.; Singh, T.R.R.; Donnelly, R.F. Microneedles for intradermal and transdermal drug delivery. *Eur. J. Pharm. Sci.* **2013**, *50*, 623–637. [CrossRef]
5. Jeong, W.Y.; Kwon, M.; Choi, H.E.; Kim, K.S. Recent advances in transdermal drug delivery systems: A review. *Biomater. Res.* **2021**, *25*, 24. [CrossRef] [PubMed]
6. Choudhary, N.; Singh, A.P. Transdermal drug delivery system: A review. *IJPP* **2021**, *8*, 5–9. [CrossRef]
7. Han, T.; Das, D.B. Potential of combined ultrasound and microneedles for enhanced transdermal drug permeation: A review. *Eur. J. Pharm. Biopharm.* **2015**, *89*, 312–328. [CrossRef]
8. Schoellhammer, C.M.; Blankschtein, D.; Langer, R. Skin permeabilization for transdermal drug delivery: Recent advances and future prospects. *Expert Opin. Drug Deliv.* **2014**, *11*, 393–407. [CrossRef]
9. Kretsos, K.; Kasting, G.B. A geometrical model of dermal capillary clearance. *Math. Biosci.* **2007**, *208*, 430–453. [CrossRef]
10. Prausnitz, M.R.; Langer, R. Transdermal drug delivery. *Nat. Biotechnol.* **2008**, *26*, 1261–1268. [CrossRef]
11. Ita, K.B. Transdermal drug delivery: Progress and challenges. *J. Drug Deliv. Sci. Technol.* **2014**, *24*, 245–250. [CrossRef]
12. Menon, G.K. New insights into skin structure: Scratching the surface. *Adv. Drug Deliv. Rev.* **2002**, *54*, S3–S17. [CrossRef] [PubMed]
13. Liu, X.; Kruger, P.; Maibach, H.; Colditz, P.B.; Roberts, M.S. Using skin for drug delivery and diagnosis in the critically ill. *Adv. Drug Deliv. Rev.* **2014**, *77*, 40–49. [CrossRef]
14. Williams, A.C.; Barry, B.W. Penetration enhancers. *Adv. Drug Deliv. Rev.* **2012**, *64*, 128–137. [CrossRef]
15. Gorzelanny, C.; Mess, C.; Schneider, S.W.; Huck, V.; Brandner, J.M. Skin Barriers in Dermal Drug Delivery: Which Barriers Have to Be Overcome and How Can We Measure Them? *Pharmaceutics* **2020**, *12*, 684. [CrossRef] [PubMed]
16. Bos, J.D.; Meinardi, M.M.H.M. The 500 Dalton rule for the skin penetration of chemical compounds and drugs: The 500 Dalton rule for skin penetration of chemical compounds and drugs. *Exp. Dermatol.* **2000**, *9*, 165–169. [CrossRef]

17. Souto, E.B.; Fangueiro, J.F.; Fernandes, A.R.; Cano, A.; Sanchez-Lopez, E.; Garcia, M.L.; Severino, P.; Paganelli, M.O.; Chaud, M.V.; Silva, A.M. Physicochemical and biopharmaceutical aspects influencing skin permeation and role of SLN and NLC for skin drug delivery. *Heliyon* **2022**, *8*, e08938. [CrossRef]
18. Atta-ur-Rahman, W.; Caldwell, G.; Iqbal Choudhary, M.; Yan, Z. (Eds.) *Frontiers in Drug Design and Discovery*; Bentham Science Publishers: Sharjah, United Arab Emirates, 2012; Volume 4, ISBN 978-1-60805-202-8.
19. Mitragotri, S. Devices for overcoming biological barriers: The use of physical forces to disrupt the barriers. *Adv. Drug Deliv. Rev.* **2013**, *65*, 100–103. [CrossRef]
20. Begum, S.G.; Chetty, C.M.; Pavithra, B.; Akhila, B.; Gayathri, C.; Ruksar, S.; Sravani, T.; Voleti, V.K. A review on emulgels—A novel approach for topical drug delivery. *Asian J. Pharm. Res. Dev.* **2019**, *7*, 70–77. [CrossRef]
21. Khullar, R.; Kumar, D.; Seth, N.; Saini, S. Formulation and evaluation of mefenamic acid emulgel for topical delivery. *Saudi Pharm. J.* **2012**, *20*, 63–67. [CrossRef]
22. Kumar, L.; Verma, R. In vitro evaluation of topical gel prepared using natural polymer. *Int. J. Drug Deliv.* **2010**, *2*, 58–63. [CrossRef]
23. Yadav, S.K.; Mishra, M.K.; Tiwari, A.; Shukla, A. Emulgel: A New Approach For Enhanced Topical Drug Delivery. *Int. J. Curr. Pharm. Sci.* **2016**, *9*, 15. [CrossRef]
24. Farjami, T.; Madadlou, A. An overview on preparation of emulsion-filled gels and emulsion particulate gels. *Trends Food Sci. Technol.* **2019**, *86*, 85–94. [CrossRef]
25. Said Dos Santos, R.; Rosseto, H.C.; Bassi Da Silva, J.; Vecchi, C.F.; Caetano, W.; Bruschi, M.L. The effect of carbomer 934P and different vegetable oils on physical stability, mechanical and rheological properties of emulsion-based systems containing propolis. *J. Mol. Liq.* **2020**, *307*, 112969. [CrossRef]
26. Kawana, M.; Miyamoto, M.; Ohno, Y.; Kihara, A. Comparative profiling and comprehensive quantification of stratum corneum ceramides in humans and mice by LC/MS/MS. *J. Lipid Res.* **2020**, *61*, 884–895. [CrossRef]
27. Bolla, P.K.; Clark, B.A.; Juluri, A.; Cheruvu, H.S.; Renukuntla, J. Evaluation of Formulation Parameters on Permeation of Ibuprofen from Topical Formulations Using Strat-M® Membrane. *Pharmaceutics* **2020**, *12*, 151. [CrossRef]
28. Janus, E.; Ossowicz, P.; Klebeko, J.; Nowak, A.; Duchnik, W.; Kucharski, Ł.; Klimowicz, A. Enhancement of ibuprofen solubility and skin permeation by conjugation with L-valine alkyl esters. *RSC Adv.* **2020**, *10*, 7570–7584. [CrossRef]
29. Ossowicz, P.; Klebeko, J.; Janus, E.; Nowak, A.; Duchnik, W.; Kucharski, Ł.; Klimowicz, A. The effect of alcohols as vehicles on the percutaneous absorption and skin retention of ibuprofen modified with L-valine alkyl esters. *RSC Adv.* **2020**, *10*, 41727–41740. [CrossRef]
30. Ossowicz-Rupniewska, P.; Klebeko, J.; Świątek, E.; Bilska, K.; Nowak, A.; Duchnik, W.; Kucharski, Ł.; Struk, Ł.; Wenelska, K.; Klimowicz, A.; et al. Influence of the Type of Amino Acid on the Permeability and Properties of Ibuprofenates of Isopropyl Amino Acid Esters. *IJMS* **2022**, *23*, 4158. [CrossRef]
31. Ossowicz-Rupniewska, P.; Nowak, A.; Klebeko, J.; Janus, E.; Duchnik, W.; Adamiak-Giera, U.; Kucharski, Ł.; Prowans, P.; Petriczko, J.; Czapla, N.; et al. Assessment of the Effect of Structural Modification of Ibuprofen on the Penetration of Ibuprofen from Pentravan® (Semisolid) Formulation Using Human Skin and a Transdermal Diffusion Test Model. *Materials* **2021**, *14*, 6808. [CrossRef]
32. Klebeko, J.; Ossowicz-Rupniewska, P.; Nowak, A.; Janus, E.; Duchnik, W.; Adamiak-Giera, U.; Kucharski, Ł.; Prowans, P.; Petriczko, J.; Czapla, N.; et al. Permeability of Ibuprofen in the Form of Free Acid and Salts of L-Valine Alkyl Esters from a Hydrogel Formulation through Strat-M™ Membrane and Human Skin. *Materials* **2021**, *14*, 6678. [CrossRef] [PubMed]
33. Rozwadowski, Z. Deuterium isotope effects on 13C chemical shifts of lithium salts of Schiff bases amino acids. *J. Mol. Struct.* **2005**, *753*, 127–131. [CrossRef]
34. Song, Y.H.; Gwak, H.S.; Chun, I.K. The effects of terpenes on the permeation of lidocaine and ofloxacin from moisture-activated patches. *Drug Deliv.* **2009**, *16*, 75–81. [CrossRef] [PubMed]
35. Khiao In, M.; Richardson, K.C.; Loewa, A.; Hedtrich, S.; Kaessmeyer, S.; Plendl, J. Histological and functional comparisons of four anatomical regions of porcine skin with human abdominal skin. *Anat. Histol. Embryol.* **2019**, *48*, 207–217. [CrossRef] [PubMed]
36. Jacobi, U.; Kaiser, M.; Toll, R.; Mangelsdorf, S.; Audring, H.; Otberg, N.; Sterry, W.; Lademann, J. Porcine ear skin: An in vitro model for human skin. *Skin Res. Technol.* **2007**, *13*, 19–24. [CrossRef]
37. Kopečná, M.; Macháček, M.; Nováčková, A.; Paraskevopoulos, G.; Roh, J.; Vávrová, K. Esters of terpene alcohols as highly potent, reversible, and low toxic skin penetration enhancers. *Sci. Rep.* **2019**, *9*, 14617. [CrossRef]
38. Davies, D.J.; Ward, R.J.; Heylings, J.R. Multi-species assessment of electrical resistance as a skin integrity marker for in vitro percutaneous absorption studies. *Toxicol. Vitr.* **2004**, *18*, 351–358. [CrossRef]
39. Ossowicz-Rupniewska, P.; Bednarczyk, P.; Nowak, M.; Nowak, A.; Duchnik, W.; Kucharski, Ł.; Rokicka, J.; Klimowicz, A.; Czech, Z. Sustainable UV-Crosslinkable Acrylic Pressure-Sensitive Adhesives for Medical Application. *IJMS* **2021**, *22*, 11840. [CrossRef]
40. Fick, A. On Liquid Diffusion. *Philos. Mag.* **1855**, *10*, 30–39. [CrossRef]
41. Higuchi, T. Physical chemical analysis of percutaneous absorption process from creams and ointments. *J. Soc. Cosmet. Chem.* **1960**, *11*, 85–97.
42. Scheuplein, R.J.; Blank, I.H. Permeability of the skin. *Physiol. Rev.* **1971**, *51*, 702–747. [CrossRef] [PubMed]
43. Kunita, R.; Nishijima, T.; Todo, H.; Sugibayashi, K.; Sakaguchi, H. A Mathematical Approach Using Strat-M® to Predict the Percutaneous Absorption of Chemicals under Finite Dose Conditions. *Pharmaceutics* **2022**, *14*, 1370. [CrossRef] [PubMed]

44. R Bhatt, T.; Golwala, D. In-Vitro Study on Permeation of different Semi-solid dosage forms of Timolol Maleate using Franz cell. *RJPT* **2022**, *15*, 2721–2726. [CrossRef]
45. Banaee, S.; Hee, S.S.Q. Glove permeation of chemicals: The state of the art of current practice, Part 1: Basics and the permeation standards. *J. Occup. Environ. Hyg.* **2019**, *16*, 827–839. [CrossRef]

gels

MDPI

Article

Response Surface Methodology for Optimization of Hydrogel-Forming Microneedles as Rapid and Efficient Transdermal Microsampling Tools

Jiraporn Leanpolchareanchai [1] and Nantana Nuchtavorn [2,*]

[1] Department of Pharmacy, Faculty of Pharmacy, Mahidol University, 447 Sri Ayudhaya Rd., Rajathevee, Bangkok 10400, Thailand; jiraporn.lea@mahidol.ac.th
[2] Department of Pharmaceutical Chemistry, Faculty of Pharmacy, Mahidol University, 447 Sri Ayudhaya Rd., Rajathevee, Bangkok 10400, Thailand
* Correspondence: nantana.nuc@mahidol.edu; Tel.: +66-2-644-8695

Abstract: Microneedles (MNs) have shown a great potential for the microsampling of dermal interstitial fluid (ISF) in a minimally invasive manner for point-of-care testing (POCT). The swelling properties of hydrogel-forming microneedles (MNs) allow for passive extraction of ISF. Surface response approaches, including Box-Behnken design (BBD), central composite design (CCD), and optimal discrete design, were employed for the optimization of hydrogel film by studying the effects of independent variables (i.e., the amount of hyaluronic acid, Gantrez^{TM} S-97, and pectin) on the swelling property. The optimal discrete model was selected to predict the appropriate variables, due to the good fit of the experimental data and the model validity. The analysis of variance (ANOVA) of the model demonstrated *p*-value < 0.0001, R^2 = 0.9923, adjusted R^2 = 0.9894, and predicted R^2 = 0.9831. Finally, the predicted film formulation containing 2.75% *w*/*w* hyaluronic acid, 1.321% *w*/*w* Gantrez^{TM} S-97, and 1.246% *w*/*w* pectin was used for further fabrication of MNs (525.4 ± 3.8 μm height and 157.4 ± 2.0 μm base width), which possessed 1508.2 ± 66.2% swelling, with 124.6 ± 7.4 μL of collection volume, and could withstand thumb pressure. Moreover, almost 50% of MNs achieved a skin insertion depth of approx. 400 μm, with 71.8 ± 3.2% to 78.3 ± 2.6% recoveries. The developed MNs show a promising prospect in microsample collection, which would be beneficial for POCT.

Keywords: response surface methodology; hydrogel; microneedles; interstitial fluid; transdermal; microsampling

Citation: Leanpolchareanchai, J.; Nuchtavorn, N. Response Surface Methodology for Optimization of Hydrogel-Forming Microneedles as Rapid and Efficient Transdermal Microsampling Tools. *Gels* **2023**, *9*, 306. https://doi.org/10.3390/gels9040306

Academic Editors: Ying Huang, Zhengwei Huang and Xuanjuan Zhang

Received: 16 March 2023
Revised: 1 April 2023
Accepted: 3 April 2023
Published: 6 April 2023

1. Introduction

The utilization of hypodermic needles for drug administration and sample extraction has a long-standing history. Despite their widespread usage, however, this traditional method has notable drawbacks, including pain, invasiveness, psychological distress, biohazardous waste, and the necessity for trained healthcare professionals. In recent times, the development of microneedle (MN) systems has been gaining increasing interest as a more effective approach for sample extraction (blood or interstitial fluid (ISF)) and drug delivery, due to their various advantages over hypodermic needle-based systems. MN systems are characterized by micron-sized needles that offer a minimally invasive and more efficient method of drug delivery and sample extraction [1]. Furthermore, MN-based point-of-care testing (POCT) devices have been receiving considerable attention owing to their high potential for extracting and detecting target analytes present in ISF. Skin ISF is located under the stratum corneum layer and abundant in the dermis layer with a thickness from 1500 to 3000 μm [2], which contains exogenous drugs and clinically relevant biomarkers [3]. Therefore, ISF could be used for drug monitoring or disease diagnosis. Microsampling methods in the invasive forms for ISF collection, including suction blistering, iontophoresis, sonophoresis, and microdialysis, require a long extraction time, expensive equipment, and the assistance of medical professionals. To overcome these limitations, MNs with heights of 50–1500 μm can penetrate through an epidermal layer with a thickness of 50–200 μm to extract ISF from the dermis, with advantages such as being self-applicable with minimal pain and invasiveness.

MNs have been investigated for the delivery of drug molecules or the monitoring of biomarkers [4,5], which can be categorized into solid, hollow, coated, dissolving, porous, and hydrogel-forming MNs [6]. Among them, hydrogel-forming MNs are fabricated by covalently or non-covalently crosslinked three-dimensional networks of hydrophilic polymers. The hydrogel-forming MNs can absorb and swell in water or bodily fluids without dissolving, resulting in the formation of a gel-like substance. This property allows for sustained and controlled drug release. The drug can be loaded into the hydrogel matrix, and it slowly releases after the MNs swell and degrade. The swelling ability of hydrogel-forming MNs has been leveraged for other biomedical applications, such as biosensing. For example, hydrogel-forming MNs have been used as minimally invasive sensors for monitoring glucose levels in diabetic patients [7], alcohol monitoring, carcinoembryonic antigen detection for early breast cancer detection [8], and psoriasis therapy and diagnosis [9]. Moreover, the hydrogel-forming MNs overcome the disadvantages of solid and dissolving MNs, which require a long extraction time, repeated application times, and the assistance of external vacuum or pressure [10,11]. Solid MNs are designed to puncture the skin, creating a micro-incision that can be utilized for the collection of bodily fluids and the delivery of active substances to a targeted area. Dissolvable MNs dissolve upon insertion into tissues, leading to the creation of small openings on the tissue surface, which allow ISF to escape. While this unique property presents significant potential for a wide range of medical applications, particularly in precise drug delivery through the rapid dissolution of polymers, it may limit the applicability of ISF extraction [12,13]. Hollow MNs have the possibility of needle hole blockage. Additionally, porous MNs need a backing layer as a reservoir to store the fluids, and increasing the porosity can decrease the mechanical strength of the MNs. In contrast, coated MNs feature a core that is coated with a surface layer. The coating of the MN typically comprises a blend of drugs and polymers that are utilized for drug delivery purposes. Conversely, the coating resembles a semi-permeable membrane and serves for the collection of bodily fluids for drug/biomarker assay purposes [12]. Moreover, the coated layers on the needles are limited because thickly coated needles limit needle insertion capability into the skin.

The development of hydrogel-forming MNs for health monitoring assay purposes should approach a collection volume of $\geq$1 μL, with a collection time of $\leq$20 min; thus, the method should be applied in a simple and rapid manner [14]. Statistical designs in particular response surface methodology (RSM) are effective experimental design methods to assist the optimization of MNs that can be performed with multiple factors and levels. RSM (e.g., Box-Behnken design (BBD), central composite design (CCD)) is commonly employed for the optimization of the formulations, which offers benefits for the study of the interaction between factors and responses [15]. Consequently, the model provides a prediction of optimal experimental conditions. Moreover, the statistical designs substantially reduce the resources, time, and effort needed to develop the formulations [16].

Hydrogels are commonly formulated from a variety of natural polymers, including chitosan, alginate, cellulose, hyaluronic acid, pectin, and starch, as well as synthetic polymers such as polyvinyl alcohol and polyethylene glycol. However, the irreparable impact of synthetic polymers on the environment has driven a recent increase in the use of natural polymers. This is due to their biodegradability, nontoxicity, widespread availability, and cost-effectiveness [17]. Hyaluronic acid and Gantrez^{TM} S-97 have been utilized for hydrogel-forming MNs. Hyaluronic acid is a naturally occurring linear polysaccharide made up of repeating N-acetyl-d-glucosamine and d-glucuronic acid units, which has been widely used for hydrogel formulations in biomedical applications due to its excellent hydrophilicity, biocompatibility, biodegradability, nonimmunogenicity, and nontoxicity [18,19]. Gantrez^{TM} S-97 is a synthetic copolymer of methyl vinyl ether and maleic anhydride (free acid), which is a non-toxic and biocompatible crosslinker with good mechanical strength. It has been widely used in denture adhesives, toothpastes, and mouthwash formulations because of its excellent film-forming ability [20]. Furthermore, pectin could be an additional ingredient for altering the properties of MNs. It is a natural polymer composed of galacturonic acid that has been utilized as a gelling agent in the food industry and biomedical and pharmaceutical fields due to their biocompatibility, biodegradability, and nontoxicity [21].

This study aimed to develop hydrogel-forming MNs for the rapid collection of skin ISF in sufficient volumes by a formulation containing hydrogel with the main polymers of hyaluronic acid and Gantrez^{TM} S-97 with the addition of pectin. A variety of RSM (i.e., BBD, CCD, and optimal discrete model) were employed to study the effects of formulation variables on the significant characteristic of hydrogel film (i.e., swelling ability). Suitable models were evaluated by the analysis of variance (ANOVA). Next, the optimal hydrogel film was prepared according to the predicted values, and hydrogel-forming MNs were further fabricated by a micromolding method. Finally, the MNs were assessed by examining the morphology, mechanical strength, swelling and skin permeation abilities, and recoveries.

2. Results and Discussion

2.1. Experimental Design

The previous report demonstrated the preparation of hydrogel films using 5% *w/w* hyaluronic acid and 0.5% *w/w* Gantrez^{TM} S-97 as the main polymer and crosslinker, respectively, which yielded a good swelling degree. However, the swelling degree dropped due to incomplete crosslinking of the film caused by an insufficient amount of Gantrez^{TM} S-97. Moreover, Gantrez^{TM} S-97 above 3% *w/w* resulted in a decrease in swelling. Moreover, the report suggested the use of 2% and 4% *w/w* pectin to enhance the film's strength, but the results showed a decrease in swelling [22,23]. Therefore, RSM was employed for the optimization of hydrogel film formulations in order to obtain the appropriate swelling degree and strength; RSM generated the mathematical relationship between the independent variables (i.e., 2–8% *w/w* hyaluronic acid, 1–2% *w/w* Gantrez^{TM} S-97, and 0.5–1.5% *w/w* pectin) and the dependent variable (i.e., swelling degree). The experimental data obtained from the hydrogel film formulations listed in Table S1 (in Supplementary Materials) were utilized to create three models, including BBD, CCD, and the optimal discrete models. The appropriate model was chosen by the evaluation of different parameters obtained from regression analysis such as *p*-value, adjusted R^2, predicted R^2, and predicted residual sum of square (PRESS) value. Moreover, the ANOVA was applied to examine the significance of the model. All models were significant for the swelling degree with p-values < 0.05. The greater adjusted R^2, predicted R^2, and low PRESS value indicated the good fit of the model; therefore, the optimal (discrete) model was selected (Table 1). The quadratic equation generated from the models (Figure 1) is as follows: Y (% swelling) = $2400.08714 - 45.88457X_1 + 76.75263X_2 - 1090.60741X_3 + 63.23516X_1X_2 + 100.12782X_1X_3 + 8.75302X_2X_3 - 25.37784(X_1)^2 - 330.53100(X_2)^2 + 153.67881(X_3)^2$, where Y = % swelling, X_1 = hyaluronic acid concentration (% *w/w*), X_2 = Gantrez^{TM} S-97 concentration (% *w/w*), and X_3 = pectin concentration (% *w/w*). In addition, the coefficient values in Figure 1 illustrated the synergistic (positive sign) and antagonistic (negative sign) effects of variables. The alteration of hyaluronic acid and pectin concentrations were the most influential factors on the swelling degree. The optimized hydrogel film formulation predicted from the model consisted of 2.75% *w/w* hyaluronic acid, 1.321% *w/w* Gantrez^{TM} S-97, and 1.246% *w/w* pectin, offering the swelling degree of 958.76. The model was a good fit for the range of independent variables used to establish the model. However, the relationship between the independent and dependent variables can change outside of this range.

Table 1. Fit statistics of the models.

Model	Variables and Level	F-Value	p-Value	R^2	Adjusted R^2	Predicted R^2	Adeq Precision	PRESS
BBD	2–8% w/w Hyaluronic acid,	120.34	<0.0001	0.9722	0.9641	0.9505	32.71	2.35×10^5
CCD	1–2% w/w Gantrez™ S-97,	258.73	<0.0001	0.9894	0.9856	0.9784	49.46	8.81×10^4
Optimal discrete	and 0.5–1.5% w/w Pectin	341.94	<0.0001	0.9923	0.9894	0.9831	59.12	6.83×10^4

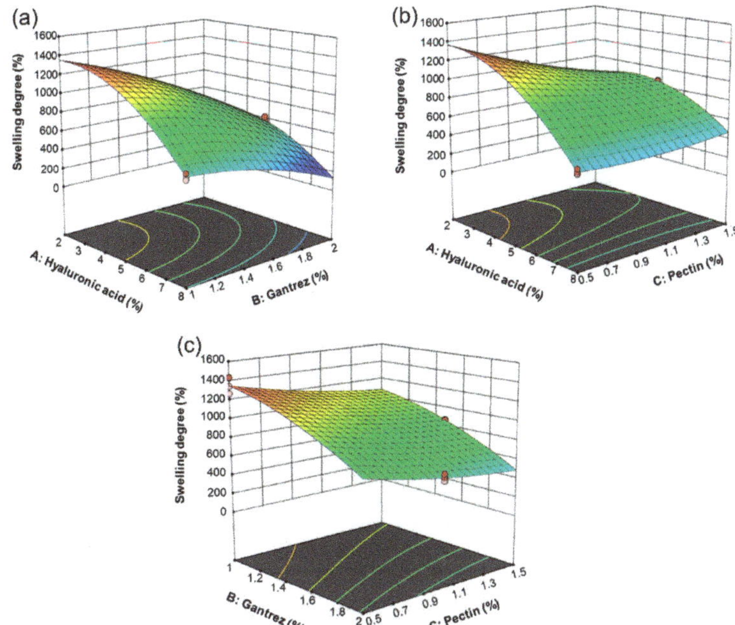

Coefficients

	Intercept	A	B	C	AB	AC	BC	A²	B²	C²
Swelling	958.756	–314.047	–294.956	–134.741	94.8527	150.192	2.18825	–228.401	–82.6328	38.4197
p-values		< 0.0001	< 0.0001	< 0.0001	< 0.0001	< 0.0001	0.8424	< 0.0001	< 0.0001	0.0102

Figure 1. The optimal RSM plots of hydrogel film formulation optimization showing the effect of variables (**a**) hyaluronic acid and Gantrez™ (**b**) hyaluronic acid and pectin and (**c**) Gantrez ™ and pectin on the swelling degree.

2.2. Hydrogel Films

The predicted value of each variable was used to formulate hydrogel films, which demonstrated 8.27 ± 0.32 mm length, 8.30 ± 0.17 mm width, and 0.20 ± 0.00 mm thickness. The Fourier transform infrared (FTIR) spectra of hydrogel ingredients and corresponding non-crosslinked and crosslinked films are shown in Figure 2. Hyaluronic acid, Gantrez™ S-97, and pectin were utilized as the main polymer, crosslinking agent, and film-strength enhancer, respectively. The crosslinked film by esterification showed peaks at 1721 and 1150 cm^{-1}, which are assigned to the C=O stretch and the C-O stretch, respectively. The peak at 3450 cm^{-1} is associated with free -OH groups of ingredients. The amide groups of hyaluronic acid show the peak of C = O stretch at 1555 and 1621 cm^{-1} and C-N stretch at 1403 cm^{-1}. The peaks representing the C=O stretch and the C-O stretch were found to decrease when the film was not crosslinked. The swelling degree of the hydrogel films was 1033.5 ± 25.7% after placing them in the artificial ISF for 5 min (Figure 3), which was slightly greater than the predicted value of 958.76. This film formulation was further utilized for the fabrication of MNs.

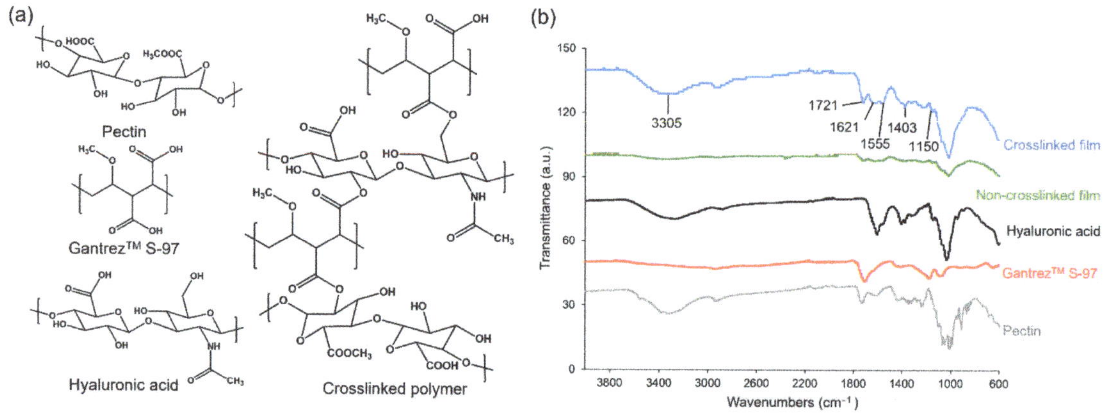

Figure 2. (**a**) The structures and (**b**) IR spectra of hydrogel ingredients, non-crosslinked, and crosslinked hydrogel films.

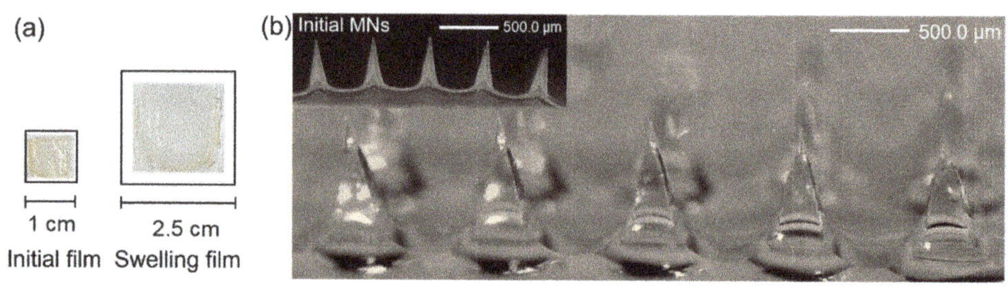

Figure 3. The dry and swelling forms in the artificial ISF for 5 min of (**a**) hydrogel films and (**b**) hydrogel-forming MNs.

2.3. Fabrication and Characterization of Hydrogel-Forming MNs

The hydrogel-forming MNs featuring 100 pyramidal-shaped needles were fabricated by a simple micromolding method. The height and base width of MNs were 525.4 ± 3.8 μm and 157.4 ± 2.0 μm, respectively, which would penetrate through the epidermis (50–200 μm thickness) and reach the dermis (1500–3000 μm thickness) for the microsampling of ISF [2]. The reduction in the height of the MNs, as compared to the micromold (600 μm), can be attributed to the viscosity of the hydrogel formulation. This viscosity may limit the flowability and distribution of the hydrogel in the mold cavity. Furthermore, hydrogel is a viscous formulation that generates a high-contact angle, resulting in low wettability [24]. Therefore, the hydrogel may not entirely fill the mold tip, which is the smallest part of the mold. Although a reduction in microneedle height occurred, it remains adequate for effective skin penetration. Furthermore, it is possible to optimize the centrifugation speed and time to overcome the limitations imposed by the viscous formulation.

Then, the MNs were placed in the artificial ISF for 5 min, which demonstrated the swelling degree of 1508.2 ± 66.2% and the collected volume of 124.6 ± 7.4 μL (Table 2). After removal from the ISF, the MNs remained in the pyramidal shape (Figure 3).

Table 2. Characteristics of the optimal hydrogel-forming MNs.

Characteristics	Value
Shape	Pyramidal
Dimension (height × base width)	525.4 ± 3.8 × 157.4 ± 2.0 µm
Swelling ability	1508.2 ± 66.2%
Collection volume	124.6 ± 7.4 µL
Recovery	71.8 ± 3.2% to 78.3 ± 2.6%
Skin insertion	Up to approx. 400 µm
Mechanical strength	Withstands pressure from a finger

In general, MN patches can be adhered to the skin by applying pressure with the thumb. Therefore, the mechanical strength test of MNs was performed in the compression mode using the constant force of 32 N per array for 30 s, mimicking thumb pressure. The bluntness of the MN tips was associated with a steep slope in the force-displacement curve (Figure 4) [25]. However, the MNs were strong enough to withstand the pressure [26,27]. Next, the Parafilm® M (PF) membrane was used for MN testing as a stratum corneum simulant, which is the outermost layer of the epidermis. In order to investigate the influence of the manual application force in the MN insertion, the created holes on an eight-stacked PF layer were examined. The created holes of 100.0 ± 0.0%, 97.7 ± 3.2% and 45.3 ± 4.2% were observed for the first, second, and third layer, respectively (Figure 5). After compression studies, the percentage height reduction was 11.0 ± 0.5%. The results suggested that the MNs were capable of skin insertion into the dermis layer since 100% of MNs achieved an insertion depth of 268 µm, and almost 50% of MNs achieved an insertion depth of 402 µm. Although the PF had significantly lower insertion depths than neonatal porcine skin, the differences were typically less than the 10% of the total needle length [28]. The developed MNs would achieve deeper insertion in the ex vivo test.

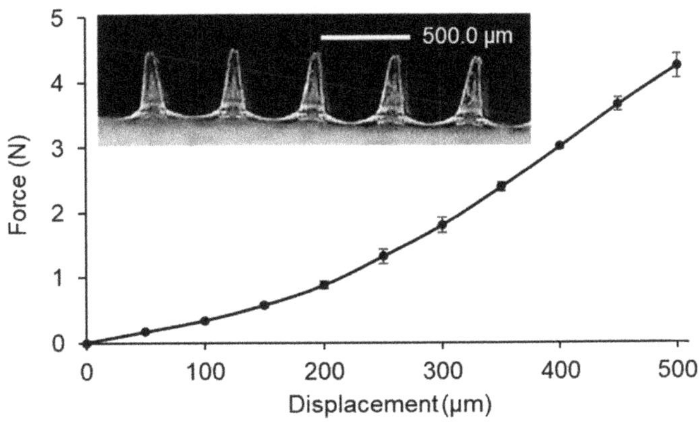

Figure 4. Force-displacement curve from mechanical strength test of hydrogel-forming MNs.

To study in vitro recovery of the extracted ISF extraction, the calibration curve of methylene blue in ISF was established in the range of 0.3125–5 µg/mL. The linear equation was y = 3459.5x + 3230 with correlation coefficient (*r*) of 0.9966. The MNs were soaked with 0.3125, 1.25, and 5 µg/mL of the methylene blue. The percent recovery of the extracted ISF was in the range of 71.8 ± 3.2 to 78.3 ± 2.6%, indicating that the MNs were able to absorb the sufficient ISF volume.

The model suggested the appropriate mixture of ingredients for the formulation of hydrogel-forming MNs, which assisted to shorten the experimental time. Compared to the previous reports, hydrogel-forming MNs exhibited enhanced swelling properties attributed to the incorporation of optimal Gantrez™ S-97. This inclusion facilitated the preservation of swelling degree, precluding their decrease during ISF collection caused by MN dissolution [22]. In addition, the model-predicted value for the addition of pectin improved the strength of the hydrogel-forming MNs, while maintaining excellent swelling properties. Also, the MNs were able to withstand thumb pressure and effectively penetrate the skin [23].

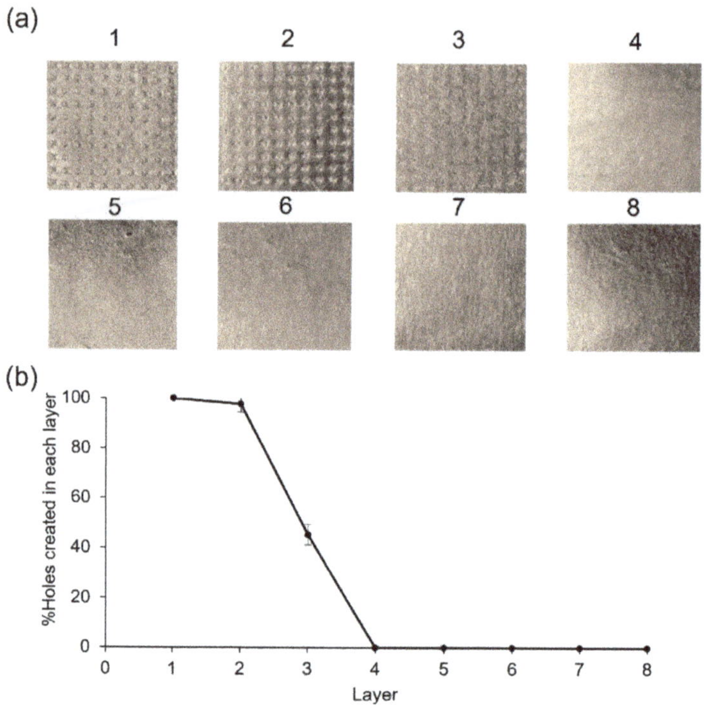

Figure 5. (**a**) Skin insertion test of hydrogel-forming MNs on (**b**) model skin (PF) layers.

3. Conclusions

Experimental design using the mathematical model RSM was efficient for the optimization of hydrogel film. The optimal discrete model was chosen for the prediction of the appropriate value of each ingredient in the formulation according to high R^2 and low PRESS. Consequently, the effects of the hydrogel formulation variables on the film characteristic can be easily predicted and precisely interpreted by using mathematical equations. The suggested hydrogel film formulation for further MN fabrication contained 2.75% *w*/*w* hyaluronic acid, 1.321% *w*/*w* Gantrez™ S-97, and 1.246% *w*/*w* pectin. The MNs displayed excellent mechanical properties for adequate insertion without breaking. All MNs exhibited the potential to penetrate the epidermis layer with an insertion depth of approx. 270 µm. Additionally, approx. 50% of the MNs achieved a skin insertion depth of approx. 400 µm. Also, the MNs showed good swelling with ISF upon insertion and rapid and passive extraction within 5 min. The proposed hydrogel-forming MNs could potentially be employed for minimally invasive microsampling and extraction of dermal ISF, which is a source of biomarkers or exogenous drugs.

4. Materials and Methods

4.1. Materials and Chemicals

Hyaluronic acid (MW 420 kDa) was purchased from Nanjing Gemsen International Co., Ltd., Nanjing, China. Gantrez™ S-97, a copolymer obtained from the free acid of methyl vinyl ether and maleic anhydride (PMVE/MA) (MW 1500 kDa), was provided by Ashland, Worcestershire, UK. Pectin from red apples was purchased from Chemipan, Bangkok, Thailand. MN molds (600 μm height, 200 μm base width, 500 μm tip to tip interspace, 10 × 10 pyramidal arrays) were purchased from Micropoint Technologies Pte Ltd., Singapore. Other chemicals used were analytical grade.

4.2. Experimental Design

A randomized RSM, including CCD, BBD, and optimal (discrete) design, was used to study the effect of three variables on hydrogel films' swelling properties. The dependent and independent variables, along with their levels, are listed in Table 1. A suitable model was chosen by assessing various parameters from the regression analysis and ANOVA such as *p*-value, adjusted R^2, predicted R^2, and PRESS values. The predicted optimal condition was tested experimentally in triplicate to validate the results.

4.3. Hydrogel Films

4.3.1. Preparation Procedure

Ten grams of each aqueous formulation (Table S1) (in Supplementary Materials) were prepared by homogenous mixing at 300 rpm. The mixtures were centrifuged (Universal 320, Andreas Hettich GmbH & Co. KG, Tuttlingen, Germany) at 4500 rpm for 15 min to remove air bubbles. Then, each formulation (0.30 ± 0.03 g) was separately cast into polydimethylsiloxane (PDMS) square-shaped molds (10 × 10 mm) and dried at room temperature for 18 h prior to peeling off the dried films. After that, the dried films were crosslinked in the oven at 80 °C for 24 h. Finally, the crosslinked films were stored in the desiccator for further analysis.

4.3.2. Characterization

The morphology, including length, width, and thickness of hydrogel films, was measured by the digital vernier caliper. Then, FTIR analysis of all ingredients and non-crosslinked and crosslinked films were carried out in the range of 4000–600 cm^{-1} (iS5 FTIR spectrometer with iD7 attenuated total reflection), Thermo Fisher Scientific, Waltham, MA, USA).

Next, the swelling test of the crosslinked hydrogel films was performed in artificial ISF (pH 7.4) [29]. An empty weighing boat was placed on the analytical balance (LAB 214i, Adam Equipment, Oxford, CT, USA) before transferring the hydrogel film, and the total weight was recorded. Then, 3 mL artificial ISF was added and the total weight recorded. At 5 min, the ISF was taken out, and the weight of the swollen film and the weighing boat was recorded. The swelling degree (%) of the hydrogel films was calculated by the following equation [30]:

$$\text{Swelling degree} = (m_t - m_0)/m_0 \times 100$$

where m_0 is the weight of the initial film, and m_t is weight of the swollen film at 5 min.

4.4. Fabrication Method of Hydrogel-Forming MNs

All the PDMS molds with 100 MNs obtained from the commercial molds were cleaned with the mixture of ethanol and water (1:1, *v*/*v*) by sonication for 20 min. The optimal formulation (0.04–0.05 g) was filled into the molds and centrifuged at 4500 rpm for 15 min at room temperature. Next, the molds were filled up to the weight of 0.150 ± 0.015 g. The MNs were dried at room temperature for 18 h, crosslinked at 80 °C for 24 h, and kept in the desiccator for further analysis.

4.5. Characterization of Hydrogel-Forming MNs

4.5.1. Morphology

The images of MNs were taken by a digital microscope (Dino-Lite Edge AM4115T-YFGW, AnMo Electronics Corporation, Taipei, Taiwan) that was capable of variable magnification of 20–220×. The accurate measurement of height and base width of the MNs was carried out by adjusting the magnification and focal height and setting the calibration.

4.5.2. Swelling Test

The swelling test procedure of the hydrogel-forming MNs was performed in the same manner as that of the hydrogel films. Then, the volume of ISF absorbed into each MN was also determined using the artificial ISF (density of 1.005 ± 0.001 g/mL). Also, the morphology of the MNs was evaluated before and after the swelling test.

4.5.3. Mechanical Strength Test

The mechanical strength of hydrogel-forming MNs was investigated by the TA.XTplusC texture analyzer (Stable Micro Systems, Godalming, UK) in the compression mode. The MNs were attached to the cylindrical probe (P/0.5, 2.45 cm diameter cylinder) by double-sided tape. The probe was lowered at a speed of 0.5 mm/s (pretest speed 1 mm/s and post-test speed 10 mm/s, trigger force 0.049 N) until a pre-set force of 32 N, mimicking thumb pressure, was achieved. Subsequently, the force was analyzed to obtain a force-displacement curve. Furthermore, the height of MNs before (H_a) and after (H_b) the application of compressive force was measured, using the digital microscope. The percentage of height reduction of MNs was calculated by the following equation [30,31]:

$$\% \text{ Height reduction} = (H_a - H_b)/H_a \times 100\%$$

4.5.4. Skin Insertion Test

The skin insertion test was performed using the validated skin model, PF, and the texture analyzer. A stack of 8 layers total of PF (134 ± 15 μm thickness for each layer) was placed on the base of the test machine, and the hydrogel-forming MNs were attached to the cylindrical probe. The procedure and all parameters were the same as mentioned in Section 4.5.3. The inserted MN arrays were removed after holding for 30 s. Then, the PF layers were unfolded, and the number of holes produced in each layer were counted and evaluated using the digital microscope. The percentage of holes in each layer generated by MNs was calculated by the following equation [28]:

$$\% \text{ Holes in PF layer} = \text{number of holes}/\text{number of MNs in the array} \times 100$$

4.5.5. In Vitro ISF Extraction and Recovery Test

A calibration curve was established by plotting the fluorescence intensity against five different concentrations of methylene blue in the artificial ISF (0.3125–5 μg/mL). Then, the hydrogel-forming MNs were weighed and placed in the ISF containing 3 concentration levels (low, medium and high) of methylene blue for 5 min. The MNs were taken out and weighed before transferring to a centrifuge tube containing 3 mL ISF. After centrifugation at 5000 rpm for 5 min, the supernatant was spectrofluorimetrically measured at the excitation wavelength of 650 nm and the emission wavelength of 685 nm. The percentage of recovery was calculated with the following equation:

$$\% \text{ Recovery} = \text{found concentration}/\text{added concentration} \times 100$$

4.6. Statistical Analysis

The RSM models employed Design-Expert software (Version 13, Stat-Ease Inc., Minneapolis, MN, USA). All experiments were performed in triplicate. The data were expressed as mean ± standard deviation.

Supplementary Materials: The following supporting information can be downloaded at: https://www.mdpi.com/article/10.3390/gels9040306/s1, Table S1: Hydrogel film formulations.

Author Contributions: Conceptualization, J.L. and N.N.; methodology, J.L. and N.N.; software, J.L. and N.N.; validation, J.L. and N.N.; formal analysis, J.L. and N.N.; investigation, J.L. and N.N.; resources, J.L. and N.N.; data curation, J.L. and N.N.; writing—original draft preparation, J.L. and N.N.; writing—review and editing, J.L. and N.N.; visualization, J.L. and N.N.; supervision, N.N.; project administration, N.N.; funding acquisition, J.L. and N.N. All authors have read and agreed to the published version of the manuscript.

Funding: This research paper was supported by Specific League Funds from Mahidol University. The APC was funded by Mahidol University.

Institutional Review Board Statement: Not applicable.

Informed Consent Statement: Not applicable.

Data Availability Statement: The data presented in this study are available on request from the corresponding author.

Acknowledgments: The authors would like to acknowledge Department of Pharmaceutical Chemistry, Department of Pharmacy and Central Laboratory Unit, Faculty of Pharmacy, Mahidol University for the equipment and facilities support.

Conflicts of Interest: The authors declare no conflict of interest.

References

1. Gowda, B.H.J.; Ahmed, M.G.; Sanjana, A. Can microneedles replace hypodermic needles? *Resonance* **2022**, *27*, 63–85. [CrossRef]
2. Makvandi, P.; Kirkby, M.; Hutton, A.R.J.; Shabani, M.; Yiu, C.K.Y.; Baghbantaraghdari, Z.; Jamaledin, R.; Carlotti, M.; Mazzolai, B.; Mattoli, V.; et al. Engineering microneedle patches for improved penetration: Analysis, skin models and factors affecting needle insertion. *Nanomicro Lett.* **2021**, *13*, 93. [CrossRef] [PubMed]
3. Kashaninejad, N.; Munaz, A.; Moghadas, H.; Yadav, S.; Umer, M.; Nguyen, N.-T. Microneedle arrays for sampling and sensing skin interstitial fluid. *Chemosensors* **2021**, *9*, 83. [CrossRef]
4. Liu, G.-S.; Kong, Y.; Wang, Y.; Luo, Y.; Fan, X.; Xie, X.; Yang, B.-R.; Wu, M.X. Microneedles for transdermal diagnostics: Recent advances and new horizons. *Biomaterials* **2020**, *232*, 119740. [CrossRef]
5. Chen, S.; Miyazaki, T.; Itoh, M.; Matsumoto, H.; Moro-oka, Y.; Tanaka, M.; Miyahara, Y.; Suganami, T.; Matsumoto, A. A porous reservoir-backed boronate gel microneedle for efficient skin penetration and sustained glucose-responsive insulin delivery. *Gels* **2022**, *8*, 74. [CrossRef] [PubMed]
6. Vora, L.K.; Moffatt, K.; Tekko, I.A.; Paredes, A.J.; Volpe-Zanutto, F.; Mishra, D.; Peng, K.; Raj Singh Thakur, R.; Donnelly, R.F. Microneedle array systems for long-acting drug delivery. *Eur. J. Pharm. Biopharm.* **2021**, *159*, 44–76. [CrossRef]
7. Liu, Y.; Yu, Q.; Luo, X.; Yang, L.; Cui, Y. Continuous monitoring of diabetes with an integrated microneedle biosensing device through 3D printing. *Microsyst. Nanoeng.* **2021**, *7*, 75. [CrossRef]
8. Xie, L.; Zeng, H.; Sun, J.; Qian, W. Engineering microneedles for therapy and diagnosis: A survey. *Micromachines* **2020**, *11*, 271. [CrossRef]
9. Gowda, B.H.J.; Ahmed, M.G.; Hani, U.; Kesharwani, P.; Wahab, S.; Paul, K. Microneedles as a momentous platform for psoriasis therapy and diagnosis: A state-of-the-art review. *Int. J. Pharm.* **2023**, *632*, 122591. [CrossRef]
10. Dugam, S.; Tade, R.; Dhole, R.; Nangare, S. Emerging era of microneedle array for pharmaceutical and biomedical applications: Recent advances and toxicological perspectives. *Future J. Pharm. Sci.* **2021**, *7*, 19. [CrossRef]
11. Miura, S.; Yamagishi, R.; Miyazaki, R.; Yasuda, K.; Kawano, Y.; Yokoyama, Y.; Sugino, N.; Kameda, T.; Takei, S. Fabrication of high-resolution fine microneedles derived from hydrolyzed hyaluronic acid gels in vacuum environment imprinting using water permeable mold. *Gels* **2022**, *8*, 785. [CrossRef]
12. Ma, S.; Li, J.; Pei, L.; Feng, N.; Zhang, Y. Microneedle-based interstitial fluid extraction for drug analysis: Advances, challenges, and prospects. *J. Pharm. Anal.* **2023**, *13*, 111–126. [CrossRef] [PubMed]
13. Nasiri, M.I.; Vora, L.K.; Ershaid, J.A.; Peng, K.; Tekko, I.A.; Donnelly, R.F. Nanoemulsion-based dissolving microneedle arrays for enhanced intradermal and transdermal delivery. *Drug Deliv. Transl. Res.* **2022**, *12*, 881–896. [CrossRef] [PubMed]
14. Samant, P.P.; Niedzwiecki, M.M.; Raviele, N.; Tran, V.; Mena-Lapaix, J.; Walker, D.I.; Felner, E.I.; Jones, D.P.; Miller, G.W.; Prausnitz, M.R. Sampling interstitial fluid from human skin using a microneedle patch. *Sci. Transl. Med.* **2020**, *12*, eaaw0285. [CrossRef] [PubMed]
15. Suriyaamporn, P.; Opanasopit, P.; Ngawhirunpat, T.; Rangsimawong, W. Computer-aided rational design for optimally Gantrez® S-97 and hyaluronic acid-based dissolving microneedles as a potential ocular delivery system. *J. Drug Deliv. Sci. Technol.* **2021**, *61*, 102319. [CrossRef]
16. Anis, N.; Ahmed, D. Modelling and optimization of polyphenol and antioxidant extraction from *Rumex hastatus* by green glycerol-water solvent according to response surface methodology. *Heliyon* **2022**, *8*, e11992. [CrossRef]
17. Soleimani, S.; Heydari, A.; Fattahi, M. Swelling prediction of calcium alginate/cellulose nanocrystal hydrogels using response surface methodology and artificial neural network. *Ind. Crops Prod.* **2023**, *192*, 116094. [CrossRef]

18. Mero, A.; Campisi, M. Hyaluronic acid bioconjugates for the delivery of bioactive molecules. *Polymers* **2014**, *6*, 346–369. [CrossRef]
19. Aragona, P.; Papa, V.; Micali, A.; Santocono, M.; Milazzo, G. Long term treatment with sodium hyaluronate-containing artificial tears reduces ocular surface damage in patients with dry eye. *Br. J. Ophthalmol.* **2002**, *86*, 181. [CrossRef]
20. Gardner, C.M.; Burke, N.A.D.; Chu, T.; Shen, F.; Potter, M.A.; Stöver, H.D.H. Poly(methyl vinyl ether-alt-maleic acid) polymers for cell encapsulation. *J. Biomater. Sci. Polym. Ed.* **2011**, *22*, 2127–2145. [CrossRef]
21. Freitas, C.M.P.; Coimbra, J.S.R.; Souza, V.G.L.; Sousa, R.C.S. Structure and applications of pectin in food, biomedical, and pharmaceutical industry: A review. *Coatings* **2021**, *11*, 922. [CrossRef]
22. Larrañeta, E.; Henry, M.; Irwin, N.J.; Trotter, J.; Perminova, A.A.; Donnelly, R.F. Synthesis and characterization of hyaluronic acid hydrogels crosslinked using a solvent-free process for potential biomedical applications. *Carbohydr. Polym.* **2018**, *181*, 1194–1205. [CrossRef]
23. Demir, Y.K.; Metin, A.Ü.; Şatıroğlu, B.; Solmaz, M.E.; Kayser, V.; Mäder, K. Poly (methyl vinyl ether-co-maleic acid)—pectin based hydrogel-forming systems: Gel, film, and microneedles. *Eur. J. Pharm. Biopharm.* **2017**, *117*, 182–194. [CrossRef]
24. Lin, Y.-H.; Lee, I.C.; Hsu, W.-C.; Hsu, C.-H.; Chang, K.-P.; Gao, S.-S. Rapid fabrication method of a microneedle mold with controllable needle height and width. *Biomed. Microdevices* **2016**, *18*, 85. [CrossRef]
25. Gittard, S.D.; Chen, B.; Xu, H.; Ovsianikov, A.; Chichkov, B.N.; Monteiro-Riviere, N.A.; Narayan, R.J. The effects of geometry on skin penetration and failure of polymer microneedles. *J. Adhes. Sci. Technol.* **2013**, *27*, 227–243. [CrossRef]
26. Courtenay, A.J.; McAlister, E.; McCrudden, M.T.C.; Vora, L.; Steiner, L.; Levin, G.; Levy-Nissenbaum, E.; Shterman, N.; Kearney, M.-C.; McCarthy, H.O.; et al. Hydrogel-forming microneedle arrays as a therapeutic option for transdermal esketamine delivery. *J. Control. Release* **2020**, *322*, 177–186. [CrossRef] [PubMed]
27. Caffarel-Salvador, E.; Brady, A.J.; Eltayib, E.; Meng, T.; Alonso-Vicente, A.; Gonzalez-Vazquez, P.; Torrisi, B.M.; Vicente-Perez, E.M.; Mooney, K.; Jones, D.S.; et al. Hydrogel-forming microneedle arrays allow detection of drugs and glucose *in vivo*: Potential for use in diagnosis and therapeutic drug monitoring. *PLoS ONE* **2016**, *10*, e0145644. [CrossRef]
28. Larrañeta, E.; Moore, J.; Vicente-Pérez, E.M.; González-Vázquez, P.; Lutton, R.; Woolfson, A.D.; Donnelly, R.F. A proposed model membrane and test method for microneedle insertion studies. *Int. J. Pharm.* **2014**, *472*, 65–73. [CrossRef] [PubMed]
29. Bretag, A.H. Synthetic interstial fluid for isolated mammalian tissue. *Life Sci.* **1969**, *8*, 319–329. [CrossRef]
30. Donnelly, R.F.; McCrudden, M.T.C.; Zaid Alkilani, A.; Larrañeta, E.; McAlister, E.; Courtenay, A.J.; Kearney, M.-C.; Singh, T.R.R.; McCarthy, H.O.; Kett, V.L.; et al. Hydrogel-forming microneedles prepared from "super swelling" polymers combined with lyophilised wafers for transdermal drug delivery. *PLoS ONE* **2014**, *9*, e111547. [CrossRef]
31. Donnelly, R.F.; Singh, T.R.R.; Garland, M.J.; Migalska, K.; Majithiya, R.; McCrudden, C.M.; Kole, P.L.; Mahmood, T.M.T.; McCarthy, H.O.; Woolfson, A.D. Hydrogel-forming microneedle arrays for enhanced transdermal drug delivery. *Adv. Funct. Mater.* **2012**, *22*, 4879–4890. [CrossRef] [PubMed]

gels

MDPI

Article

Design and Evaluation of Paeonol-Loaded Liposomes in Thermoreversible Gels for Atopic Dermatitis

Yu Wang [1], Yan Yue [1,†], Ruoyang Jia [1,†], Xinyi Liu [1], Zhiqing Cheng [1], Yongfeng Cheng [2,3,*], Yinxiang Xu [4], Zili Xie [5] and Hongmei Xia [1,*]

1 College of Pharmacy, Anhui University of Chinese Medicine, Hefei 230012, China
2 Clinical College of Anhui Medical University, Hefei 230031, China
3 School of Life Science, University of Science and Technology of China, Hefei 230027, China
4 Zhaoke (Hefei) Pharmaceutical Co., Ltd., Hefei 230088, China
5 Anhui Institute for Food and Drug Control, Hefei 230051, China
* Correspondence: yfcheng@ustc.edu.cn (Y.C.); sm_shine@ahtcm.edu.cn (H.X.); Tel./Fax.: +86-13965033210 (H.X.)
† These authors contributed equally to this work.

Abstract: Paeonol (PAE) is a hydrophobic drug. In this study, we encapsulated paeonol in a lipid bilayer of liposomes (PAE-L), which delayed drug release and increased drug solubility. When PAE-L was dispersed in gels (PAE-L-G) based on a poloxamer matrix material for local transdermal delivery, we observed amphiphilicity, reversible thermal responsiveness, and micellar self-assembly behavior. These gels can be used for atopic dermatitis (AD), an inflammatory skin disease, to change the surface temperature of the skin. In this study, we prepared PAE-L-G at an appropriate temperature for the treatment of AD. We then assessed the gel's relevant physicochemical properties, in vitro cumulative drug release, and antioxidant properties. We found that PAE-loaded liposomes could be designed to increase the drug effect of thermoreversible gels. At 32 °C, PAE-L-G could change from solution state to gelatinous state at 31.70 ± 0.42 s, while the viscosity was 136.98 ± 0.78 MPa.S and the free radical scavenging rates on DPPH and H_2O_2 were 92.24 ± 5.57% and 92.12 ± 2.71%, respectively. Drug release across the extracorporeal dialysis membrane reached 41.76 ± 3.78%. In AD-like mice, PAE-L-G could also relieve skin damage by the 12th day. In summary, PAE-L-G could play an antioxidant role and relieve inflammation caused by oxidative stress in AD.

Keywords: paeonol; thermosensitive gel; liposome–hydrogel; antioxidant; atopic dermatitis

Citation: Wang, Y.; Yue, Y.; Jia, R.; Liu, X.; Cheng, Z.; Cheng, Y.; Xu, Y.; Xie, Z.; Xia, H. Design and Evaluation of Paeonol-Loaded Liposomes in Thermoreversible Gels for Atopic Dermatitis. *Gels* **2023**, *9*, 198. https://doi.org/10.3390/gels9030198

Academic Editors: Esmaiel Jabbari and Kummara Madhusudana Rao

Received: 9 February 2023
Revised: 24 February 2023
Accepted: 27 February 2023
Published: 5 March 2023

1. Introduction

Atopic dermatitis (AD), also known as atopic eczema, is a chronic relapsing inflammatory disease that is often associated with other atopic conditions such as asthma, rhinitis, and food allergies [1,2]. The main manifestations are skin barrier dysfunction, skin itching, etc. [3–5]. Its onset is usually concentrated in early childhood, but 2–10% of adults have AD, leading to a significant decrease in quality of life and increased psychological burdens [6–9]. The factors causing AD are not clear at present, but are mainly attributed to filaggrin (FLG), resulting in impaired skin barrier function [10] and changes in immune system function [11,12]. There are also possible environmental, genetic, and other factors. Currently, local corticosteroid drugs are used to treat moderate to severe AD patients clinically, but long-term use will lead to skin thinning, epidermal atrophy, vascular dilation, and other adverse reactions [13,14], and can even lead to patients' poor compliance with the drugs. Therefore, to develop a natural medicine as an alternative therapy, Komal [15] took inspiration from curcumin and encapsulated tetrahydrocoumarin in a solid lipid nanoparticle gel, which enhanced the targeting of the drug to the skin in a controlled release manner, overcoming the skin barrier and improving bioavailability. Gallic acid, which has been shown to be effective for AD treatment, can be combined with a pH/temperature responsive gel for enhanced transdermal delivery [16]. Therefore, using Chinese herbal ingredients to treat AD has become a new trend.

Paeonol (PAE) is an active ingredient that is isolated from the dried roots or whole grass of *Cynanchum paniculatum* (Bunge) *Kitagawa, Paeonia fruticosa Andr,* and *P. lactiflora Pall* in the Ranunculaceae family [17]. A large number of studies have shown that paeonol has good anti-inflammatory [18], anti-allergy [19], anti-infection [20], and antioxidant [21] effects. However, paeonol has poor solubility in water, is volatile at room temperature, and has poor bioavailability, and so it is necessary to design smart polymer materials to deliver the drug. In addition to their ability to encapsulate hydrophilic and hydrophobic drugs, liposomes also provide better permeability and locally increased drug concentration, thereby reducing the stimulation potential [22]. Liposomes are an effective drug delivery system for the treatment of skin inflammatory diseases, because compared with traditional preparations, liposomes can carry a higher amount of drug to the affected area [23]. However, liposomes have poor stability, so they are modified in a variety of ways to increase it.

Hydrogels have been widely used in drug delivery, regenerative medicine, cosmetic technology, and other fields. Due to their three-dimensional network structure, combined with weak cohesion in the form of cross-linked covalent bonds and hydrogen or ionic bonds [24], hydrogels can swell in aqueous solution and have high water absorption. Since the human body is composed of a large amount of water, hydrogel polymers can penetrate deep into the skin [24]. In cosmetic applications, hydrogels are mainly applied to the scalp and skin and used in oral care [25,26]. In recent years, it has been found that these hydrogels, as transport carriers of natural plants, are beneficial for promoting hair growth [27] and protecting the scalp via pH value changes [28]. In the treatment of atopic dermatitis, damage to the skin barrier caused by inflammatory symptoms leads to increased cuticle transepidermal water loss, and it has been shown that the skin epidermal barrier and homeostasis are affected by age [29]. Hydrogels are breathable and moisturizing [30–32] and they reduce water loss, alleviate dry skin symptoms, and maintain skin elasticity. As a new type of topical preparation that has emerged in recent years, temperature-sensitive gels have the advantages of easy application and good biocompatibility. Poloxamers are ABA-type triblock copolymers consisting of PEO (A) and PPO (B) units. Poloxamer 407 (P407) requires a lower concentration for gel formation at 25 °C than other poloxamer series. At room temperature (25 °C), the solution appears as a flowing viscous liquid; at the gelling temperature, it changes into a semi-solid transparent gel [33,34]. At temperatures above the low-critical solution temperature, poloxamer will lose its PPO block, resulting in micellar formation and solution conversion to gels. Unfortunately, the low gelation temperature of poloxamer 407 (P407) as an in situ gel carrier was reported to impede drug administration [35]. Therefore, the addition of poloxamer 188 (P188) as a polymer additive can reduce prescription dosage and achieve heat sensitivity to the physiological environment. Poloxamer can also be used as a surfactant [36], emulsifier, and slow-release material [37,38]. It is easily absorbed after local administration and has good stability. However, poloxamers can form a hydrogel containing a large amount of water, which can cause a blasting reaction such that drugs are released in large quantities in the initial stage [39], preventing drugs from having a long-term therapeutic effect.

Liposome-in-gel systems combine the advantages of liposomes and gels; liposomes can reduce the explosive release of hydrogels [40], and gels can maintain the integrity and function of liposomes, thus improving the bioavailability of drugs [41]. Liposomes in thermosensitive gels have previously been shown to achieve slow-release functionality [42,43].

In this study, paeonol-loaded liposomes were prepared using the thin-film dispersion method. P407 and P188 were used to design heat-sensitive in situ gels to achieve the double delivery of a paeonol-loaded liposome-in-gel for transdermal drug delivery [44]. We produced the solution at room temperature, which can rapidly undergo phase transformation to form gels after being applied to the inflammatory site (32–37 °C), to prolong the retention time of PAE on the skin surface and increase the drug action time, and to provide a platform for the development of transdermal drug delivery. In this study, we investigated the antioxidant capacity of paeonol solution, paeonol-loaded liposomes (PAE-L), paeonol-loaded gels (PAE-G), paeonol-loaded liposomes in temperature-sensitive gels (PAE-L-G), and their effect on AD-like skin of mice induced by 2,4-dinitrochlorobenzene (DNCB).

2. Results and Discussion

2.1. Standard Curve of Paeonol

We selected the corresponding concentration value in the appropriate range (0–0.015 mg/mL) to draw the standard curve and take the concentration–absorbance as the coordinates. The linear regression equation is written as follows: $A = 85.126 \times C + 0.0117$, and the regression coefficient was $R^2 = 0.9992$, indicating that the linearity was good in this concentration range.

2.2. Characterization of PAE-Loaded Liposomes

Because liposomes are composed of a kind of nanoscale artificial vesicle and consist of phospholipids and cholesterol, PAE-L appears to the naked eye as a thick, yellowish liquid. The shape of PAE-L under an electron microscope (Glacios-200KV, Thermo Fisher Scientific, Waltham, MA, USA) appears to be spherical. The hydration particle sizes of B-L and PAE-L were 127 ± 8.1 nm and 132.6 ± 11.5 nm, respectively. Zeta potential was negative (−19.4 ± 0.8 mV and −17.9 ± 2.1 mV). The encapsulation rate of PAE-L was 86.47 ± 8.22%.

2.3. Physicochemical Characterization of Formulations

The surface temperature of the skin was 32–35 °C [45]. When the body suffers from skin damage, its temperature will slightly rise by 0.5–1 °C [46]. Therefore, when the temperature-sensitive in situ gels act on the skin surface, the minimum gelling temperature is 32.5–34 °C, and the maximum is 36–37 °C. Therefore, the gelling temperature of the prescriptions should be between 32 °C and 37 °C. In order to obtain the most suitable carrier for the skin surface of AD mice, poloxamer gels (Table 1), which are temperature-sensitive at 32 °C, 35 °C, and 37 °C, respectively, were selected to check their physical and chemical properties. In the case of the same polymers and similar concentration compositions, both PAE and PAE-L have less influence on the gelling temperature than the blank in situ gels. For temperature-sensitive gels, it is necessary to quickly change from solution to gel form in a short time. Although the gelling ability of PAE-L-G was weakened at 32 °C, 35 °C, and 37 °C, requiring 51.80 ± 5.05 s, 62.92 ± 7.67 s, and 82.58 ± 2.34 s, respectively, there was no significant difference compared to blank gels. In addition, the gelling temperature and gelling capacity were mainly affected by the prescription amounts of P407 and P188 in the gels. The viscosity had been reported to be affected by increasing polymer concentration or use [47,48]. Therefore, viscosity levels need to be within the appropriate range of values. There was no significant difference between the viscosity of blank gels and PAE-G, but the viscosity of the formula was significantly improved after the addition of liposomes, and the viscosity could even be higher than 100 MPa.S at the gelling temperature. The ratio of prescription dosage had little influence on the viscosity change.

Table 1. Physicochemical properties of poloxamer gels at different temperatures.

	Group	Gelation Temperature (°C)	Gelation Capacity (s)	Viscosity 4 °C (MPa·S)	Viscosity 32–37 °C (MPa·S)	pH 4 °C	pH 32–37 °C
	G	31.25 ± 0.78	44.51 ± 2.60	19.32 ± 1.44	91.99 ± 0.65	7.25 ± 0.03	7.04 ± 0.02
32 °C	PAE-G	31.80 ± 0.42	44.79 ± 2.46	19.06 ± 0.55	92.80 ± 0.09	7.38 ± 0.01	7.16 ± 0.05
	PAE-L-G	31.70 ± 0.42	51.80 ± 5.05	30.34 ± 0.06	136.98 ± 0.78	7.33 ± 0.04	7.18 ± 0.03
	G	34.45 ± 0.49	58.29 ± 5.85	28.73 ± 0.39	94.44 ± 0.13	7.27 ± 0.02	7.11 ± 0.12
35 °C	PAE-G	34.70 ± 0.14	59.61 ± 0.86	29.58 ± 0.02	94.38 ± 0.51	7.48 ± 0.02	7.23 ± 0.27
	PAE-L-G	34.60 ± 0.57	62.92 ± 7.67	36.36 ± 6.32	149.28 ± 0.22	7.47 ± 0.01	7.23 ± 0.01
	G	36.90 ± 0.14	73.41 ± 1.35	50.07 ± 0.22	120.94 ± 1.41	7.47 ± 0.05	7.27 ± 0.17
37 °C	PAE-G	36.70 ± 0.28	69.66 ± 0.95	46.22 ± 7.13	123.43 ± 1.69	7.55 ± 0.06	7.34 ± 0.03
	PAE-L-G	36.70 ± 0.71	82.58 ± 2.34	56.39 ± 5.76	197.80 ± 7.14	7.52 ± 0.03	7.38 ± 0.14

In this study, temperature-responsive gels were prepared using the cold method. The solutions of gels were seen clearly by visual observation under black background. The pH of the skin surface was first determined by Heuss in 1892 [49]. The main reason for this is that both the lipid tissue and acidic environment can provide better conditions for lipid metabolism. Studies have shown that the pH level rises in patients with atopic dermatitis [50]. Therefore, the pH value of the skin carrier is an important parameter because differentiation of the skin pH can regulate adverse reactions such as redness or itching [51]. As can be seen from the pH data in Table 1, the pH values of blank gels were measured as 7.04 ± 0.02, 7.11 ± 0.12, and 7.27 ± 0.17 at the gelling temperature. For PAE-L-G, the pH values were 7.18 ± 0.03, 7.23 ± 0.01, and 7.38 ± 0.14, and the pH value decreased with the increase in temperature. The dosage of prescription and the loading of PAE had little effect on the pH value. In addition, skin carriers with alkaline had been reported to improve skin health in some conditions [52]. Therefore, the medicinal carrier in this study can play a role in the treatment of atopic dermatitis.

Combined with the above results, it can be seen that although the temperature-sensitive gels can form semi-solid medicinal carriers on AD-like skin within the range of 32–37 °C, there is no significant difference, and the encapsulated PAE or PAE-L-G would not have a great impact on their physical and chemical properties. However, at the gelling temperature of 32 °C, the concentration of P407 and P188 required for prescription is small, which can reduce the cost of medicine and achieve the ideal therapeutic effect. Therefore, the thermosensitive gels at 32 °C were selected for the follow-up study on AD inflammation diseases.

2.4. In Vitro Permeation Study

PAE is a lipophilic drug, but its solubility and skin accumulation can be improved by the carrier of liposomes and gels. According to the analysis of the dialysis release curve (Figure 1A), the cumulative release rates of the PAE group were 72.32 ± 1.68% within 12 h, while the maximum cumulative release rate of the PAE-L group was only 59.94 ± 4.88%. Therefore, the PAE solution can release the drug quickly in a short time, while the liposome group plays a sustained-release role. As Figure 1B,C show, PAE-L and PAE-L-G could achieve their effect in the local long-term due to sustained and controlled release in the treatment process. Among them, the cumulative release rate (R%) from PAE-L-G was the lowest, mainly since the drug needed to traverse not only the lipid bilayer but also the three-dimensional network structure of the gels. However, in the same formulation, when changing the concentration ratio of P407 and P188, it could be seen that blank gels at 32 °C had the highest drug release in a certain time. For PAE-L-G, the final R% of the three different temperatures were 41.76 ± 3.78%, 46.14 ± 1.04%, and 43.47 ± 0.68%, respectively. There is no significant difference. In summary, the thermosensitive gels not only played the function of drug storage, but also had the ability to help drugs penetrate the skin, so they can be used as a carrier for local skin delivery. Meanwhile, it was further indicated that PAE-L-G at 32 °C was more suitable for the treatment of symptoms caused by AD.

At the same time, OriginPro 2021 software was used to simulate the release curve, and common models such as Zero-order kinetics, First-order kinetics, Higuchi, Korsmeyer–Peppas, Hixson–Crowell, and WeibullCDF were used to analyze the drug release kinetics. Among them, Table 2 lists the equations of the fitting curve whose correlation coefficients are greater than 0.95 and the worst fitting effect, and also shows the corresponding kinetic constants and exponential parameters. The WeibullCDF model can make $R^2 > 0.97$, indicating that it is more suitable for this model and has feasibility and a linear relationship, while the Hixson–Crowell model has negative R^2, indicating that the time-cumulative drug release curve does not conform to the trend of this model.

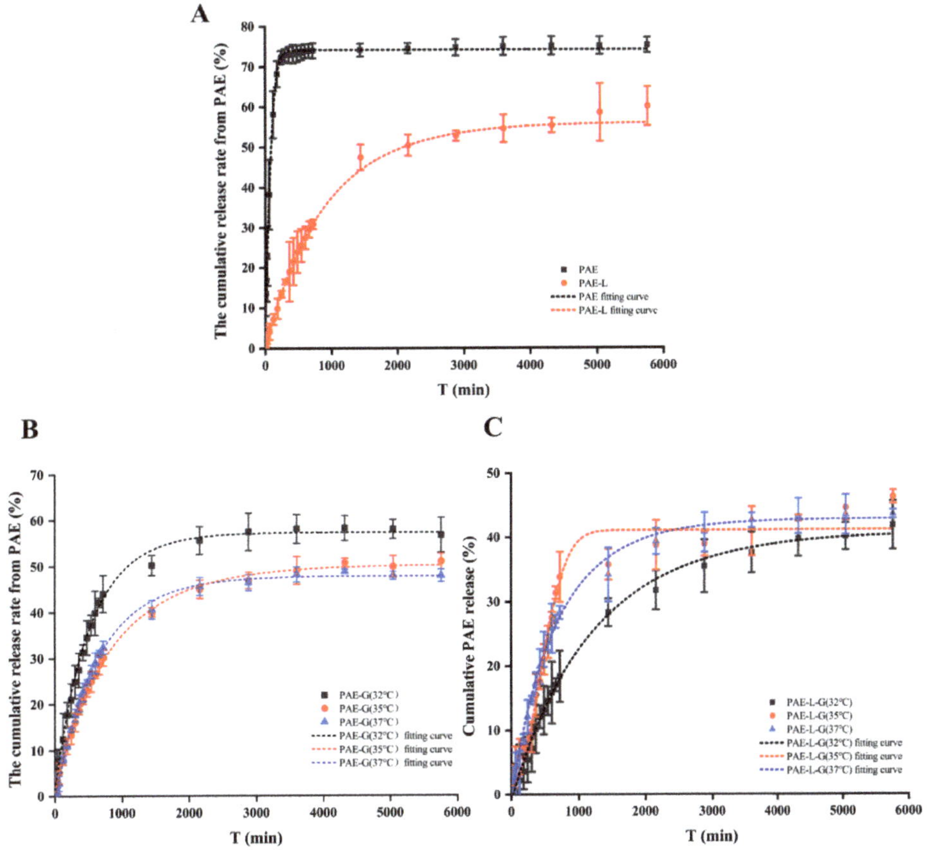

Figure 1. WeibullCDF fitting curve of different formulations. (**A**) PAE and PAE-L. (**B**) PAE-G at 32 °C, 35 °C, and 37 °C. (**C**) PAE-L-G at 32 °C, 35 °C, and 37 °C.

Table 2. The fitting curve values under different mathematical models.

Group		Relevant Data	First-Order	WeibullCDF	Hixson–Crowell
PAE		Equation	$Q = 74.1414\,[1 - \exp(-0.0152t)]$	$Q = 10.3580 + 63.8181\,[1 - e^{-(t/94.2344)^{1.3040}}]$	$Q = 100\,[1 - (1 - 2.7546t)^3]$
		R^2	0.9931	0.9984	−3.9777
PAE-L		Equation	$Q = 55.2900\,[1 - \exp(-0.0017t)]$	$Q = -1.1750 + 5.2263\,[1 - e^{-(t/881.6552)^{0.9298}}]$	$Q = 100\,[1 - (1 - 8.7042t)^3]$
		R^2	0.9983	0.9988	0.6961
PAE-G	32 °C	Equation	$Q = 57.0866\,[1 - \exp(-0.0020t)]$	$Q = 0.7776 + 56.4113\,[1 - e^{-(t/515.9190)^{0.9777}}]$	$Q = 100\,[1 - (1 - 1.9744t)^3]$
		R^2	0.9957	0.9961	−0.1147
	35 °C	Equation	$Q = 49.7029\,[1 - \exp(-0.0013t)]$	$Q = 0.0281 + 50.8791\,[1 - e^{-(t/864.5301)^{0.8868}}]$	$Q = 100\,[1 - (1 - 6.6639t)^3]$
		R^2	0.9978	0.9995	0.6268
	37 °C	Equation	$Q = 48.6397\,[1 - \exp(-0.0014t)]$	$Q = -0.6052 + 48.9826\,[1 - e^{(t/646.9570)^{1.0363}}]$	$Q = 100\,[1 - (1 - 5.6325t)^3]$
		R^2	0.9973	0.9996	0.6506
PAE-L-G	32 °C	Equation	$Q = 40.6142\,[1 - \exp(-8.0875t)]$	$Q = -1.0828 + 41.0662\,[1 - e^{(t/1209.4106)^{0.9575}}]$	$Q = 100\,[1 - (1 - 1.1847t)^3]$
		R^2	0.9962	0.9971	0.6384
	35 °C	Equation	$Q = 43.0370\,[1 - \exp(-0.0016t)]$	$Q = 1.3481 + 39.6793\,[1 - e^{-(t/565.3188)^{2.0324}}]$	$Q = 100\,[1 - (1 - 5.6873t)^3]$
		R^2	0.9574	0.9791	−0.0549
	37 °C	Equation	$Q = 42.4447\,[1 - \exp(-0.0015t)]$	$Q = -1.7778 + 44.5795\,[1 - e^{-(t/662.4938)^{0.9005}}]$	$Q = 100\,[1 - (1 - 5.4228t)^3]$
		R^2	0.9930	0.9946	0.0349

2.5. Antioxidant Capacity via Scavenging Free Radicals

2.5.1. Experiment of the Oxidation Resistance on H_2O_2

Hydrogen peroxide is a kind of reactive oxygen species (ROS). In the human body environment, it can be produced from many biological membranes, such as the mitochondrial membrane, microsomal membrane, and cytoplasm [53]. When the body's internal REDOX balance system is broken, ROS accumulates too much, especially in skin cells, which can accelerate skin aging, induce a cellular inflammatory response, and inhibit the immune function of skin cells.

According to the curve of inhibition rate changing with the concentration in Figure 2A, the antioxidant rate was positively correlated with the concentration of PAE at five different concentrations (0.05, 0.1, 0.15, 0.2, and 0.25 mg/mL). The relationship between the drug concentration and the antioxidant rate was in line with the logarithmic curve, and the free radical inhibition rate showed an upward trend under the change in paeonol concentration, gradually approaching 100%. In Figure 2B, the inhibition rates of PAE, PAE-L, PAE-G, and PAE-L-G on free radicals were 86.89 ± 4.07%, 89.61 ± 5.49%, 99.49 ± 3.49%, and 92.12 ± 2.71%. At the same concentration, the free radical scavenging rate of PAE should be the maximum, because its release and reaction were direct, so it could have a better effect. However, in the experimental results, the inhibition rate of PAE-G was the best, possibly because poloxamer itself could be used as a solvent, increasing the surface solubility of paeonol. In addition, it had been proved that poloxamer 188 could protect cell membranes by eliminating free radicals [54]. Although paeonol in the PAE-L-G was coated with the lipid layer of liposomes and the triblock copolymer of gels, it still had a good antioxidant capacity due to the synergism of drugs and excipients of liposomes and gels.

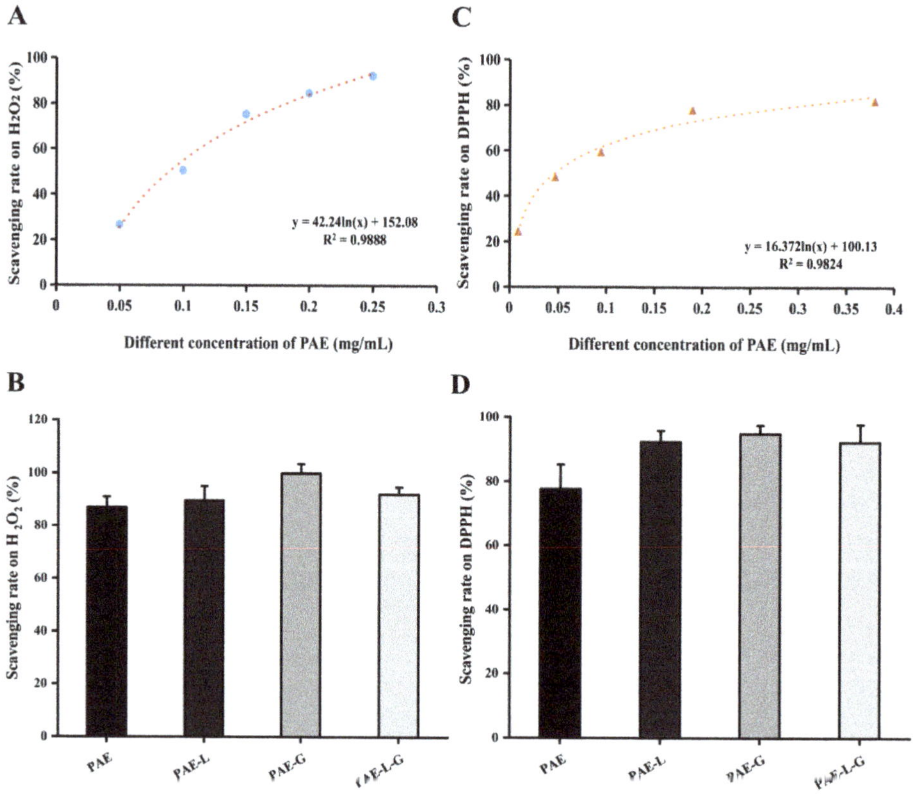

Figure 2. Free radical scavenging capacity. (**A**) Scavenging ability of paeonol at different concentrations on H_2O_2 free radicals. (**B**) Expression of H_2O_2 free radical scavenging ability of different preparations of PAE. (**C**) Scavenging rate on DPPH by different concentrations of PAE. (**D**) The scavenging ability of PAE, PAE-L, PAE-G, and PAE-L-G to DPPH free radicals, respectively.

2.5.2. Scavenging DPPH Free Radical Effect

The DPPH radical is a synthetic, single-electron, stable, nitrogen-centered paramagnetic compound. When the free radical scavenger is present, DPPH accepts an electron or hydrogen atom to form a stable DPPH-H compound, which will fade the solution and reduce absorbance. The degree of discoloration is quantitatively related to the number of electrons received (free radical scavenger activity) [55,56]. Due to the simple structure of DPPH free radicals and the easy control of reaction, DPPH free radicals can be used as an evaluation index of antioxidant performance.

During the experiment, shown as Figure 2C,D, the scavenging rate on free radicals increased with the difference in drug concentration and reaction time with DPPH. Because the scavenging rate of free radicals except PAE can reach more than 90% within 2 h of each prescription, the experimental results showed the same trend as the H_2O_2 experiment, and the effect of PAE-G was the best, mainly because poloxamer gels played a synergistic role in this reaction process. The results showed that the PAE-L-G could fully exert the scavenging free radical effect. There was no significant difference between the groups.

2.6. Pharmacodynamic Studies of Atopic Dermatitis

2.6.1. Reduce AD-like Skin Symptoms

The atopic dermatitis model was performed on the skin of mice except for the blank group in the first four days. As can be seen from Figure 3A, no obvious symptoms of skin injury were observed on the first day of DNCB administration. It can be seen in Figure 3C,D that the skin thickness and injury severity of each group reached a peak on the fourth day. Therefore, dry skin and peeling symptoms began at five days. The secondary sensitization on the ear was performed on the 8th, 12th, and 16th days, respectively, but the corresponding skin symptoms had been alleviated to some extent. The skin status of the positive control group and PAE-L-G group was restored to 70% on the 8th day; there were still scabs and skin damage in PAE, PAE-L, and PAE-G groups, and only the skin of the PAE-L-G group recovered on the last day. As scratching times were recorded on the last day, Figure 3B showed that p values between the administration group and the model group were all greater than 0.05, with no significant difference. In this experiment, compound dexamethasone acetate cream was used as a positive control, but it was found that the mice showed emaciation and slow skin recovery after the 8th day, which even led to the death of the mice. The main reason was that dexamethasone belongs to the hormone class, which can not only achieve anti-inflammatory, anti-sensitization, antipruritic effects, as well as reduce exudation, but it can also lead to skin atrophy and fungal infection, causing physical damage. Therefore, it is necessary to find more effective alternatives.

The spleen is an important immune organ of the body, which plays a role in indicating the strength of immune function [57]. The spleen index of AD-like mice induced by DNCB is shown in Table 3. The spleen in the model group accounted for the largest proportion, indicating spleen enlargement caused by inflammatory symptoms of atopic dermatitis. In terms of the experimental results, the degree of spleen enlargement in the PAE-L group was slightly lower than that in the model group, and the group with temperature-sensitive gel could achieve the same effect as that in the blank group, expressing that the addition of gel could reduce the body's immunity and inhibit dermatitis. The difference was not statistically significant ($p > 0.05$). As the ear was sensitized three times, but the ear was not treated with drugs, the measurement of ear swelling degree on the 17th day showed that, compared with the model group, PAE-L-G could better inhibit the inflammation damage, hypertrophy, and swelling of the ear, and even showed better efficacy than the positive control group, and that the inhibition rate of paeonol solution was lower. Combined with these results, PAE-L-G could be effective not only for topical use but also for systemic inflammation.

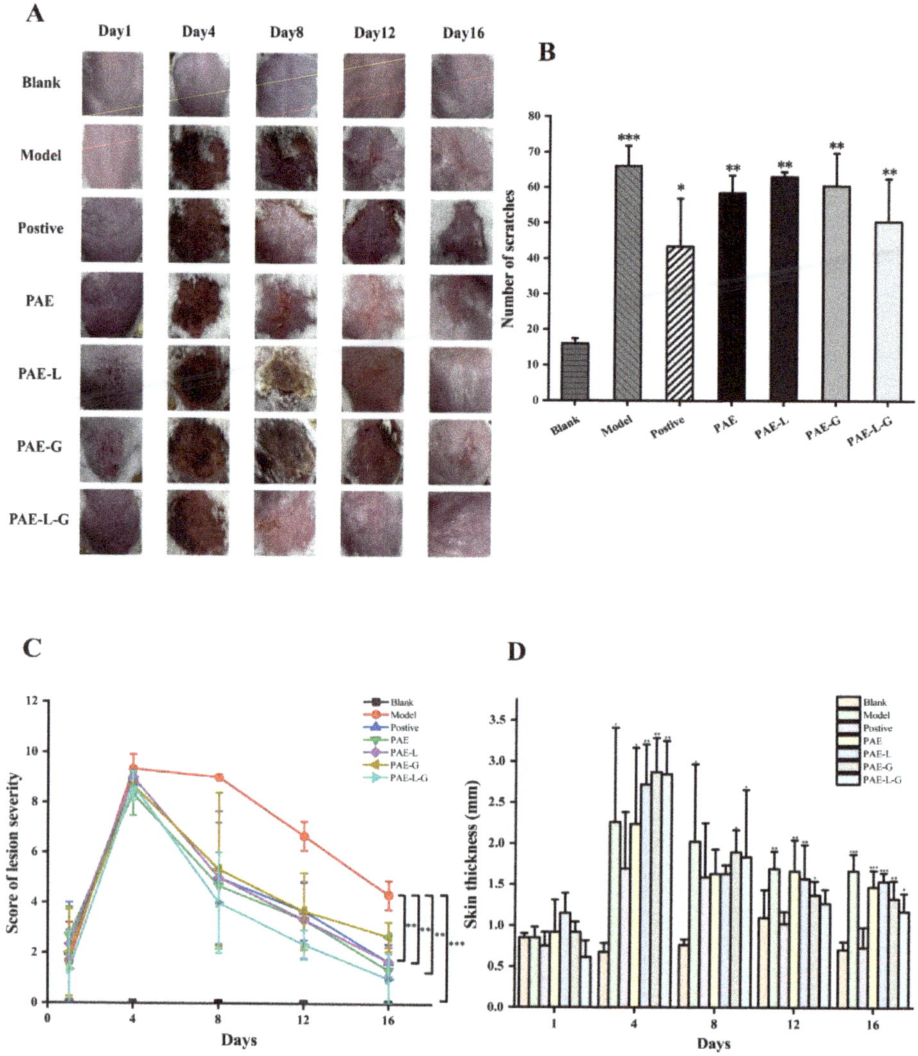

Figure 3. Skin conditions and behavioral manifestations in AD-like mice. (**A**) Skin surface symptoms of mice. (**B**) Number of scratches. (**C**) Score of lesion severity. (**D**) Skin thickness. This difference is statistically significant. (*** $p < 0.001$; ** $p < 0.01$ and * $p < 0.05$ vs. normal group).

Table 3. Spleen index and ear swelling. The data are means ± SD (n = 3). Means with different letters (a,b) are significantly different ($p < 0.05$) via Duncan's test.

Group	Weight (g)	Splenic Weight (g)	Spleen Index (%)	Swelling Degree (mg)	Swelling Inhibition Rate (%)
Blank	28.14 ± 1.50	0.1077 ± 0.0187	0.38 ± 0.05	0.57 ± 0.64 [b]	83.17
Model	21.31 ± 5.55	0.0921 ± 0.0096	0.46 ± 0.18	3.37 ± 0.70 [a]	-
Positive	22.85 ± 1.57	0.0852 ± 0.0694	0.34 ± 0.07	1.25 ± 1.63 [ab]	62.87
PAE	30.29 ± 2.82	0.1161 ± 0.0034	0.39 ± 0.05	2.50 ± 2.12 [ab]	25.74
PAE-L	27.69 ± 0.97	0.1184 ± 0.0430	0.43 ± 0.17	2.90 ± 0.95 [ab]	13.86
PAE-G	27.82 ± 2.18	0.1034 ± 0.0325	0.37 ± 0.09	2.47 ± 1.25 [ab]	26.73
PAE-L-G	31.52 ± 1.17	0.1151 ± 0.0009	0.37 ± 0.01	1.23 ± 1.23 [ab]	63.37

2.6.2. PAE Content in the Skin of AD-like Mice

To detect the drug content in the skin of AD experimental mice, the standard curve of skin homogenate (Figure 4) was first established according to the concentration of the standard curve established previously. The results show that: A = 0.1737 × C + 0.9911 (R^2 = 0.9975), and the absorbance has a linear relationship with this concentration range. According to the measurement of mice in the administration group (Table 4), the drug content in the PAE group was the highest, up to 1.89 ± 0.59 μg/mL, while that of PAE-L-G was slightly higher than that of the PAE-L and PAE-G groups.

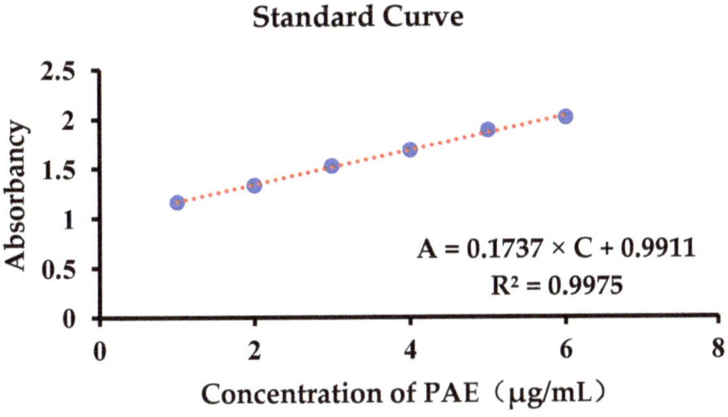

Figure 4. Standard curve of PAE in the skin homogenate.

Table 4. Drug content of skin in different groups of AD-like mice.

Group	Drug Concentration (μg/mL)
PAE	1.89 ± 0.59
PAE-L	1.11 ± 0.33
PAE-G	1.15 ± 0.36
PAE-L-G	1.50 ± 0.41

2.6.3. Inhibit the Production of MDA in Tissue Homogenate

Oxygen free radicals act on the unsaturated fatty acids of lipids to produce lipid peroxide. The latter is gradually decomposed into a series of compounds, including malondialdehyde (MDA). Lipid oxidation levels can be detected by detecting MDA levels. Under acidic and high temperature conditions, MDA can be condensed with thiobarbituric acid (TBA) to produce brown–red trimethylecylate, which can be detected at 532 nm. Studies have shown that the detection of MDA levels could be used as a test index for the symptoms of photoaged skin caused by ultraviolet exposure [58]. Fardin [59] used MDA as an antioxidant parameter to study oxidative stress degree in atopic dermatitis.

In the AD experiment, MDA levels were detected in all organs of mice (Figure 5). There was no significant difference between PAE-L-G and positive control in the supernatant of organ tissue homogenate, indicating that PAE-L-G had a high inhibitory ability of lipid peroxidation in AD mice and could well inhibit oxidative stress response. The inhibition rates of PAE-L-G in the skin, liver, and brain were 50.03 ± 13.25%, 89.15 ± 2.43%, and 95.38 ± 6.97%, respectively, while those in the positive control were 93.99 ± 8.50%, 74.19 ± 4.46%, and 139.54 ± 6.79%. In the heart, lung, and kidney, PAE-L-G had no statistical significance compared to PAE, PAE-L, and PAE-G ($p > 0.05$).

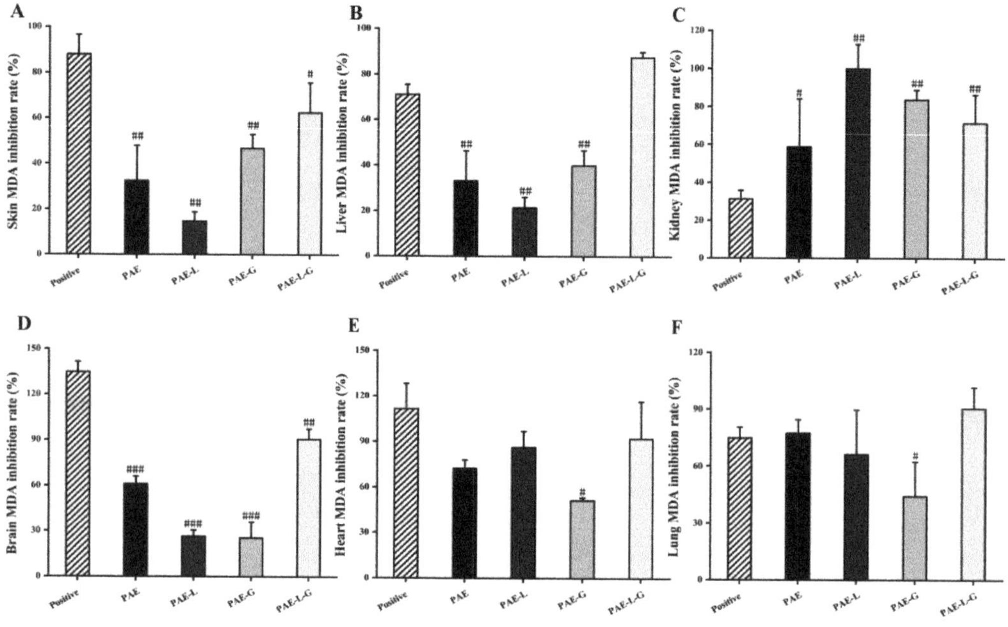

Figure 5. Inhibitory rate of different samples on MDA production in different organ homogenates. (**A**) Skin; (**B**) Liver; (**C**) Brain; (**D**) Kidney; (**E**) Heart; (**F**) Lung. The data are means ± SD (n = 3). $^{###}$ $p < 0.001$; $^{##}$ $p < 0.01$ and $^{\#}$ $p < 0.05$ vs. control group.

2.6.4. Blocking the Damage of AD-like Mouse Skin

Since AD is the primary manifestation of type 2 inflammatory disorder in keratinocytes, it leads to the increase in reactive oxygen species (ROS) in the cells [60,61]. The skin is exposed to the environment, oxygen, and bacteria for a long time, which induces oxidative stress response. ROS is produced by the immune system and has a good defense effect, but when ROS levels are too high in the body, the number of genes expressing inflammatory cells increases, leading to the persistence and deterioration of the inflammatory reaction. In order to prevent the loss caused by ROS, antioxidant enzymes such as superoxide dismutase (SOD), catalase (CAT), and glutathione peroxidase (GPx) are released as a protective mechanism [62]. SOD has the function of dismutating two superoxide radicals into hydrogen peroxide and oxygen. Thus, it has anti-aging and antioxidant effects on the skin, which can be used as an indicator to detect AD-like symptoms.

In Figure 6, except for the PAE solution, SOD activity content in other drug administration groups was significantly higher than that in the model group ($p < 0.01$), showing statistical significance. Among them, PAE-L-G could achieve the same effect as the positive control, and could even make SOD activity reach a non-oxidation level. Combined with the results, PAE-L-G could relieve inflammatory symptoms caused by ROS-induced oxidative stress and it played an antioxidant role.

Due to the limited clinical application of paeonol, the encapsulation substrates are currently divided into conventional dosage forms, polymer delivery systems, and lipid-based delivery systems [63]. Tablets are common dosage forms on the market, which are used for oral administration, and the quality is stabilized through the coating technology. In Guo group's study [64], paeonol and gastroretention tablets of paeonol were applied to gastric ulcer models at the same time. Changes in dosage forms can increase the residence time of drugs in the stomach and give full play to the advantages of local administration. However, rapid drug absorption and first-pass elimination effects necessary for oral administration still exist, reducing the bioavailability of the drug. The transethosomes in the lipid system show the ability to deform into cellular vesicles. The fluidization of stratum corneum (SC) lipid is induced by adding ethanol, which is conducive to drug penetration through the skin barrier and increases drug penetration. The application of paeonol transethosome to pig ear skin improved the stability of the drug and skin deposition [65]. The bulk of the preparations in polymer delivery systems consists of micron and nanoscale particles, such as microparticles, nanocapsules, and nanospheres, which can improve drug efficacy, safety, and controllability and reduce toxic side effects. At the same time, it has great potential for application in skin diseases. However, the preparation process is complicated and the research cost is high. In our study, by preparing paeonol-loaded liposomes in heat-sensitive gels, on the one hand, the intelligent responsiveness improved its targeting; on the other hand, diffusion experiment can find that the liposome gel system enhances its continuous drug release time, and the local drug concentration increases. The detection of antioxidant properties reflects the observable treatment for AD-like immune skin diseases.

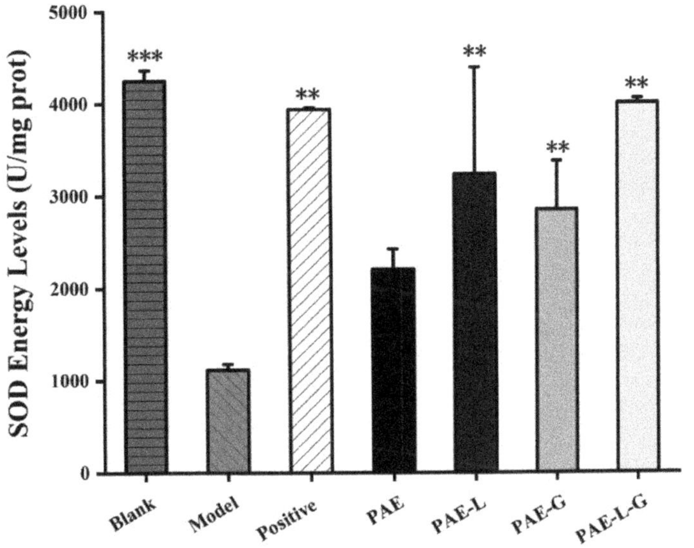

Figure 6. SOD activity value of mice skin. Values represent mean ± SD (n = 3, standard one-way ANOVA, *** $p < 0.001$ and ** $p < 0.01$ vs. model group).

3. Conclusions

In this study, PAE-L-G suitable for the treatment of AD was prepared through film dispersion and cold expansion, and its characterization and physicochemical properties were evaluated, which proved that its platform based on transdermal drug delivery was feasible. The results of in vitro diffusion experiments proved that, compared to the PAE solution, PAE-L-G had a better ability of controlled release and sustained efficacy for a longer time. In addition, different mathematical models were used to fit the release curve, which conforms to the first-order release kinetics and WeibullCDF. Free radical scavenging studies showed that liposomes and the network structure of gels might affect drug release. The inhibition rate of PAE-L-G with antioxidant properties was lower than that of other preparation groups, but there was no significant difference. Meanwhile, MDA experiments on various organs also proved that PAE-L-G could be used as an antioxidant. The results of behavioral and histological experiments on the AD mouse model indicated that PAE-L-G could relieve skin dryness, damage, and itching, even without the side effects of the positive control group, and so PAE-L-G could pass across the SC to work. T-SOD was used as an antioxidant enzyme; higher enzyme activity could be used as an index to evaluate the amount of ROS production, and PAE-L-G could reach a level similar to that of the blank group. The hydrogel itself can be used as the loading material of emollients. At the same time, the local therapy for AD can improve the skin condition, enhance the moisturizing ability, and can play a synergistic effect with paeonol. Poloxamer gels can respond to the change in character according to the change in temperature. The high antioxidant properties of paeonol can alleviate the symptoms of skin damage caused by immune reactions and reduce the generation of toxic side effects. In summary, we believe that paeonol-loaded liposomes in thermally reversible hydrogels are promising materials for skin delivery and have broad application prospects.

4. Materials and Methods

4.1. Materials and Reagents

Paeonol (purity 99%) and Hydrogen peroxide (H_2O_2) were obtained from Shanghai Suyi Chemical Reagent Co., Ltd. (Shanghai, China) and soybean phospholipids (purity 99%), cholesterol (purity 99%), NaOH, and Trichloroacetic acid (TCA) were purchased from Sinopharm Chemical Reagent Co., Ltd. (Shanghai, China), including sulfate heptahydrate. Poloxamer 407 and 188 were purchased from BAST Company. KH_2PO_4 was purchased from Guangxilong Science Co., Ltd. (Guangxi, China) and 1,1-diphenyl-2-picrylhydrazyl (DPPH, 98%), thiobarbituric acid (TBA), and 2,4-Dinitrochlorobenzene (DNCB) were provided by Shanghai Yuanye Biology Science and Technology Co., Ltd. (Shanghai, China). A T-SOD Test kit was acquired from Jiancheng Bioengineering Institute (Nanjing, China). All the reagents were AR grade.

4.2. Animals

Healthy Kunming mice (females weighing 20 ± 2 g) were purchased from the Animal Laboratory Center of Anhui University of Traditional Chinese Medicine (Hefei, China). The animal experiment protocol involved in this experiment conforms to the review and approval of the Ethics Committee of Anhui University of Chinese Medicine (Hefei, China). The indoor temperature was kept at 25 ± 2 °C, the relative humidity was kept at 50–70%, the light rhythm was 12 h, the feeding environment was quiet and ventilated, and the experimental animals were given ordinary feed and clean water.

4.3. Establishment of Standard Curves

The 0.0119 g of paeonol was accurately weighed with a balance, dissolved by adding a small amount of anhydrous ethanol, and then fixed into a 100 mL volumetric flask with phosphate buffer (PBS, pH 7.4). The mother liquor was completely dissolved by ultrasound to obtain 119 μg/mL. It was then diluted into 1.19 μg/mL, 3.57 μg/mL, 5.95 μg/mL, 8.33 μg/mL, 10.71 μg/mL, 13.09 μg/mL solution. Absorbance was measured at 274 nm with an ultraviolet spectrophotometer (UV-1000, AOE Instrument Co., Ltd., Shanghai, China). Values were recorded, and standard curves were established.

4.4. Preparation of Paeonol-Loaded Liposomes (PAE-L)

PAE-L was prepared using the thin-film dispersion method. Firstly, according to the ratio of phospholipid and cholesterol prescribed in our laboratory [58], 0.3 g phospholipid, 0.1 g cholesterol, and 0.0028 g paeonol were accurately weighed and placed into a 250 mL beaker, and then 5 mL anhydrous ethanol was added and placed into the ultrasonic cleaner (HS3120, Tianjin Hengao Technology Development Co., Ltd., Tianjin, China) to dissolve completely and quickly spread on the beaker wall rotary evaporator to form liposome film, and an appropriate amount of PBS was added to shake it with magnetic stirrers.

4.5. Characterization of PAE-L

The particle size and potential of PAE-L were measured by the Malvern nano-particle size potential analyzer (ZEN3690, US Malvern Instrument Co., Ltd., Malvern, UK) [66]. The encapsulation rate (ER%) of PAE-L was determined as follows: the 10 mL volumetric bottle was washed, dried, and set aside. A total of 1 mL PAE-L was placed in a 10 mL volumetric bottle, and PBS buffer was added for constant volume. The bottle stopper was pressed and mixed evenly. Then, 4 mL of the mixture was precisely absorbed, centrifuged at 4500 rpm for 15 min, and the supernatant was taken. The absorbance was measured at the maximum absorption wavelength of paeonol at 274 nm. According to the standard curve of paeonol drawn previously, the concentration of the drug that failed to be encapsulated by liposome was calculated. Next, 1 mL anhydrous ethanol was added to 1 mL PAE-L taken out again from a 10 mL volumetric bottle, ultrasonic dissolution was completed, and the concentration of all drugs in the supernatant was measured. The formula of drug encapsulation rate (ER%) is shown in (1):

$$ER\% = \frac{[C_w - C_n]}{C_w} \times 100\% \quad (1)$$

where "C_n" is the concentration of the drug that has not been successfully encapsulated and "C_w" is the concentration of the drug encapsulated in the liposome.

4.6. Preparation of Paeonol-Loaded Liposomes in Thermoreversible Gels (PAE-L-G)

The prescribed amount of Poloxamer 407 and Poloxamer 188 were mixed and stirred evenly and the volume was fixed into the final preparation system and swelled overnight in a 4 °C refrigerator. PAE-L was evenly dispersed in the thermosensitive gels to finally prepare the PAE-L-G [67].

4.7. Characterization of PAE-L-G

4.7.1. Visual Appearance and Clarity

We visually observed the appearance and transparency of formulations against a black background.

4.7.2. Gelling Temperature and Gelling Time

The change in gel from a flowing solution state to a non-flowing solid state can be observed by the tube inversion method [68,69]. Therefore, the gel temperature of temperature-sensitive gel can be obtained by this method. We washed and dried multiple test tubes to avoid interference from impurities. A total of 2 mL of gel with different prescription concentrations were prepared and placed in a controlled, constant temperature, water bath environment. The temperature was adjusted at the rate of 0.1 °C per minute, and the gel was kept in it for 1 min. Then, the tube was vertically raised to 90° to detect whether the gel flowed. According to the ratio of different concentrations of P407 and P188, the temperature control range was set between 20 °C and 40 °C, and three parallel measurements were made to take the mean value.

Detection of gelation time is also called gelation capacity, and we measured this value using the same test tube method as the temperature measurement, with a stopwatch to record the time of hydrogel gelation, that is, the gelation capacity.

4.7.3. Determination of Viscosity

The viscosity was determined using an Ostwald viscometer (1831, Ring Light Glass Instrument Co., Ltd., Taizhou, China) [68]. Before measurement, the Ostwald viscometer was cleaned with anhydrous alcohol, and then the alcohol was volatilized. We fixed the viscometer on an iron frame and kept the liquid level balanced (Figure 7). We then poured 5 mL of distilled water into the viscometer, use the ear wash ball to draw the liquid until it reached score 1, started timing, and stopped timing when the liquid level dropped to score 2.

Each group of samples was measured three times repeatedly, and the recorded decline time of the solution was calculated using the viscosity Formula (2), and, finally, the viscosity of the sample was obtained:

$$\eta_2 = \frac{\rho_2 t_2}{\rho_1 t_1} \times \eta_1 \quad (2)$$

In the above formula, "η" represents viscosity (MPa·S), "ρ" represents density (g/mL), and "t" represents the time (s). The digits "1" and "2" refer to water and sample, respectively.

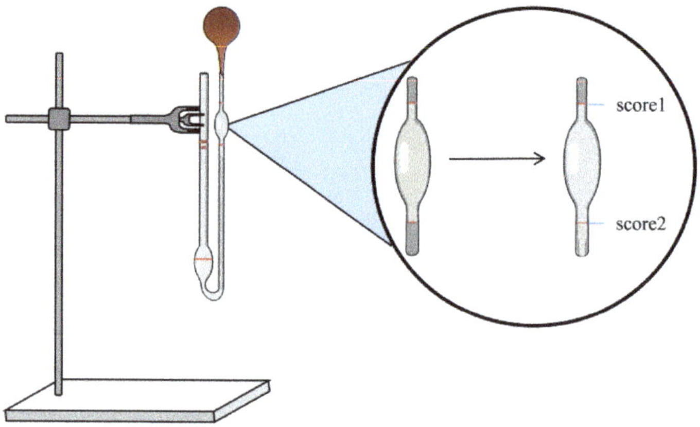

Figure 7. Method of measurement for viscometers.

4.7.4. pH Value

The pH of the liposomes-in-gels was measured using a calibrated pH meter (PHS-3E, Shanghai Lei Instrument Co., Ltd., Shanghai, China). The samples were made in triplicate and the parallel measurements were repeated three times to calculate the mean value of the preparation.

4.8. Antioxidant Ability

4.8.1. Scavenging H_2O_2 Assay

Hydrogen peroxide solution was prepared by dissolving 0.1 mL hydrogen peroxide in 50 mL PBS. The samples of the three groups were prepared at the same time. The blank group was treated with 0.6 mL PBS and 1.8 mL H_2O_2 solution, and the sample group was treated with 0.6 mL different preparations and 1.8 mL H_2O_2 solution, and the mixed solution of 0.6 mL different preparations and 1.8 mL PBS was used as the control group. After preparation, the three groups of preparations were incubated at room temperature for 10 min and protected from light, and the absorbance was measured at the detection wavelength (230 nm), and the clearance rate was obtained according to Formula (3):

$$\text{RSA of antioxidant activity} = \left[1 - \frac{A_S - A_X}{A_O}\right] \times 100\% \quad (3)$$

where "A_S" represents absorbance of the sample group, "A_X" represents absorbance of the control group, and "A_O" represents absorbance of the blank group.

4.8.2. Scavenging DPPH Free Radical

A total of 8 mg of DPPH (1,1-diphenyl-2-trinitrophenylhydrazine) was weighed into a 100 mL weighing bottle with ethanol absolute. The experiment was divided into the blank group, sample group, and control group. The blank group was mixed with 1 mL PBS and 2 mL DPPH solution, the sample group was mixed with 1 mL of different preparations prepared and 2 mL DPPH solution, and the control group was mixed with 2 mL PBS and 1 mL preparation. After being mixed evenly and gently shaken, they were allowed to react for 2 h in the environment while avoiding light. Finally, the absorbance was measured at 517 nm, and the result was obtained by using the clearance Formula (3).

4.9. In Vitro Drug Release across the Dialysis Membrane

The improved Franz diffusion well was adopted [70], with a diffusion area of 1.825 cm^2 and the receiving cell volume was 15 mL. We prepared three sets of controllable temperature magnetic stirrers (85-2, Changzhou Yineng Experimental Instrument Factory, Changzhou, China), and set the temperatures to 32 °C, 35 °C, and 37 °C, respectively, which were stable between ±0.5 °C. Samples of different gelling temperatures were placed on it, while PBS buffers (pH = 7.4) were prepared as a supplement to the receiving pool. The same volume of PBS was added for every 2 mL of sample solution collected. 5 min, 10 min, 20 min, 30 min, 1 h, 2 h, 3 h, 4 h, 5 h, 6 h, 7 h, 8 h, 9 h, 10 h, 11 h, 12 h, 24 h, 36 h, 48 h, 60 h, 72 h, 84 h, and 96 h were used as sampling nodes, and the absorbance of samples was timely measured at 274 nm. This step was repeated three times in parallel, and the cumulative drug release rate Q_n (%) was obtained according to Formula (4):

$$Q_n(\%) = \frac{\left(C_n \times V_n + \sum_{i=1}^{n-1} C_i \times V_i\right)}{Q_T} \times 100\% \quad (4)$$

where "C_n" is the concentration of the drug sampled each time, "V_n" is the volume of dissolved liquid sampled each time, "C_i" and "V_i" are the accumulated concentration and volume of the drug sampled before, and "Q_T" is the theoretical drug amount.

The function of the fitting curve is to use a mathematical model to fit a series of data into a smooth curve, explore the internal relationship between the two groups of data, and understand the change trend between the data. Therefore, in order to study the drug release mechanism of each sample group, a variety of fitting models were selected to determine the best fitting curve.

4.10. Preliminary Antioxidant Activity in Atopic Dermatitis

4.10.1. Establishment of the Atopic Dermatitis Model

The mice were fed adaptively for several days, and the hair in a 2 × 2 cm^2 area was shaved on the back of the mice on the day before the start of the experiment. 5% DNCB solution was prepared (dissolved in a mixture according to the ratio of acetone: olive oil = 3:1), and 0.5% DNCB was diluted on a 5% basis. 100 μL of 5% DNCB solution was used to sensitize the back skin for 2 days. The blank group was given the same volume of acetone and olive oil carrier. After 5 days of intervals, the 0.5% DNCB solution was applied to ear skin 3 times, once every 3 days to sensitize. On the fourth day of the experiment, the blank group and the model group were treated with PBS, the positive control group was smeared with the same volume of compound dexamethasone acetate cream, and the administration group was treated with PAE, PAE-L, PAE-G, and PAE-L-G, respectively, for 12 consecutive days. The mice were fasted for 12 h after the final treatment and sacrificed on the 17th day of the experiment. During the experiment, the skin thickness of the mice's backs and ears were collected for measurement. Organs were used for follow-up experiments. Spleen index was calculated by the ratio of the spleen to body weight, and the ear swelling degree was calculated by weight reduction of the left and right ears. Ear swelling inhibition rate was also recorded using Formula (5):

$$Swelling\ inhibition\ rate(\%) = \frac{A_m - A_s}{A_m} \times 100\% \quad (5)$$

where "A_m" is the ear swelling degree of the model group and "A_s" is the ear swelling degree of sample groups.

4.10.2. Evaluation of the Severity of Dermatitis and Observation of Scratching

Scratching behavior was recorded for 20 min after the last sensitization with 0.5% DNCB (on day 16). A mouse raised its paw and continued to scratch for a long time until it returned to the floor. This was recorded one time.

4.10.3. Scoring of Dermatitis

The skin lesion score was mainly divided into four aspects: (I) skin flaring and hemorrhage, (II) crust formation and xerosis cutis, (III) edema, and (IV) excoriation and erosion, which were scored on three scales: 0 = no symptoms, 1 = mild, 2 = moderate, and 3 = severe. Finally, the total score of the four groups was used as the dermatitis score.

4.10.4. Determination of Drug Content

The back skin of mice in the blank group was cut up and added to anhydrous ethanol (1:10 mL) to homogenate (FSH-2A, Changzhou Putian Experimental Instrument Factory, Changzhou, China) in three batches. The homogenate was wrapped with plastic wrap to prevent the volatilization of ethanol. Then, the skin was ultrasonic for 30 min and transferred to a centrifuge tube, and centrifuged at the speed of 3500 r/min for 5 min. Because the absorbance value of the supernatant measured directly was too large, the supernatant diluted 10 times with ethanol was selected. Six test tubes were added with 1mL of supernatant diluted 10 times, and then PBS and five drug concentrations were added, respectively. The *A* value was determined under the characteristic wavelength of PAE (274 nm), and the drug standard curve of skin homogenate was established. We then took 2 mL of the back skin homogenate supernatant of the drug administration group (diluted 10 times) to obtain the *A* value required by the absorbance-drug concentration equation and calculated the drug content.

4.10.5. Inhibition of the Production of Malonedione (MDA)

The organs and skin tissues of the mice were removed on the 17th day after they were killed by neck removal. After repeated rinsing with physiological saline (pH = 7) to remove the residual blood, they were dried with filter paper. The tissue was weighed and 1:9 cold physiological saline was added for full homogenization (physiological saline was added three times), the samples were collected in test tubes, and the speed of the overspeed centrifuge (TG16-W, Hunan Xiangli Experimental Instrument Factory, Hunan, China) was set at 4000 rpm. After the time was kept for 15 min, the supernatant was taken. A total of 1 mL supernatant + 9 mL dilution medium was used to make 10% analytical solution.

After mixing 0.375% TBA and 5.6% TCA at a ratio of 2:1, 3 mL was added into 1 mL tissue homogenate. The thermostatic water bath was heated to 95 °C in advance, and the mixture was incubated in the test tube for 40 min. After the water cooled to room temperature and centrifuge spun at 4000 r/min for 8 min, we measured the absorbance value of the supernatant at wavelength 532 nm and calculated the inhibition rate (IR%):

$$IR(\%) = \frac{A_m - A_s}{A_m - A_b} \times 100\% \quad (6)$$

where "*A*" is the absorbance value, "*m*" is the model group, "*s*" is the sample group, and "*b*" is the blank group.

4.10.6. Evaluate the Antioxidant Effect of Superoxide Dismutase (SOD) in AD Mice Skin

The animal tissue weight was accurately weighed by adding a homogenate medium (0.9% physiological saline) of 9 times the volume at the ratio of weight (g): volume (mL) = 1:9. The homogenate was prepared into 10% homogenate using mechanical homogenization under the conditions of an ice water bath. The supernatant was taken for determination after being centrifuged at the speed of 2500–3000 rpm for 10 min. We then added the application solution of color removal agent into the monitoring and measuring tube and kept a constant temperature (37 °C) water bath for 40 min. Next, we added color-developing agent and mixed well, left the mixture at room temperature for 10 min, and, at the wavelength of 550 nm in zero distilled water, performed color comparison. Specific operations were conducted according to the experimental procedure of the SOD kit.

4.11. Statistical Analysis

All experiments in this paper were repeated three times, and the results were expressed by mean ± SEM. Data analysis charts were drawn by OriginPro 2021 (OriginLab Corp., Northampton, MA, USA) and statistical results were analyzed by one-way analysis of variance (ANOVA) with SPSS software (IBM, Armonk, NY, USA). A p value less than 0.05 was considered to be statistically different between the two groups of data.

Author Contributions: Conceptualization, H.X.; methodology, Y.W.; software, R.J.; validation, Y.W., Y.Y. and R.J.; formal analysis, Y.W.; investigation, Y.W., Y.Y., X.L., Y.C., Z.C. and R.J.; resources, X.L., Y.C. and Z.C.; data curation, Y.W., Y.Y. and R.J.; writing—original draft preparation, Y.W.; writing—review and editing, H.X.; visualization, Y.X. and Z.X.; supervision, Z.X. and H.X.; project administration, Y.X. and H.X.; funding acquisition, Y.C. and H.X. All authors have read and agreed to the published version of the manuscript.

Funding: This research was funded by the Department of Education of Anhui Province of China (KJ2019A0470, KJ2020A0222, KJ2019A0943, KJ2019A0946, KJ2021A1227), the Research and development project entrusted by the enterprise: research on antioxidant biological activity of quercetin (2022HZ049), the Quality Engineering Project of the Anhui Provincial Department of Education in 2021 (2021jyxm0824),the Natural Science Foundation of Anhui Province of China (1608085MH227), the Quality Engineering Project of Anhui University of Chinese Medicine in 2021 (2021zlgc042), and the Anhui University of Chinese Medicine annual innovation training program for College Students (2017142, 2017171).

Institutional Review Board Statement: This study was conducted according to the guidelines of the Laboratory Animal Center of Anhui University of Chinese Medicine and was approved by the Animal Ethical Committee of Anhui University of Chinese Medicine (SYXK (wan) 2020-001).

Informed Consent Statement: Not applicable.

Data Availability Statement: Not applicable.

Conflicts of Interest: The authors declare no conflict of interest.

References

1. Weidinger, S.; Novak, N. Atopic dermatitis. *Lancet* **2016**, *387*, 1109–1122. [CrossRef]
2. Na, C.H.; Baghoomian, W.; Simpson, E.L. A Therapeutic Renaissance—Emerging Treatments for Atopic Dermatitis. *Acta Derm. Venereol.* **2020**, *100*, 367–379. [CrossRef]
3. Sparavigna, A.; Setarol, M.; Gualandri, V. Cutaneous pH in children affected by atopic dermatitis and in healthy children: A multicenter study. *Ski. Res. Technol.* **1999**, *5*, 221–227. [CrossRef]
4. Boothe, W.D.; Tarbox, J.A.; Tarbox, M.B. Atopic Dermatitis: Pathophysiology. *Adv. Exp. Med. Biol.* **2017**, *1027*, 21–37.
5. Thyssen, J.P.; Rinnov, M.R.; Vestergaard, C. Disease Mechanisms in Atopic Dermatitis: A Review of Aetiological Factors. *Acta Derm. Venereol.* **2020**, *100*, 340–348. [CrossRef] [PubMed]
6. Borok, J.; Matiz, C.; Goldenberg, A.; Jacob, S.E. Contact Dermatitis in Atopic Dermatitis Children-Past, Present, and Future. *Clin. Rev. Allergy Immunol.* **2019**, *56*, 86–98. [CrossRef]
7. Bylund, S.; von Kobyletzki, L.B.; Svalstedt, M.; Svensson, Å. Prevalence and Incidence of Atopic Dermatitis: A Systematic Review. *Acta Derm. Venereol.* **2020**, *100*, 320–329. [CrossRef]
8. Tracy, A.; Bhatti, S.; Eichenfield, L.F. Update on pediatric atopic dermatitis. *Cutis* **2020**, *106*, 143–146. [CrossRef]
9. Reich, K.; Thyssen, J.P.; Blauvelt, A.; Eyerich, K.; Soong, W.; Rice, Z.P.; Hong, H.C.-H.; Katoh, N.; Valenzuela, F.; DiBonaventura, M.; et al. Efficacy and safety of abrocitinib versus dupilumab in adults with moderate-to-severe atopic dermatitis: A randomised, double-blind, multicentre phase 3 trial. *Lancet* **2022**, *400*, 273–282. [CrossRef]
10. Sroka-Tomaszewska, J.; Trzeciak, M. Molecular Mechanisms of Atopic Dermatitis Pathogenesis. *Int. J. Mol. Sci.* **2021**, *22*, 4130. [CrossRef]
11. Liu, F.-T.; Goodarzi, H.; Chen, H.-Y. IgE, mast cells, and eosinophils in atopic dermatitis. *Clin. Rev. Allergy Immunol.* **2011**, *41*, 298–310. [CrossRef] [PubMed]
12. Corrêa, M.P.; Areias, L.L.; Correia-Silva, R.D.; D'Ávila, S.; Leopoldino, A.M.; Greco, K.V.; Gil, C.D. The Role of Galectin-9 as Mediator of Atopic Dermatitis: Effect on Keratinocytes. *Cells* **2021**, *10*, 947. [CrossRef] [PubMed]
13. Barnes, L.; Kaya, G.; Rollason, V. Topical Corticosteroid-Induced Skin Atrophy: A Comprehensive Review. *Drug Saf.* **2015**, *38*, 493–509. [CrossRef] [PubMed]
14. Padula, C.; Machado, I.P.; Vigato, A.A.; de Araujo, D.R. New Strategies for Improving Budesonide Skin Retention. *Pharmaceutics* **2021**, *14*, 30. [CrossRef]

15. Saini, K.; Modgill, N.; Singh, K.K.; Kakkar, V. Tetrahydrocurcumin Lipid Nanoparticle Based Gel Promotes Penetration into Deeper Skin Layers and Alleviates Atopic Dermatitis in 2,4-Dinitrochlorobenzene (DNCB) Mouse Model. *Nanomaterials* **2022**, *12*, 636. [CrossRef] [PubMed]
16. Chatterjee, S.; Hui, P.C.-L.; Kan, C.-W.; Wang, W. Dual-responsive (pH/temperature) Pluronic F-127 hydrogel drug delivery system for textile-based transdermal therapy. *Sci. Rep.* **2019**, *9*, 11658. [CrossRef]
17. Zhang, L.; Li, D.-C.; Liu, L.-F. Paeonol: Pharmacological effects and mechanisms of action. *Int. Immunopharmacol.* **2019**, *72*, 413–421. [CrossRef]
18. Hu, Y.S.; Han, X.; Yu, P.J.; Jiao, M.M.; Liu, X.H.; Shi, J.B. Novel paeonol derivatives: Design, synthesis and anti-inflammatory activity in vitro and in vivo. *Bioorg. Chem.* **2020**, *98*, 103735. [CrossRef]
19. Wang, Y.; Tang, Z.; Guo, X.; Zhao, Y.; Ren, S.; Zhang, Z.; Lv, H. Hyaluronic acid-cyclodextrin encapsulating paeonol for treatment of atopic dermatitis. *Int. J. Pharm.* **2022**, *623*, 121916. [CrossRef] [PubMed]
20. Tang, H.; Yang, D.; Zhu, L.; Shi, F.; Ye, G.; Guo, H.; Deng, H.; Zhao, L.; Xu, Z.; Li, Y. Paeonol Interferes With Quorum-Sensing in Pseudomonas aeruginosa and Modulates Inflammatory Responses In Vitro and In Vivo. *Front. Immunol.* **2022**, *13*, 896874. [CrossRef] [PubMed]
21. Guo, S.; Zhang, Q. Paeonol protects melanocytes against hydrogen peroxide-induced oxidative stress through activation of Nrf2 signaling pathway. *Drug Dev. Res.* **2021**, *82*, 861–869. [CrossRef] [PubMed]
22. Guimarães, D.; Cavaco-Paulo, A.; Nogueira, E. Design of liposomes as drug delivery system for therapeutic applications. *Int. J. Pharm.* **2021**, *601*, 120571. [CrossRef]
23. Jain, S.; Kale, D.P.; Swami, R.; Katiyar, S.S. Codelivery of benzoyl peroxide & adapalene using modified liposomal gel for improved acne therapy. *Nanomedicine* **2018**, *13*, 1481–1493.
24. Mitura, S.; Sionkowska, A.; Jaiswal, A. Biopolymers for hydrogels in cosmetics: Review. *J. Mater. Sci. Mater. Med.* **2020**, *31*, 50. [CrossRef] [PubMed]
25. Costa, E.M.; Silva, S.; Veiga, M.; Tavaria, F.K.; Pintado, M.M. A review of chitosan's effect on oral biofilms: Perspectives from the tube to the mouth. *J. Oral Biosci.* **2017**, *59*, 205–210. [CrossRef]
26. Teng, Y.-Y.; Zou, M.-L.; Liu, S.-Y.; Jia, Y.; Zhang, K.-W.; Yuan, Z.-D.; Wu, J.-J.; Ye, J.-X.; Yu, S.; Li, X.; et al. Dual-Action Icariin-Containing Thermosensitive Hydrogel for Wound Macrophage Polarization and Hair-Follicle Neogenesis. *Front. Bioeng. Biotechnol.* **2022**, *10*, 902894. [CrossRef]
27. Kim, H.S.; Kwon, H.-K.; Lee, D.H.; Le, T.N.; Park, H.-J.; Kim, M.I. Poly(gamma-Glutamic Acid)/Chitosan Hydrogel Nanoparticles For Effective Preservation And Delivery Of Fermented Herbal Extract For Enlarging Hair Bulb And Enhancing Hair Growth. *Int. J. Nanomed.* **2019**, *14*, 8409–8419. [CrossRef]
28. Cavallaro, G.; Caruso, M.R.; Milioto, S.; Fakhrullin, R.; Lazzara, G. Keratin/alginate hybrid hydrogels filled with halloysite clay nanotubes for protective treatment of human hair. *Int. J. Biol. Macromol.* **2022**, *222*, 228–238. [CrossRef]
29. Montero-Vilchez, T.; Cuenca-Barrales, C.; Rodriguez-Pozo, J.-A.; Diaz-Calvillo, P.; Tercedor-Sanchez, J.; Martinez-Lopez, A.; Molina-Leyva, A.; Arias-Santiago, S. Epidermal Barrier Function and Skin Homeostasis in Atopic Dermatitis: The Impact of Age. *Life* **2022**, *12*, 132. [CrossRef]
30. Lee, S.G.; Kim, S.R.; Cho, H.I.; Kang, M.H.; Yeom, D.W.; Lee, S.H.; Lee, S.; Choi, Y.W. Hydrogel-Based Ultra-moisturizing Cream Formulation for Skin Hydration and Enhanced Dermal Drug Delivery. *Biol. Pharm. Bull.* **2014**, *37*, 1674–1682. [CrossRef]
31. Barbosa, A.I.; Torres, T.; Lima, S.A.C.; Reis, S. Hydrogels: A Promising Vehicle for the Topical Management of Atopic Dermatitis. *Adv. Ther.* **2021**, *4*, 2100028. [CrossRef]
32. Ramos Campos, E.V.; Proença, P.L.D.F.; Doretto-Silva, L.; Andrade-Oliveira, V.; Fraceto, L.F.; de Araujo, D.R. Trends in nanoformulations for atopic dermatitis treatment. *Expert Opin. Drug. Deliv.* **2020**, *17*, 1615–1630. [CrossRef]
33. García-Couce, J.; Schomann, T.; Chung, C.K.; Que, I.; Jorquera-Cordero, C.; Fuentes, G.; Almirall, A.; Chan, A.; Cruz, L.J. Thermosensitive Injectable Hydrogels for Intra-Articular Delivery of Etanercept for the Treatment of Osteoarthritis. *Gels* **2022**, *8*, 488. [CrossRef] [PubMed]
34. Ilić-Stojanović, S.; Nikolić, L.; Nikolić, V.; Ristić, I.; Cakić, S.; Petrović, S.D. Biomedical Applications of Thermosensitive Hydrogels for Controlled/Modulated Piroxicam Delivery. *Gels* **2023**, *9*, 70. [CrossRef] [PubMed]
35. Hirun, N.; Kraisit, P.; Tantishaiyakul, V. Thermosensitive Polymer Blend Composed of Poloxamer 407, Poloxamer 188 and Polycarbophil for the Use as Mucoadhesive In Situ Gel. *Polymers* **2022**, *14*, 1836. [CrossRef]
36. Abdeltawab, H.; Svirskis, D.; Hill, A.G.; Sharma, M. Increasing the Hydrophobic Component of Poloxamers and the Inclusion of Salt Extend the Release of Bupivacaine from Injectable In Situ Gels, While Common Polymer Additives Have Little Effect. *Gels* **2022**, *8*, 484. [CrossRef]
37. Jaquilin, P.J.R.; Oluwafemi, O.S.; Thomas, S.; Oyedeji, A.O. Recent advances in drug delivery nanocarriers incorporated in temperature-sensitive Pluronic F-127–A critical review. *J. Drug Deliv. Sci. Technol.* **2022**, *72*, 103390. [CrossRef]
38. Cui, N.; Dai, C.-Y.; Mao, X.; Lv, X.; Gu, Y.; Lee, E.-S.; Jiang, H.-B.; Sun, Y. Poloxamer-Based Scaffolds for Tissue Engineering Applications: A Review. *Gels* **2022**, *8*, 360. [CrossRef]
39. Yao, H.; Lu, H.; Zou, R.; Chen, X.; Xu, H. Preparation of Encapsulated Resveratrol Liposome Thermosensitive Gel and Evaluation of Its Capability to Repair Sciatic Nerve Injury in Rats. *J. Nanomater.* **2020**, *2020*, 2840162. [CrossRef]
40. Lin, Y.-C.; Gao, M.-Y.; Wu, Y.-J.; Fang, Y.-P. Lipid-enveloped PLGA as a hybrid carrier for sustained delivering camptothecin in ovarian cancer. *IET Nanobiotechnol.* **2017**, *11*, 797–802. [CrossRef]

41. Zhang, Q.; Yang, X.; Wu, Y.; Liu, C.; Xia, H.; Cheng, X.; Cheng, Y.; Xia, Y.; Wang, Y. In Vitro Evaluation of Kaempferol-Loaded Hydrogel as pH-Sensitive Drug Delivery Systems. *Polymers* **2022**, *14*, 3205. [CrossRef] [PubMed]
42. Ding, W.; Li, Y.; Hou, X.; Li, G. Bleomycin A6-loaded anionic liposomes with in situ gel as a new antitumoral drug delivery system. *Drug Deliv.* **2016**, *23*, 88–94. [CrossRef] [PubMed]
43. Nie, S.; Hsiao, W.L.; Pan, W.; Yang, Z. Thermoreversible Pluronic F127-based hydrogel containing liposomes for the controlled delivery of paclitaxel: In vitro drug release, cell cytotoxicity, and uptake studies. *Int. J. Nanomed.* **2011**, *6*, 151–166.
44. Wang, W.; Hui, P.C.L.; Wat, E.; Ng, F.S.F.; Kan, C.W.; Wang, X.; Wong, E.C.W.; Hu, H.; Chan, B.; Lau, C.B.S.; et al. In vitro drug release and percutaneous behavior of poloxamer-based hydrogel formulation containing traditional Chinese medicine. *Colloids Surf. B Biointerfaces* **2016**, *148*, 526–532. [CrossRef] [PubMed]
45. Okur, N.Ü.; Hökenek, N.; Okur, M.E.; Ayla, Ş.; Yoltaş, A.; Siafaka, P.I.; Cevher, E. An alternative approach to wound healing field; new composite films from natural polymers for mupirocin dermal delivery. *Saudi Pharm. J.* **2019**, *27*, 738–752. [CrossRef] [PubMed]
46. Hornstein, O.P.; Boissevain, F.; Wittmann, H. Non-invasive Measurement of the Vascular Dynamics of Dermographism. *J. Dermatol.* **1991**, *18*, 79–85. [CrossRef]
47. Chaudhary, B.; Verma, S. Preparation and evaluation of novel in situ gels containing acyclovir for the treatment of oral herpes simplex virus infections. *Sci. World J.* **2014**, *2014*, 280928. [CrossRef]
48. Wei, Y.; Li, C.; Zhu, Q.; Zhang, X.; Guan, J.; Mao, S. Comparison of thermosensitive in situ gels and drug-resin complex for ocular drug delivery: In vitro drug release and in vivo tissue distribution. *Int. J. Pharm.* **2020**, *578*, 119184. [CrossRef] [PubMed]
49. Heuss, E. Die Reaktion des Schweisses beim gesunden Menschen. *Monatschr. Prakt. Dermatol.* **1892**, *14*, 341–343.
50. Rizi, K.; Green, R.J.; Donaldson, M.X.; Williams, A.C. Using pH abnormalities in diseased skin to trigger and target topical therapy. *Pharm. Res.* **2011**, *28*, 2589–2598. [CrossRef]
51. Erol, İ.; Üstündağ Okur, N.; Orak, D.; Sipahi, H.; Aydin, A.; Özer, Ö. Tazarotene-loaded in situ gels for potential management of psoriasis: Biocompatibility, anti-inflammatory and analgesic effect. *Pharm. Dev. Technol.* **2020**, *25*, 909–918. [CrossRef]
52. Schmid-Wendtner, M.H.; Korting, H.C. The pH of the skin surface and its impact on the barrier function. *Skin Pharmacol. Physiol.* **2006**, *19*, 296–302. [CrossRef]
53. Yang, B.; Chen, Y.; Shi, J. Reactive Oxygen Species (ROS)-Based Nanomedicine. *Chem. Rev.* **2019**, *119*, 4881–4985. [CrossRef]
54. Paprikar, A.; Soni, A.; Kaushal, N.; Lin, S. Sublingual insulin administration: Application of hydroxypropyl beta-cyclodextrin and poloxamer188 as permeation enhancers. *Pharm. Dev. Technol.* **2021**, *26*, 233–242. [CrossRef]
55. Li, X. Comparative Study of 1,1-Diphenyl-2-picryl-hydrazyl Radical (DPPH•) Scavenging Capacity of the Antioxidant Xanthones Family. *Chem. Sel.* **2018**, *3*, 13081–13086. [CrossRef]
56. Sirivibulkovit, K.; Nouanthavong, S.; Sameenoi, Y. Paper-based DPPH Assay for Antioxidant Activity Analysis. *Anal. Sci. July* **2018**, *34*, 795–800. [CrossRef]
57. Nam, J.H.; Choi, J.; Monmai, C.; Rod-In, W.; Jang, A.Y.; You, S.; Park, W.J. Immune-Enhancing Effects of Crude Polysaccharides from Korean Ginseng Berries on Spleens of Mice with CyclophosphamideInduced Immunosuppression. *J. Microbiol. Biotechnol.* **2022**, *32*, 256–262. [CrossRef]
58. Liu, C.; Xia, Y.; Li, Y.; Cheng, Y.; Xia, H.; Wang, Y.; Yue, Y.; Wu, Y.; Cheng, X.; Xu, Y.; et al. Ligustrazine as an Extract from Medicinal and Edible Plant Chuanxiong Encapsulated in Liposome-Hydrogel Exerting Antioxidant Effect on Preventing Skin Photoaging. *Polymers* **2022**, *14*, 4778. [CrossRef] [PubMed]
59. Nosratabadi, R.; Khajepour, F.; Zangouyee, M.R.; Khosravimashizi, A.; Afgar, A.; Abdollahi, V.; Dabiri, S. Caraway extract alleviates atopic dermatitis by regulating oxidative stress, suppressing Th2 cells, and upregulating Th1 cells in mice. *Asian Pac. J. Trop. Biomed.* **2022**, *12*, 421–429. [CrossRef]
60. Pyeon, S.; Kim, O.-K.; Yoon, H.-G.; Kim, S.; Choi, K.-C.; Lee, Y.-H.; Lee, J.; Park, J.; Jun, W. Water Extract of Rubus coreanus Prevents Inflammatory Skin Diseases In Vitro Models. *Plants* **2021**, *10*, 1230. [CrossRef] [PubMed]
61. Khan, A.Q.; Agha, M.V.; Sheikhan, K.S.A.M.; Younis, S.M.; Tamimi, M.A.; Alam, M.; Ahmad, A.; Uddin, S.; Buddenkotte, J.; Steinhoff, M. Targeting deregulated oxidative stress in skin inflammatory diseases: An update on clinical importance. *Biomed. Pharmacother.* **2022**, *154*, 113601. [CrossRef]
62. Nguyen, C.T.; Sah, S.K.; Zouboulis, C.C.; Kim, T.-Y. Inhibitory effects of superoxide dismutase 3 on Propionibacterium acnes-induced skin inflammation. *Sci. Rep.* **2018**, *8*, 4024. [CrossRef] [PubMed]
63. Adki, K.M.; Kulkarni, Y.A. Chemistry, pharmacokinetics, pharmacology and recent novel drug delivery systems of paeonol. *Life Sci.* **2020**, *250*, 117544. [CrossRef]
64. Guo, Y.; Du, Y.; Xie, L.; Pu, Y.; Yuan, J.; Wang, Z.; Zhang, T.; Wang, B. Effects of Paeonol and Gastroretention Tablets of Paeonol on Experimental Gastric Ulcers and Intestinal Flora in Rats. *Inflammation* **2020**, *43*, 2178–2190. [CrossRef]
65. Chen, Z.X.; Li, B.; Liu, T.; Wang, X.; Zhu, Y.; Wang, L.; Wang, X.H.; Niu, X.; Xiao, Y.; Sun, Q. Evaluation of paeonol-loaded transethosomes as transdermal delivery carriers. *Eur. J. Pharm. Sci.* **2017**, *99*, 240–245. [CrossRef]
66. Wang, M.; Xia, H.; Yang, X.; Zhang, Q.; Li, Y.; Wang, Y.; Xia, Y.; Xie, Z. Berberine Hydrochloride-loaded Liposomes Gel: Preparation, Characterization and Antioxidant Activity. *Indian J. Pharm. Educ. Res.* **2023**, *57*, 74–82. [CrossRef]
67. Arslan, A.; Kose Ozkan, C.; Sig, A.K.; Dogan, E.; Esim, O.; Cetinkaya, S.; Atalay, F.; Tas, C.; Savaser, A.; Ozkan, Y. Evaluation of a novel oxiconazole nitrate formulation: The thermosensitive gel. *Saudi Pharm. J.* **2018**, *26*, 665–672. [CrossRef] [PubMed]

68. Xia, H.; Jin, H.; Cheng, Y.; Cheng, Z.; Xu, Y. The Controlled Release and Anti-Inflammatory Activity of a Tetramethylpyrazine-Loaded Thermosensitive Poloxamer Hydrogel. *Pharm. Res.* **2019**, *36*, 52. [CrossRef] [PubMed]
69. Praveen, C.; Ujwala, D. Synthesis and Evaluation of Water Insoluble but Swellable Bioadhesive Polymer for Ocular Drug Delivery. *Indian J. Pharm. Educ. Res.* **2019**, *53*, 225–235. [CrossRef]
70. Wu, Y.; Wang, M.; Li, Y.; Xia, H.; Cheng, Y.; Liu, C.; Xia, Y.; Wang, Y.; Yue, Y.; Cheng, X.; et al. The Fabrication of Docetaxel-Containing Emulsion for Drug Release Kinetics and Lipid Peroxidation. *Pharmaceutics* **2022**, *14*, 1993. [CrossRef]

gels

MDPI

Article

pH Sensitive Drug Delivery Behavior of Palmyra Palm Kernel Hydrogel of Chemotherapeutic Agent

Kummara-Madhusudana Rao [1,2], Kummari Subba Venkata Krishna Rao [3], Ramasubba-Reddy Palem [4], Uluvangada-Thammaiah Uthappa [2], Chang-Sik Ha [5,*] and Sung-Soo Han [1,2,*]

1 School of Chemical Engineering, Yeungnam University, 280 Daehak-Ro, Gyeongsan 38541, Republic of Korea
2 Research Institute of Cell Culture, Yeungnam University, 280 Daehak-Ro, Gyeongsan 38541, Republic of Korea
3 Department of Chemistry, Yogi Vemana University, Kadapa 516 003, India
4 Department of Medical Biotechnology, Dongguk University, 32 Dongguk-ro, Ilsandong-gu, Goyang 10326, Republic of Korea
5 Department of Polymer Science and Engineering, School of Chemical Engineering, Pusan National University, Busan 46241, Republic of Korea
* Correspondence: csha@pnu.edu (C.-S.H.); sshan@yu.ac.kr (S.-S.H.); Tel.: +82-53-810-2773 (S.-S.H.); Fax: +82-53-810-4686 (S.-S.H.)

Abstract: This study examined the gel behavior of naturally-occurring palmyra palm kernel (PPK). Due to the presence of polysaccharide in PPK hydrogels, they exhibit excellent swelling behavior in response to pH. Chemotherapeutic drug 5-fluorouracil (5-FU) was encapsulated in these gels using an equilibrium swelling technique. It was found that 5-FU had an encapsulation efficiency of up to 62%. To demonstrate the drug stability in the gels, the PPK hydrogels were characterized using fourier transform infrared spectroscopy, differential scanning calorimetry, and X-ray diffraction. The results showed that the PPK hydrogel matrix contained molecularly dispersed 5-FU drug. The PPK hydrogel exhibited a denser structure and a rough surface, according to images obtained by scanning electron microscopy. In vitro release tests were carried out at pH 1.2 (gastric fluid) and 7.4 (intestinal fluid). The efficacy of the encapsulation and the release patterns were influenced by the network topology of the PPK hydrogel. The release patterns showed that 5-FU was released gradually over a time internal of more than 12 h. The findings suggest that naturally-occurring PPK hydrogels loaded with chemotherapeutic drugs could be employed to treat colon cancer.

Keywords: palmyra palm kernel; polysaccharide; hydrogels; pH-sensitive; 5-fluorouracil; colon cancer

Citation: Rao, K.-M.; Rao, K.S.V.K.; Palem, R.-R.; Uthappa, U.-T.; Ha, C.-S.; Han, S.-S. pH Sensitive Drug Delivery Behavior of Palmyra Palm Kernel Hydrogel of Chemotherapeutic Agent. *Gels* **2023**, 9, 38. https://doi.org/10.3390/gels9010038

Academic Editor: Akira Matsumoto

Received: 17 December 2022
Revised: 28 December 2022
Accepted: 30 December 2022
Published: 2 January 2023

1. Introduction

Polysaccharides from plant resources have a history of being included in dose formulations for effective drug administration. Significant potential has been shown for polysaccharides such as alginate, carrageenan, guar gum, gum arabic, and konjac glucomannan as drug delivery methods for controlled release [1]. Therefore, drug delivery carriers such as microparticles, beads, tablets, and crosslinked hydrogels composed of natural materials are important for drug delivery applications. Among these, hydrogels are very important because of their responsiveness, degradability, and stability [2,3].

Recently, several hydrogels were developed from a combination of natural-natural and natural-synthetic polymers. In this study, the gel behavior of palmyra palm kernel was identified. Palmyra palms are economically useful and cultivated widely in tropical regions. The plant has been used traditionally as a stimulant, anti-leprotic, diuretic, and antiphlogistic. The palmyra palm is one of the most important trees in India, and has some 800 different uses. Indian (Andhra Pradesh) fruit, known as toddy palm seeds (Taati munjalu), is a delicate halwa/jelly-like fresh fruit beloved for its sweet, soft flesh and reviving sugary water inside. The kernel has a moisture content of 92.6%, 0.46% protein, 1% fat, and 6.29% carbs [4]. The kernel has mostly polysaccharides such as pectin, pullulan, and mannose-cellulose [5–7]. However, complete swelling properties, such as pH swelling, are not reported.

Antineoplastic agent 5-FU is an acidic, water-soluble, hydrophilic medication frequently used in clinical chemotherapy to treat solid tumors. With the antimetabolite 5-fluorouracil (5-FU), a common component of cancer chemotherapy, it is possible to avoid scarring after trabeculectomy and increase the likelihood of successful long-term retinal reattachment [8,9]. The intravenously-administered 5-FU in the treatment of colon cancer results in severe systemic effects, due to its toxic effects on the normal cells in the body [10]. The primary mechanism of action of the medication is as a thymidylate synthase inhibitor, which prevents DNA synthesis, interferes with nucleic acid synthesis, and eventually stops cell proliferation. pH-sensitive hydrogels are crucial for colon cancer drug delivery using 5-FU as model chemotherapeutic agent. Over the last decade, several pH-sensitive carriers, such as nanoparticles, microspheres, hydrogels, and enteric-coated materials, have been used for colon cancer drug delivery [11–16].

In the present contribution, 5-FU is encapsulated in palmyra palm kernel (PPK) hydrogels. PPK hydrogels can potentially be applied for colon cancer drug delivery based on their swelling and pH-dependent release behavior. This study examined their release characteristics. To the best of the authors' knowledge, PPK was identified as a hydrogel for the first time because it could swell in aqueous media and show similar properties to chemically-synthesized hydrogels.

2. Results and Discussion

2.1. Development of 5-FU Loaded PPK Hydrogels and Its Characterization

For drug delivery applications, many researchers have developed hydrogels composed of natural carbohydrate polymers, such as chitosan, sodium alginate, pectin, carrageenan, and dextrin. The present study developed naturally-occurring palmyra palm fruit hydrogels (Figure 1); the PPK hydrogels have many hydrophilic units and exhibit pH-sensitive properties. The presence of polysaccharides and a number of hydrophilic functionalities in the hydrogels was thought to be the cause of the pH sensitivity. Assoi et al. developed a method for the extraction of pectin polysaccharide from repined and young sugar palms through microwave-assisted extraction [5]. The PPK hydrogels are safe and economically useful. The pH sensitivity is also expected because pectin polysaccharide is present in palm fruit. Therefore, 5-FU was successfully encapsulated in the PPK hydrogels by immersing PPK hydrogel in a known concentration of 5-FU solution. The encapsulation efficiencies were also calculated, and Table 1 lists their values. The 5-FU-PPK hydrogels could be beneficial for drug delivery in the treatment of colon cancer.

Table 1. In vitro release parameters and encapsulation efficiency values of various PPK hydrogels.

Formulation Code	n	r	Kp ($\times 10^2$)	%EE
P1 (soft PPK)	0.3592	0.9686	0.2324	62.82
P2 (hard PPK)	0.2832	0.9867	0.8512	59.42
P3 (more harder PPK)	0.2691	0.9912	0.9862	54.69

FTIR, DSC, XRD, and SEM were used to characterize the PPK hydrogels. These characterization techniques provided useful information on PPK hydrogels. Figure 2 shows the FT-IR spectra of (a) PPK hydrogel (P2), (b) 5-FU-PPK hydrogel (P2), and (c) 5-FU. The sharp band at 2924 cm^{-1} indicated C-H stretching, whereas the broadband at 3436 cm^{-1} was linked to OH stretching. Additional distinguishing signals were identified for the C-O-H bending at 1451 cm^{-1}, C-O-C stretching at 1107 cm^{-1} and C-O stretching vibrations at 1037 cm^{-1}. The carboxylic C=O group was given the band at 1656 cm^{-1}. On the other hand, a peak at 1656 cm^{-1} (carboxylic C=O) was shifted to lower frequency (1648 cm^{-1}) for 5-FU-PPK hydrogel, suggesting hydrogen bonding interactions between PPK functional groups and 5-FU.

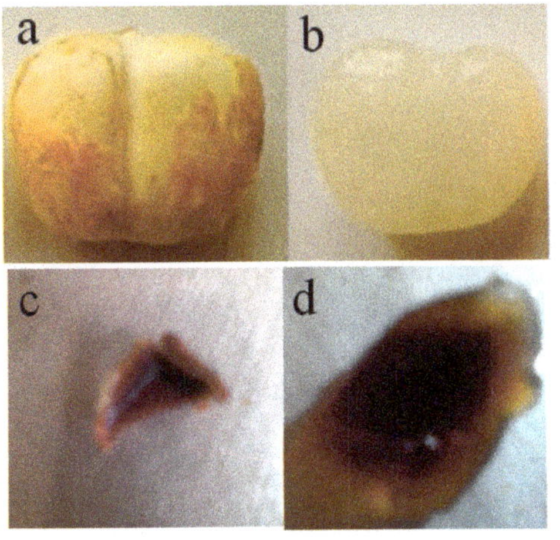

Figure 1. Digital photographs of (**a**) PPK fruit, (**b**) PPK after seed removal, (**c**) dry PPK hydrogel, and (**d**) swollen PPK hydrogel (photographs were captured using Nikon-D5200, Japan Optical Industries Co., Ltd.).

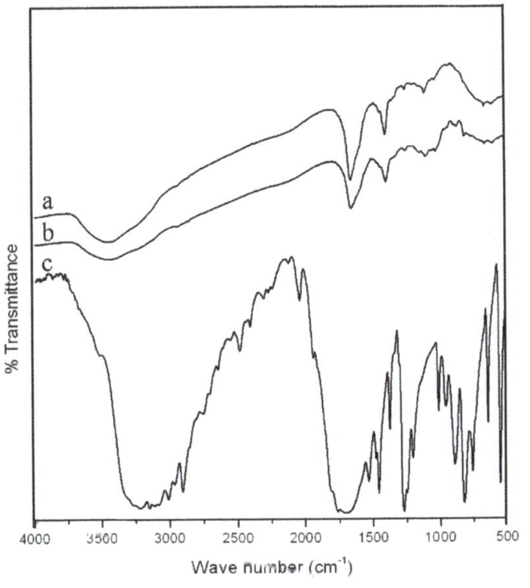

Figure 2. FTIR spectra of PPK hydrogel for P2 formulation: (a) pure PPK hydrogel, (b) 5-FU-PPK hydrogel, and (c) 5-FU.

Studies using DSC and XRD are crucial for determining the drugs' characteristics after being encapsulated in hydrogels. Figure 3 displays the DSC values for PPK hydrogel (P2), 5-FU-PPK hydrogel (P2), and 5-FU. Due to melting and polymorphism of 5-FU, the DSC curve showed a strong peak at 285 °C (Curve C) [15,16]. On the other hand, assuming that the drug is dispersed molecularly in the PPK gel, this peak did not show up in the case of the 5-FU-PPK hydrogels.

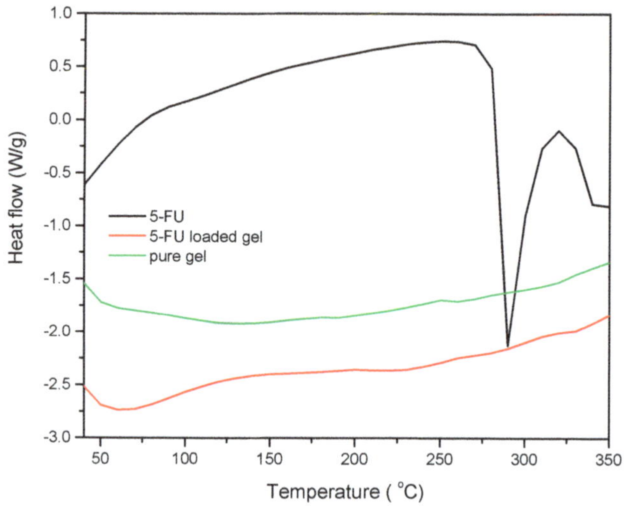

Figure 3. DSC curves of PPK hydrogel for P2 formulation: pure PPK hydrogel, 5-FU-PPK hydrogel, and 5-FU.

Figure 4 shows the XRD patterns of pure 5-FU, 5-FU-PPK hydrogel (P2), and PPK hydrogel (P2). Due to the drug's crystalline structure, pure 5-FU had peaks at 17 °C, 29 °C, and 32 °C. However, in the 5-FU loaded PPK hydrogels, the crystalline peaks of 5-FU vanished, indicating that the 5-FU molecules were dispersed throughout the hydrogel matrix [17,18]. Due to the hydrogels' amorphous structures, both drug-loaded and pure hydrogels showed peaks at 23 °C and 41.2 °C. SEM pictures of the PPK hydrogels for the P2 formulation are shown in Figure 5. The surface morphology of pure PPK hydrogel revealed a denser structure and a rough surface.

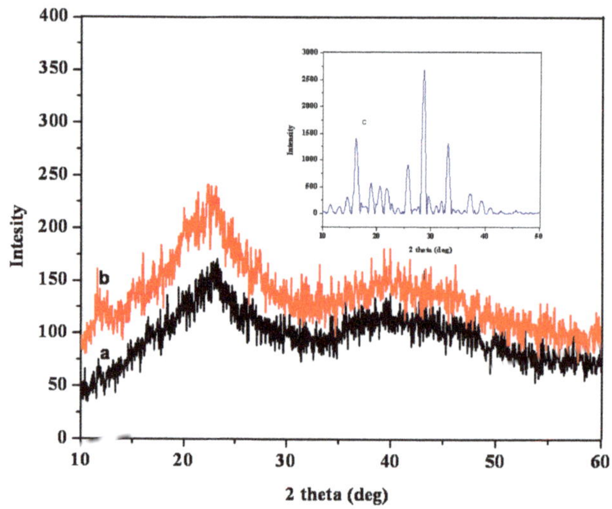

Figure 4. XRD patterns of PPK hydrogel for P2 formulation: (a) pure PPK hydrogel, (b) 5-FU-PPK hydrogel, and (c) 5-FU.

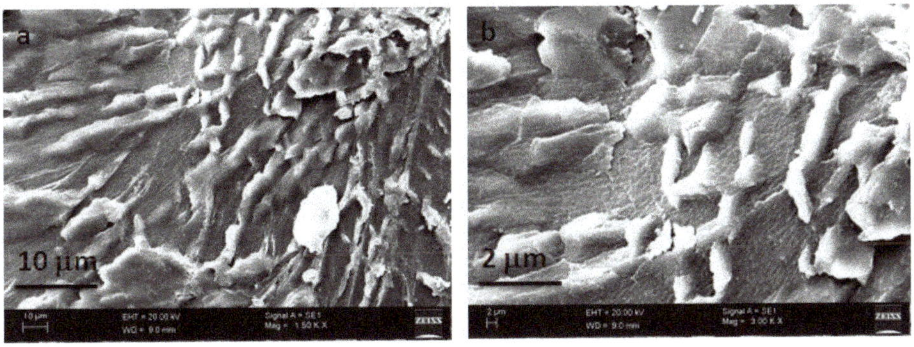

Figure 5. SEM images of PPK hydrogel for P2 formulation: (**a**) low magnification, and (**b**) high magnification ([**a**] ×1.50KX; [**b**] ×3.00KX).

2.2. Swelling Study and pH-Responsive Property of PPK Hydrogels

PPK hydrogels were permitted to swell in DDW and various pH solutions (Figure 6). The swelling of the three types of PPK hydrogels depends on their aging time. Figure 6a depicts the PPK hydrogels' swelling behavior in DDW as a function of time. The swelling order of the PPK hydrogels was P1 > P2 > P3. The pH swelling of all hydrogels was examined at different pH solutions. The %ESR values of PPK hydrogels in the pH solutions depends on the functional groups in the hydrogels. Figure 6b displays the swelling ratio of PPK hydrogel (P2) in various pH solutions. The swelling ratios for all hydrogels increased slowly up to pH 5.0 and significantly increased at higher pH. It is expected that PPK hydrogels have carboxylic functional groups, which are responsible for its ionization under different pH conditions. At lower pH, the carboxylic groups were converted to –COOH, and hydrogen bonding developed. A higher pH, on the other hand, caused the carboxylic groups to become ionic and the repelling force to become stronger, which promoted swelling. Similar to chemically-synthesized hydrogels, the newly-identified natural hydrogels exhibited excellent swelling behavior in aqueous solutions at various pH.

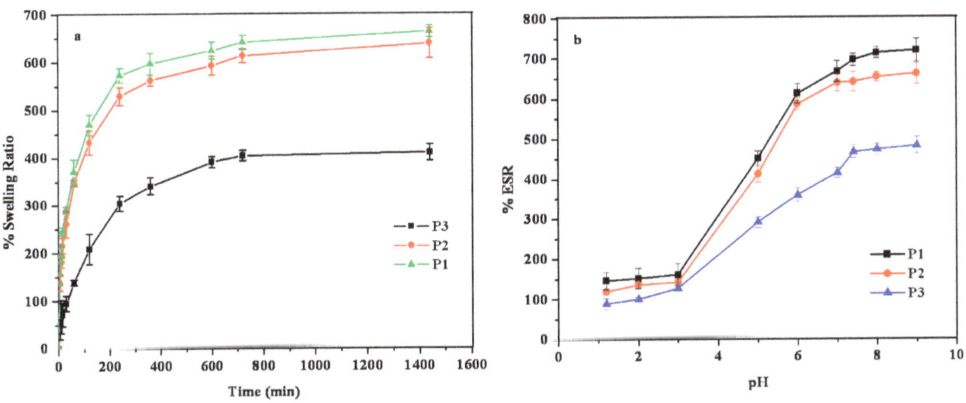

Figure 6. (**a**) Swelling studies of PPK hydrogels in DDW at 37 °C and (**b**) %ESR of PPK hydrogels at various pH conditions.

2.3. In Vitro Release Study

The in vitro release performance of different PPK hydrogels (P1, P2, P3) loaded with 5-FU drug was analyzed in pH 7.4 conditions at 37 °C. As can be seen in Figure 7, the network structure of PPK hydrogel produced from PPK fruits with various aging periods affects the percentage of 5-FU release from PPK hydrogel. As the aging time increased, the PPK hydrogels showed a more rigid structure. The order of the 5-FU release trend from the PPK hydrogels was P1 > P2 > P3. The rigid network structure with microvoid contracting was responsible for the lowest release of 5-FU from the P3 formulation. Therefore, the buffer molecules could not travel easily into the networks of the PPK hydrogels. In general, the release profile of drugs from hydrogels depends on the crosslinking between polymer chains. At a higher level of crosslinking, the drug molecules cannot easily be diffused out from the hydrogel, due to rigid network structure [17]. Similarly, the release of 5-FU sped up from the softer PPK hydrogels, while becoming slower from the harder PPK hydrogels.

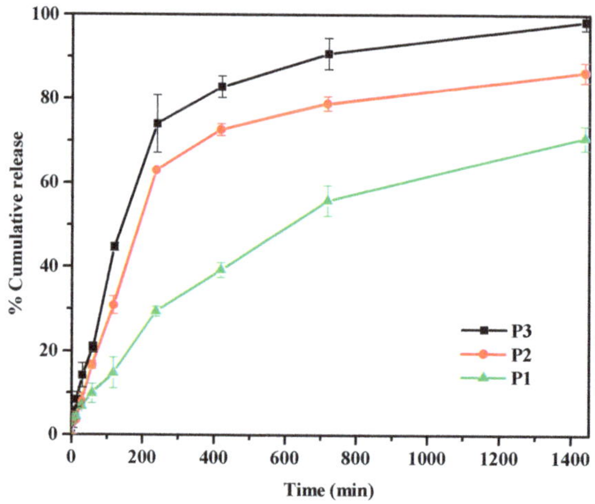

Figure 7. In vitro release behavior of 5-FU through PPK hydrogels in pH 7.4 at 37 °C for P1, P2, and P3.

The in vitro release of 5-FU from PPK hydrogels (P2 formulation) were carried out in gastro-intestinal fluids at 37 °C (Figure 8). The release patterns were found to depend on the pH condition. The % of 5-FU was higher at higher pH conditions, whereas the %5FU was lower for gastric fluid (pH 1.2). This finding can be explained on the basis of functional groups that exist in the PPK hydrogels. PPK contains pectin-based polysaccharides, which have ionizable carboxylic acid groups that can influence drug release. Negative charge repulsion forces were generated at higher pH conditions (7.4) because of the ionization of carboxylic groups, thereby resulting in maximum drug release. At pH 1.2, the ionization repulsive forces were decreased due to the protonation of carboxylate ions, which meant drug release was slower. Thus, 5-FU release was dependent on the pH-responsive behavior of PPK hydrogel, which can have benefits for drug delivery in colon cancer patients. Moreover, the release profiles were fitted using Korsmeyer-Peppas equation to corroborate the release mechanism as follows [19]:

$$\frac{M_t}{M_\infty}A = k_p t^n$$

where k_p is the rate constant, n is the release exponent, and M_t/M_∞ is the percentage of the drug released at time t. The release mechanism was evaluated using the estimated n value.

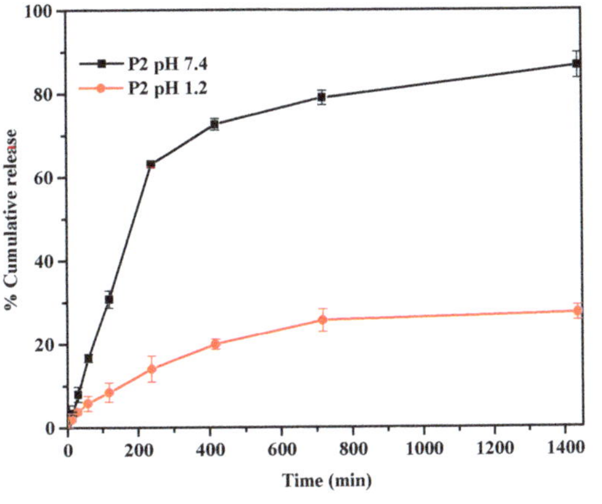

Figure 8. In vitro release behavior of the PPK hydrogel (P2) in pH 1.2 and 7.4 buffer media at 37 °C.

For all formulations, n and k were calculated using the least square method, and the results are shown in Table 1 as a summary. According to the equation, the log value of the percentage of 5-FU release from PPK hydrogel was plotted as a function of log time. Fickian diffusion governs the drug release from the polymer matrix if $n < 0.5$. For $n > 0.5$, anomalous or non-Fickian drug diffusion occurs. Release kinetics that are completely non-Fickian or case II type for $n = 1$. The n values of the current data were 0.26 to 0.35, indicating a Fickian diffusion pattern. Thus, the n values are affected by the network topology of PPK hydrogel.

3. Conclusions

The pH swelling responsive behavior of palmyra fruit hydrogels collected from palmyra palm tree was reported. It is interesting to note that in alkaline conditions, the PPK hydrogels showed greater swelling behavior. When 5-FU was introduced to the PPK hydrogels as a model drug, its encapsulation efficiency reached 62%. DSC and XRD revealed the molecular dispersion of 5-FU in the PPK hydrogels. In vitro release tests showed that PPK hydrogel released 5-FU in intestinal fluid more quickly than in stomach fluid. The naturally-occurring PPK hydrogels were stable for more than one year. The economically cheap and safe natural hydrogels were found to be an efficient carrier for colon cancer drug delivery of 5-FU.

4. Materials and Methods

Palmyra palm fruits were collected from the surroundings of the Yaganti temple, Kurnool district, Andhrapradesh, India. The fruit kernel seed was removed (Figure 1a,b show digital photographs of the palmyra palm kernel before and after removing the seed, respectively), and the white jelly kernel was then cut into small pieces and immersed repeatedly in water and aqueous methanol for three days to remove any water-soluble sugars and other materials. The resulting pieces were air-dried and then dried at 40 °C in an electronically-controlled hot air oven. The pieces were used further for swelling, drug loading, and characterization studies. In this study, three types of fruits were used according to their aging time (soft, hard, and harder fruits, named P1, P2, and P3 and listed in Table 1). Because many hydrophilic units are present in the palmyra palm fruit, they are swellable in water. Figure 1c,d present digital photographs of dry PPK gel and swollen PPK gel, respectively. 5-Fluorouracil (5-FU) was purchased from Himedia, India.

4.1. Swelling Studies

Mass measurements were used to determine the swelling capacity of PPK hydrogels when placed in DDW at 37 °C. First, the dry PPK hydrogels were accurately weighed and submerged in DDW for various time periods. After swelling, the PPK hydrogels were carefully removed and wiped to remove any surface-adhered water. The pH-sensitive properties of PTK hydrogel were measured by recording the swelling of the hydrogels at different pH levels (pH 1.2, 2.0, 3.0, 5.0, 6.0, 7.0, 7.4, 8.0, and 9.0). The percentage swelling (%SR) ratio and the equilibrium swelling (%ESR) ratio were calculated in accordance with the formulae below:

$$\%\ \mathrm{SR} = \frac{W_s - W_d}{W_d} \times 100$$

$$\%\ \mathrm{ESR} = \frac{W_e - W_d}{W_d} \times 100$$

where W_s is the weight of the gel while it is swelling at time t, W_d is the weight of the hydrogel when it is dry, and W_e is the weight of the gel after it has reached equilibrium in the water.

4.2. 5-FU Loading

5-FU was loaded into PPK hydrogels by an equilibrium swelling method in drug solutions according to previously reported work [17]. The hydrogels were then allowed to swell for 24 h at 37 °C in a known amount of drug solution. 5-FU only dissolves in water at a very low concentration (13 mg/mL). On the other hand, its sodium salt is 65 mg/mL solubilized. Gel discs were submerged in a 5-FU aqueous solution, which was neutralized with NaOH, to load the maximal amount of 5-FU into the PPK hydrogel network. The 5-FU in the solvent was adsorbed onto the hydrogels during this process.

5-FU loading efficiency of 5FU-PPK hydrogels was assessed spectrophotometrically. To extract the 5-FU from the PPK hydrogels, a 5-FU loaded PPK hydrogel disc was submerged in 20 mL of buffer solution and rapidly agitated for 48 h. After filtering, the solution was measured with a UV-Vis spectrophotometer at a wavelength of 270 nm. The following equations were utilized for the calculation of drug loading (%DL) and encapsulation efficiency (%EE):

$$\%\ \text{Drug loading} = \frac{\text{Amount of drug in hydrogel}}{\text{Amount of hydrogel}} \times 100$$

$$\%\ \text{Encapsulation Efficiency} = \frac{\text{Actual loading}}{\text{Theoritical loading}} \times 100$$

Table 1 lists the data for various formulations.

4.3. In Vitro 5-FU Release

The release of 5-FU from PPK hydrogels was carried out employing dissolution experiments using a LabIndia DS-8000 tablet dissolution tester system fitted with eight baskets. The drug release performance was measured at 37 °C with rotation speed of 100 rpm at different time intervals (5, 15, 30, 60, 120, 140, 420, 720, and 1440 min). The 5-FU release from the 5-FU-PPK hydrogels was examined in fluids with gastric pH values of 1.2 and intestinal pH values of 7.4. Aliquot samples were taken out and examined using UV spectrophotometry (UV-3092, LAB INDIA, Mumbai, India) at a maximum wavelength of 270 nm at regular intervals.

4.4. Characterization

Fourier transform infrared spectroscopy (FTIR) was performed on the PPK hydrogels, drug-loaded PPK hydrogel, and pure 5-FU using a Perkin Elmer (Impact 410, Wisconsin, MI, USA) instrument scanning between 4000 and 500 cm^{-1}. Before scanning spectra, under a hydraulic pressure of 600 dynes/m^2, the hydrogels were finely powdered with KBr to make the pellets. The differential scanning calorimetry (DSC) curves of the PPK hydrogels, 5-FU-PPK hydrogel, and 5-FU were analyzed using a sequential thermal analyzer (TA instruments-Model-SDT-Q600). The test was performed at a heating rate of 10 °C/min with 100 mL/min purging rate of N2 environment. The PPK hydrogels were observed using scanning electron microscopy (SEM) (SUPRA25, Carl Zeiss AG, Jena, Germany) at two magnifications using acceleration voltage at 20 kV.

Author Contributions: K.-M.R.: Conceptualization, methodology, validation, formal analysis, investigation, writing—original draft preparation K.S.V.K.R.: Conceptualization and investigation, review and editing, R.-R.P.: Formal analysis and validation; U.-T.U.: Formal analysis and validation; C.-S.H.: Supervision, editing, resource, and funding acquisition; S.-S.H.: Supervision, editing, resource, and funding acquisition. All authors have read and agreed to the published version of the manuscript.

Funding: The work was supported by the National Research Foundation of Korea (NRF) Grant funded by the Ministry of Science and ICT, Korea (2021R1I1A3060098 and Brain Korea 21 Plus Program (4199990414196)) and by the Korea Institute for Advancement of Technology funded by the Ministry of Trade, Industry and Energy (P0017531). This work was also supported by Korea Institute of Planning and Evaluation for Technology in Food, Agriculture and Forestry (IPET) through High Value-added Food Technology Development Program, funded by Ministry of Agriculture, Food and Rural Affairs (MAFRA)(321027-5), and this research was also supported by the National Research Foundation of Korea (NRF) (Grant No. 2020R1A6A1A03044512).

Institutional Review Board Statement: Not applicable.

Informed Consent Statement: Not applicable.

Data Availability Statement: Not applicable.

Conflicts of Interest: The authors declare no conflict of interest.

References

1. Liao, N.; Pang, B.; Jin, H.; Xu, X.; Yan, L.; Li, H.; Shao, D.; Shi, J. Potential of lactic acid bacteria derived polysaccharides for the delivery and controlled release of oral probiotics. *J. Control. Release* **2020**, *323*, 110–124. [CrossRef] [PubMed]
2. Liu, L.; Yao, W.; Rao, Y.; Lu, X.; Gao, J. pH-Responsive carriers for oral drug delivery: Challenges and opportunities of current platforms. *Drug Deliv.* **2017**, *24*, 569–581. [CrossRef] [PubMed]
3. Beneke, C.E.; Viljoen, A.M.; Hamman, J.H. Polymeric plant-derived excipients in drug delivery. *Molecules* **2009**, *14*, 2602–2620. [CrossRef] [PubMed]
4. Subrahmanyan, V.; Bains, G.S.; Natarajan, C.P.; Bhatia, D.S. The carbohydrates of tender kernel of palmyra palm (*Borassus flabellifer*, L.). *Arch. Biochem. Biophys.* **1956**, *60*, 27–34. [CrossRef] [PubMed]
5. Assoi, S.; Konan, K.; Walker, L.T.; Holser, R.; Agbo, G.N.; Dodo, H.; Wicker, L. Functionality and yield of pectin extracted from Palmyra palm (Borassus aethiopum Mart) fruit. *LWT—Food Sci. Technol.* **2014**, *58*, 214–221. [CrossRef]
6. Assoi, S.; Konan, K.; Agbo, G.N.; Dodo, H.; Holser, R.; Wicker, L. Palmyra palm (*Borassus aethiopum* Mart.) fruits: Novel raw materials for the pectin industry. *J. Sci. Food Agric.* **2017**, *97*, 2057–2067. [CrossRef] [PubMed]
7. Sugumaran, K.R.; Gowthami, E.; Swathi, B.; Elakkiya, S.; Srivastava, S.N.; Ravikumar, R.; Gowdhaman, D.; Ponnusami, V. Production of pullulan by Aureobasidium pullulans from Asian palm kernel: A novel substrate. *Carbohydr. Polym.* **2013**, *92*, 697–703. [CrossRef] [PubMed]
8. Matsuyama, H.; Teramoto, M.; Urano, H. Analysis of solute diffusion in poly (vinyl alcohol) hydrogel membrane. *J. Membr. Sci.* **1997**, *126*, 151–160. [CrossRef]
9. Fournier, E.; Passirani, C.; Colin, N.; Bretona, P.; Sagodiraa, S.; Benoitb, J.P. Development of novel 5-FU-loaded poly(methylidene malonate 2.1.2)-based microspheres for the treatment of brain cancers. *Eur. J. Pharm. Biopharm.* **2004**, *57*, 189–197. [CrossRef] [PubMed]
10. Othman, M.H.; Zayed, G.M.; El-Sokkary, G.H.; Ali, U.F.; Abdellatif, A.A.; Othman, M. Preparation and evaluation of 5-fluorouracil loaded microsponges for treatment of colon cancer. *J. Cancer Sci. Ther.* **2017**, *9*, 307–313. [CrossRef]
11. Jain, S.K.; Jain, A.; Gupta, Y.; Ahirwar, M. Design and development of hydrogel beads for targeted drug delivery to the colon. *AAPS PharmSciTech.* **2007**, *8*, E34–E41. [CrossRef] [PubMed]
12. Du, H.; Liu, M.; Yang, X.; Zhai, G. The design of pH-sensitive chitosan-based formulations for gastrointestinal delivery. *Drug Discov.* **2015**, *20*, 1004–1011. [CrossRef] [PubMed]
13. Chourasia, M.K.; Jain, S.K. Pharmaceutical approaches to colon targeted drug delivery systems. *J. Pharm. Pharm. Sci.* **2003**, *6*, 33–66. [PubMed]

14. Nazari, M.; Saljooghi, A.S.; Ramezani, M.; Alibolandi, M.; Mirzaei, M. Current status and future prospects of nanoscale metal-organic frameworks in bioimaging. *J. Mater. Chem. B* **2022**, *10*, 8824–8851. [CrossRef]
15. Rizwan, M.; Yahya, R.; Hassan, A.; Yar, M.; Azzahari, A.D.; Selvanathan, V.; Sonsudin, F.; Abouloula, C.N. pH sensitive hydrogels in drug delivery: Brief history, properties, swelling, and release mechanism, material selection and applications. *Polymers* **2017**, *9*, 137. [CrossRef] [PubMed]
16. Mirzaei, M.; Zarch, M.B.; Darroudi, M.; Sayyadi, K.; Keshavarz, S.T.; Sayyadi, J.; Fallah, A.; Maleki, H. Silica mesoporous structures: Effective nanocarriers in drug delivery and nanocatalysts. *Appl. Sci.* **2020**, *10*, 7533. [CrossRef]
17. Rao, K.M.; Nagappan, S.; Seo, D.J.; Ha, C.S. pH sensitive halloysite-sodium hyaluronate/poly (hydroxyethyl methacrylate) nanocomposites for colon cancer drug delivery. *Appl. Clay Sci.* **2014**, *97*, 33–42. [CrossRef]
18. Reddy, P.R.S.; Rao, K.K.; Rao, K.M.; Reddy, N.S. Synthesis of Novel Hydrogels based Poly(4-Hydroxyphenylazo-3-N-(4-hydroxyphenyl) maleimide) for Specific Colon Delivery of Chemotherapeutic Agent. *J. Appl. Pharm. Sci.* **2015**, *5*, 21–28. [CrossRef]
19. Ritger, P.L.; Peppas, N.A.A. Simple equation for description of solute release II. Fickian and Anomalous release from swellable devices. *J. Control. Release* **1987**, *5*, 37–42. [CrossRef]

MDPI
St. Alban-Anlage 66
4052 Basel
Switzerland
www.mdpi.com

Gels Editorial Office
E-mail: gels@mdpi.com
www.mdpi.com/journal/gels

www.ingramcontent.com/pod-product-compliance
Lightning Source LLC
Chambersburg PA
CBHW061447210926
51850CB00008B/1641
9783725800063